D0780373

Water Reuse for Irrigation

Agriculture, Landscapes, and Turf Grass

Water Reuse for Irrigation

Agriculture, Landscapes, and Turf Grass

Edited by

Valentina Lazarova

Akiça Bahri

CRC PRESS

Boca Raton London New York Washington, D.C.

Library of Congress Cataloging-in-Publication Data

Water reuse for irrigation: agriculture, landscapes, and turf grass / edited by Valentina Lazarova, Akiça Bahri
p. cm.
Includes bibliographical references.
ISBN 1-56670-649-1 (alk. paper)
1. Irrigation water – Quality. 2. Water reuse. 3. Sewage irrigation – Environmental aspects. I. Lazarova, Valentine. II. Bahir, Akiça.

S618.45W28 2004
333.91′3–dc22 2004055223

Direct all inquiries to CRC Press, 2000 N.W. Corporate Blvd., Boca Raton, Florida 33431.

Visit the CRC Press Web site at www.crcpress.com

International Standard Book Number 1-56670-649-1
Library of Congress Card Number 2004055223
Printed in the United States of America 1 2 3 4 5 6 7 8 9 0
Printed on acid-free paper

About the Authors

Valentina Lazarova graduated with a master's degree in Sanitary Engineering from the School of Civil Engineering of Leningrad. She earned a Ph.D. degree from the Technical University of Chemical Engineering of Sofia, Bulgaria, and a postdoctoral fellowship at the National Institute of Applied Sciences in Toulouse, France. She started her professional career as project manager at the Bulgarian National Institute Vodokanalproekt and assistant professor at the Center of Biotechnology of the University of Sofia. Since 1991 Dr. Lazarova has worked at the Technical and Research Center (CIRSEE) of Suez Environnement, France, as project manager and department head for wastewater treatment and reuse.

The main focus of her academic and research activities is global water-cycle management with special attention to water and wastewater treatment and water-quality management. Her efforts have focused on the transfer and application of innovative methods and findings resulting from fundamental research to industry (wastewater-treatment design and operation). A particular emphasis of her work has been to develop, promote, and apply an interdisciplinary approach to the development of innovative water- and wastewater-treatment processes and integrated water-management strategies that are more efficient, economically viable, and technically reliable. She has made major contributions in the field of advanced biofilm reactors, disinfection, water reuse, water-quality control, integrated water-resource management, and water reuse.

Akiça Bahri holds agricultural engineer and doctor-engineer degrees from the National Polytechnical Institute of Toulouse, France, and a Ph.D. from the Department of Water Resources Engineering, Lund University, Sweden. She works for the National Research Institute for Agricultural Engineering, Water and Forestry in her home country of Tunisia. She has worked in the field of agricultural use of marginal waters (brackish and wastewater), sewage sludge, and their impacts on the environment. She is involved in policy and legislative issues regarding water reuse and land application of sewage sludge. She is in charge of research management in the field of agricultural water use. Besides being involved in research and teaching, she has wide-ranging international experience. She is member of different international scientific committees.

Abstract

The purpose of this book is to provide guidelines for the use and management of recycled water for both landscape and agricultural irrigation. The target audience is water and wastewater engineers, administrators and planners, operators, and recycled water users. Multidisciplinary knowledge in this field is summarized and can be easily used by students and researchers. To bridge the gaps between fundamental science (biology, agronomy, environmental engineering, etc.) and relatively uncharted areas of economic, institutional, and liability issues and facilitate the successful planning and operation of water-reuse projects for various purposes, these guidelines are intended to provide the needed information, analysis of the existing practice, and recommendations for agricultural and landscape irrigation. Emerging issues and concerns are also covered, such as adverse effects on plants, groundwater, environment, and public health.

Preface

Growing water scarcity worldwide and stringent water-development regulations to protect the environment are two major challenges facing water professionals in implementing integrated management of water resources. The water pollution control efforts in many countries have made treated municipal and industrial wastewater suitable for economical augmentation of the existing water supply when compared to increasingly expensive and environmentally destructive new water resource developments that include dams and reservoirs. Thus, the beneficial use of treated municipal wastewater is considered as a competitive and viable water source option. Recycled water makes it possible to close the water cycle at a point closer to cities and farms by producing water from municipal wastewater and reducing wastewater discharge into the environment.

The use of reclaimed water for agriculture is a major water-reuse practice world wide. For a number of semi-arid regions and islands, water recycling provides a major portion of the irrigation water. However, despite widespread irrigation with reclaimed wastewater, water-reuse programs are still faced with a number of technical, economic, social, regulatory, and institutional challenges. Some of the water-quality concerns and evaluation of long-term environmental, agronomic, and health impacts remain unanswered. Furthermore, the economic benefits and financial performance of water reuse in irrigation are difficult to assess and demonstrate. The economics of water reuse depends on many local conditions, which are difficult to generalize. Often the costs and benefits of water reuse need to be compared with the environmental costs and costs to downstream water users.

The aim of this book is to bridge some of these gaps, providing a synthesis of comprehensive information generated by recent advances in science and practices of water reuse for irrigation. It presents guidelines, recommendations, and codes of best practice from around the world for all types of uses of recycled water for irrigation. Emerging issues and concerns are also discussed, such as adverse effects on plants, groundwater, environment, and public health.

Planning water-reuse projects for irrigation purposes can also be valuable for planners and local authorities, as it has numerous associated benefits such as providing reliable, secure, and drought-proof water sources via recycled water, closing the water cycle at a small and large scale, improving public health and the environment, and contributing to sustainable development in both rural and densely populated urban regions. This book is the most up-to-date treatise by experts on irrigation using recycled water produced by appropriate treatment of urban wastewater.

Valentina Lazarova

Paris, France

Acknowledgments

The publication of this book is a team effort and a number of persons greatly contributed directly or indirectly to the preparation of this book. A part of this work was supported by the EC Program Environment and Climate, contract ENV4-CT98-0790 "Enhancement of Integrated Water Resource management with Water Reuse at Catchment Scale" (EU project CatchWater). I would like to thank all participants in the CatchWater consortium and the CEE advisor officers for their active collaboration. In fact, over the past years substantial research activities in the field of water were initiated in the context of the European Union's environmental research and demonstration programs, with the goal of understanding the complexity of the problems of the freshwater environment, providing decision-makers with the appropriate tools to manage water resources in an integrated way, and supporting the implementation of related European Union water policies (Water Framework Directive). Within this context, attention was also given in field studies and modelling activities aiming to develop socio-technical-economic methodologies for water reclamation schemes ensuring public health and environmental protection. Public acceptance and socio economic impacts involved with water reuse have been investigated also with the goal of providing the basis for common principals and criteria for integrated water management practices with water reuse. Several other research programs funded by Suez Environnement enabled us to improve our knowledge about water reuse in the context of different developed and emerging countries.

The guidance of Prof. Takashi Asano, Jim Crook and Bahman Sheikh were decisive for the elaboration of a comprehensive and homogenous guideline book on irrigation with recycled water.

The active support of my family and especially my daughter Raya was precious to me and helped to overcome all technical, social and psychological constraints.

Alexander Franchet and Michel Hurtrez provided essential help for the improvement of the book's illustrations.

Valentina Lazarova

Contributors

A number of North American, Central American, European, and Mediterranean experts were involved in the preparation of different chapters, as well as in the reviewing and revising of other chapters. The active participation of all contributors was critical for the quality of the manuscript:

- Takashi Asano, *Consultant, retired from University of California at Davis, Davis, California, United States*
- Akiça Bahri, *National Research Institute for Agricultural Engineering, Water, and Forestry, Ariana, Tunisia*
- Herman Bouwer, *Consultant, retired from U.S. Department of Agriculture, Phoenix, Arizona, United States*
- James Crook, *Consultant, Norwell, Massachusetts, United States*
- Paul Jeffrey, *Cranfield University, Bedford, United Kingdom*
- Blanca Jiménez Cisneros, *National Autonomous University of México, Mexico, D.F., México*
- Xavier Millet, *Consorci de la Costa Brava, Girona, Spain*
- Joe Morris, *Cranfield University, Bedford, United Kingdom*
- Ioannis Papadopoulos, *Agricultural Research Institute, Nicosia, Cyprus*
- Eric Rosenblum, *South Bay Water Recycling Environmental Services Department, San Jose, California, United States*
- Lluis Sala, *Consorci de la Costa Brava, Girona, Spain*
- Bahman Sheikh, *Consultant, San Francisco, California, United States*
- Sean Tyrrel, *Cranfield University, Bedford, United Kingdom*

Contents

1. CHALLENGES OF SUSTAINABLE IRRIGATION WITH RECYCLED WATER

Valentina Lazarova and Takashi Asano

1.1 Managing water security by water reuse 1
1.2 Objectives and contents of this volume 4
1.3 Role of water reuse for irrigation 6
1.4 Benefits and constraints of irrigation with recycled water 8
1.5 Specifics of water reuse planning 11
1.6 International experience with irrigation using recycled water 14
1.6.1 Water reuse in Europe, the Mediterranean region, and the Middle East 18
1.6.2 Water reuse in the United States 20
1.6.3 Water reuse in Central and South America 26
1.6.4 Water reuse in Asia and Oceania 26
1.7 Management actions for improvement of irrigation with recycled water 28

References 29

2. WATER QUALITY CONSIDERATIONS

Valentina Lazarova, Herman Bouwer, and Akiça Bahri

2.1 Introduction 31
2.2 Parameters with health significance 36
2.2.1 Chemicals 36
2.2.2 Pathogens 40
2.3 Parameters with agronomic significance 45
2.3.1 Salinity 45
2.3.2 Toxic ions 46
2.3.3 Sodium adsorption ratio 48
2.3.4 Trace elements 49
2.3.5 pH 53
2.3.6 Bicarbonate and carbonate 53

2.3.7 Nutrients 54
2.3.8 Free chlorine 55
2.4 Sampling and monitoring strategies 55
References 58

3. INTERNATIONAL HEALTH GUIDELINES AND REGULATIONS

James Crook and Valentina Lazarova

3.1 WHO guidelines for irrigation 64
3.2 USEPA guidelines for water reuse 66
3.3 California water recycling criteria 66
3.4 Other water reuse regulations 73
3.5 Standards for urban uses of recycled water and landscape irrigation 76
3.6 Standard enforcement and perspectives 79
References 80

4. CODE OF PRACTICES FOR HEALTH PROTECTION

Valentina Lazarova and Akiça Bahri

4.1 Introduction 83
4.2 Specific wastewater treatment for reuse purposes 86
4.2.1 Typical schemes used for production of recycled water for irrigation 87
4.2.2 Main disinfection processes used in water reuse systems 89
4.2.3 Requirements for recycled water storage and distribution 92
4.2.4 Requirements for reliability of operation of water reuse systems 93
4.3 Control of recycled water application 94
4.4 Restrictions on crops and public access 96
4.5 Human exposure control 98
References 101

5. CODE OF SUCCESSFUL AGRONOMIC PRACTICES

Valentina Lazarova, Ioannis Papadopoulos, and Akiça Bahri

5.1 Amount of water used for irrigation 104
5.2 General water quality guidelines for maximum crop production 105

5.3 Choice of management strategy of irrigation with recycled water 106
5.4 Selection of irrigation method 108
5.4.1 Criteria for selection of an appropriate irrigation method 108
5.4.2 Comparison of irrigation methods 113
5.4.3 Final considerations for the choice of irrigation method 122
5.5 Crop selection and management 123
5.5.1 Code of practices to overcome salinity hazards 123
5.5.2 Code of practices to overcome boron, sodium and chloride toxicity 135
5.5.3 Code of practices to overcome trace elements toxicity 137
5.6 Code of management practices of water application 142
5.6.1 Leaching and drainage 142
5.6.2 Using other water supplies 145
5.6.3 Adjusting fertilizer applications 145
5.6.4 Management of soil structure 146
5.6.5 Management of clogging in sprinkler and drip irrigation systems 147
5.6.6 Management of storage systems 148

References 149

6. CODES OF PRACTICES FOR LANDSCAPE AND GOLF COURSE IRRIGATION

Bahman Sheikh

6.1 Benefits of and constraints on the use of recycled water for landscape irrigation 152
6.2 Effects of recycled water on turfgrass 153
6.3 Best practices for golf course irrigation 153
6.4 Prevention of adverse effects of recycled water on turfgrass 156
6.5 Management of adverse effect of water reuse on soils 157
6.6 Recommendations to avoid adverse effects of water reuse on groundwater 157
6.7 Economic and financial aspects of landscape irrigation 158
6.8 Customer acceptance of recycled water for irrigation of landscaping and golf courses 160

References 161

7. WASTEWATER TREATMENT FOR WATER RECYCLING

Valentina Lazarova

7.1 Introduction 164
7.1.1 Choice of appropriate treatment 166
7.1.2 Main treatment processes used for wastewater treatment 166
7.1.3 Influence of sewer configuration on water quality 168
7.2 Physicochemical treatment of wastewater 168
7.2.1 Screening 168
7.2.2 Primary sedimentation 171
7.2.3 Coagulation/flocculation 171
7.2.4 Flotation 172
7.3 Biological wastewater treatment processes 173
7.3.1 Activated sludge 173
7.3.2 Trickling filters 174
7.3.3 Rotating biological contactors 176
7.4 Advanced biofilm technologies 176
7.5 Nonconventional natural systems 178
7.5.1 Lagooning 178
7.5.2 Wetlands 183
7.5.3 Infiltration-percolation 184
7.5.4 Soil-aquifer treatment 186
7.6 Advanced tertiary treatment and disinfection 188
7.6.1 Tertiary filtration 189
7.6.2 Chlorination 194
7.6.3 Chlorine dioxide 198
7.6.4 UV disinfection 198
7.6.5 Ozonation 208
7.6.6 Membrane filtration 215
7.6.7 Membranes bioreactors 216
7.7 Storage and distribution of recycled water 219
7.7.1 Short-term storage 220
7.7.2 Long-term storage 220
7.7.3 Management of recycled water storage reservoirs 224
7.7.4 Control of water quality in distribution systems 225
7.8 Criteria for selection of appropriate polishing process before irrigation 226
7.8.1 Cost of additional treatment and reuse 226
7.8.2 Main criteria for selection of disinfection process 228
References 231

8. ADVERSE EFFECTS OF SEWAGE IRRIGATION ON PLANTS, CROPS, SOIL, AND GROUNDWATER

Herman Bouwer

8.1 Toward a healthy environment and sustainable development 236
8.2 Compounds with potential adverse effects on recycled water for irrigation 237
8.3 Behavior of some compounds during irrigation with sewage effluent 238
8.3.1 Salt and water relations in irrigation soils 239
8.3.2 Behavior and potential adverse effects of nutrients in irrigation soils 244
8.3.3 Effects of disinfection by-products on groundwater 247
8.3.4 Effects of pharmaceuticals and other organic contaminants 248
8.4 Salt and groundwater water-table management for sustainable irrigation 253
8.4.1 Salt loadings 254
8.4.2 Salt tolerance of plants 255
8.4.3 Management of salty water 256
8.4.4 Future aspects for salinity management in south-central Arizona 259

References 260

9. ECONOMICS OF WATER RECYCLING FOR IRRIGATION

Joe Morris, Valentina Lazarova, and Sean Tyrrel

9.1 General principles 266
9.2 Financial analysis 266
9.3 Economic analysis 267
9.4 Benefits of recycled water for irrigation 267
9.5 Factors influencing irrigation benefits 268
9.6 Components of recycling systems for irrigation 269
9.7 Irrigation water supply options 269
9.8 Water-recycling options 271
9.9 Costs of water-recycling options 272
9.10 Prices for recycled water 275
9.11 Function of water prices 275
9.12 Criteria for setting prices for recycled water 277
9.13 Pricing instruments 278
9.14 Examples of recycled water prices 279
9.15 Conclusions 282

References 282

10. COMMUNITY AND INSTITUTIONAL ENGAGEMENT IN AGRICULTURAL WATER REUSE PROJECTS

Paul Jeffrey

10.1 Introduction 285
10.2 Public perceptions of water reuse for agricultural production 289
10.3 Institutional barriers 292
10.4 Models for participative planning 298
10.5 Participative planning processes for water-reuse projects 301

References 305

11. INSTITUTIONAL ISSUES OF IRRIGATION WITH RECYCLED WATER

Eric Rosenblum

11.1 Introduction 310
11.2 Ownership of water, wastewater, and recycled water 311
11.2.1 Water rights 311
11.2.2 Water use limits 316
11.2.3 Rights to recycled water 318
11.3 Wastewater regulations 319
11.3.1 Effluent regulations 320
11.3.2 Pretreatment to protect recycled water quality 322
11.4 Planning and implementation issues 324
11.4.1 Land use planning 324
11.4.2 Environmental regulations 326
11.4.3 Construction issues 328
11.4.4 Wholesaler/retailer issues 329
11.4.5 Customer agreements 332
11.5 Program management 333
11.5.1 Integrated planning 333
11.5.2 Matrix analysis of institutional issues 333
11.5.3 Summary of institutional guidelines 337

References 339

12. CASE STUDIES OF IRRIGATION WITH RECYCLED WATER

12.1 El Mezquital, Mexico: The Largest Irrigation District Using Wastewater

Blanca Jimenez

12.1.1 General description 345
12.1.2 Wastewater quality 347

12.1.3 Effects of wastewater reuse on agriculture and health 348
12.1.4 Mexican legislation for agricultural irrigation 348
12.1.5 Helminthiasis 349
12.1.6 Removal of helminth eggs 351
12.1.7 Filtration step 352
12.1.8 Bacteria removal 353
12.1.9 Sludge treatment and disposal 355
12.1.10 Costs of recycled water 355
12.1.11 Unplanned aquifer recharge by irrigation with wastewater 356

References 361

12.2 Water Reuse for Golf Course Irrigation in Costa Brava, Spain

Lluís Sala and Xavier Millet

12.2.1 History 363
12.2.2 Tips for adequate recycled water management in golf course irrigation 364
12.2.3 Electrical conductivity (EC) 364
12.2.4 Nutrients 365
12.2.5 Maturation pond design and management 368
12.2.6 Conclusions 372

References 373

12.3 Monterey County Water Recycling Projects: A Case Study in Irrigation Water Supply for Food Crop Irrigation

Bahman Sheikh

12.3.1 History and motivation 374
12.3.2 Project overview 374
12.3.3 Public perception 377
12.3.4 Current project status and operation 378
12.3.5 Conclusions, lessons learned and recommendations 378

References 379

13. CONCLUSIONS AND SUMMARY OF PRACTICES FOR IRRIGATION WITH RECYCLED WATER

Valentina Lazarova

13.1 Assessment of the feasibility of using recycled water for irrigation 383
13.2 Good agronomic practices for irrigation with recycled water 383

13.3 Negative impacts of irrigation water on plants and main corrective actions 388
13.4 Management practices and corrective actions for improvement of the operation of water-reuse treatment schemes 388
13.5 Successful participation programmes improved public acceptance 389
13.6 Successful initiatives to address legal and institutional issues 390

1

Challenges of Sustainable Irrigation with Recycled Water

Valentina Lazarova and Takashi Asano

CONTENTS

1.1 Managing water security by water reuse 1
1.2 Objectives and contents of this volume.................................. 4
1.3 Role of water reuse for irrigation 6
1.4 Benefits and constraints of irrigation with recycled water 8
1.5 Specifics of water reuse planning 11
1.6 International experience with irrigation using recycled water.......... 14
   1.6.1 Water reuse in Europe, the Mediterranean region, and the Middle East 18
   1.6.2 Water reuse in the United States 20
   1.6.3 Water reuse in Central and South America 26
   1.6.4 Water reuse in Asia and Oceania 26
1.7 Management actions for improvement of irrigation with recycled water.. 28
References 29

1.1 MANAGING WATER SECURITY BY WATER REUSE

Many factors will challenge water professionals in the new millennium. Growing water scarcity, rapid increase in population, rapid urbanization and megacity development, increasing competition among water users, and growing concerns for health and environmental protection are examples of important issues. Despite improvements in the efficiency of water use in many

1-56670-649-1/05/$0.00 + $1.50

developed countries, the demand for fresh water has continued to climb as the world's population and economic activity have expanded. According to the International Water Management Institute (IWMI[1]), by 2025, 1.8 billion people will live in countries or regions with absolute water scarcity. The term "absolute water scarcity" means water availability of less than the 100 m^3/inhabitant/year that is necessary for domestic and industrial use. This water availability level is not sufficient to maintain the current level of per capita food production from irrigated agriculture. Today, most countries in the Middle East and North Africa can be classified as having absolute water scarcity. By 2025, these countries will be joined by Pakistan, South Africa, large parts of India and China, and a number of other regions. These data suggest that many countries will have to manage water resources far more efficiently than they do now if they are to meet their future needs.

Water for agriculture is critical for food security. Agriculture remains the largest water user, with about 70% of the world's freshwater consumption. According to recent Food and Agriculture Organization (FAO) data,[2] only 30 to 40% of the world's food comes from irrigated land comprising 17% of the total cultivated land. In the future, water availability for agriculture will be threatened by the increasing domestic and industrial demand. Use of water for irrigation in 45 countries accounting for 83% of the world's 1995 population is forecast to increase by 22% between 1995 and 2025.[3]

The demand and pressure for irrigation are increasing to satisfy the required growth of food production, because there is little growth in cultivated areas worldwide (0.1%/year). Between 1961 and 1999, a twofold increase of the total irrigated area in the world was observed, up to 274 million ha, whereas irrigated area per capita remained almost constant at 460.7 ha/1000 people (Figure 1.1[4]).

One of the broad strategies to address this challenge to satisfy irrigation demand under conditions of increasing water scarcity in both developed and emerging countries is to conserve water and improve the efficiency of water use through better water management and policy reforms. In this context, water reuse becomes a vital alternative resource and key element of the integrated water resource management at the catchment scale.[5,6] New management strategies of irrigation must be developed and well integrated in the global water cycle.

Figure 1.2 shows the main sources of irrigation water and its interactions with the global water cycle. These sources can be classified in two major groups with specific advantages and constraints:

1. Natural sources of irrigation water:

   Rainwater accounts for an important portion of the water used to satisfy irrigation demand. However, its contribution could be considered important only in temperate climates under specific climate conditions and, predominantly, at small scale.

   Surface water (lakes and rivers) plays a major role for irrigation in both temperate and dry climates. However, surface water resources

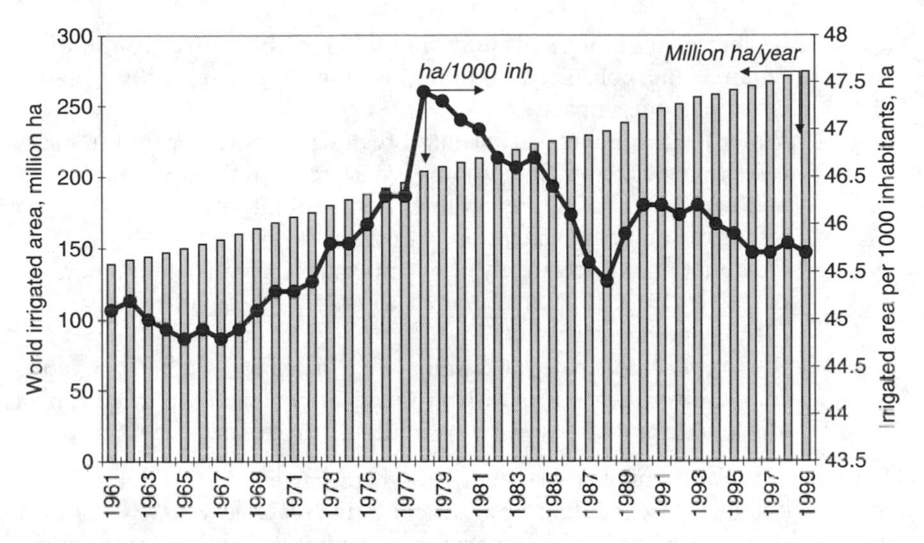

Figure 1.1 Evolution of the irrigated areas and worldwide specific irrigated areas per 1000 inhabitants (1961–1999).

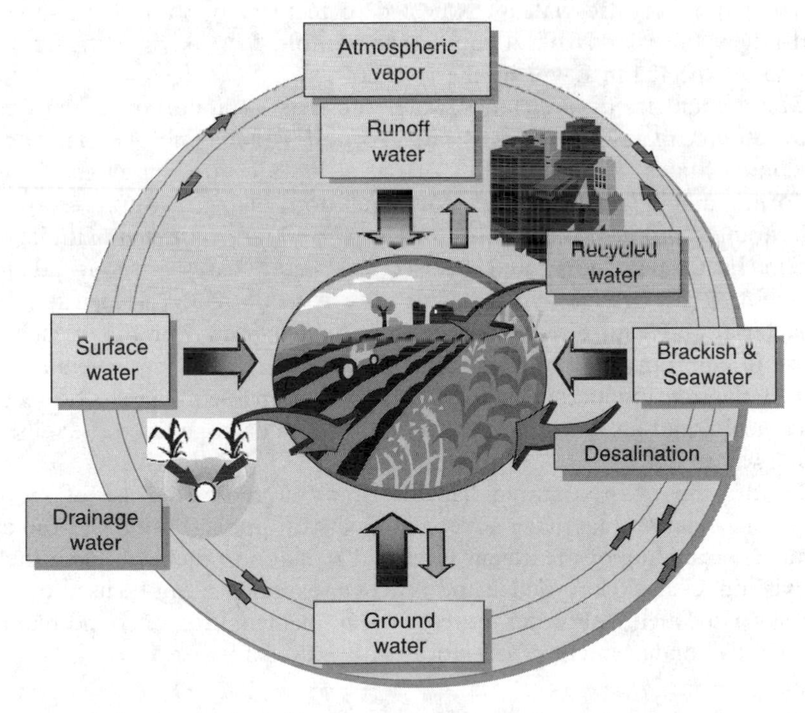

Figure 1.2 Irrigation cycle and its integration into the natural water cycle.

are becoming more and more limited, and their effective use often requires the construction of dams and reservoirs with negative environmental impact.

Water from aquifers has local and regional importance, but in many cases is associated with a progressive decrease in the water table level and withdrawal of nonrenewable fossil groundwater.

2. Alternative sources of irrigation water:

   Desalination has relatively low importance for irrigation because of its high cost; thus, it is limited to only a few small-scale cases in islands and coastal areas.

   Reuse of municipal wastewater and drainage water is a cost-competitive alternative, with growing importance for irrigation in all climatic conditions at both small and large scales.

Consequently, for a number of countries where current freshwater reserves are or will be in the near future at a critical limit, recycled water is the only significant low-cost alternative resource for agricultural, industrial, and urban nonpotable purposes. Water reuse satisfies 25% of the water demand in Israel. The contribution of water reuse is expected to reach 10 to 13% of the water demand in the next few years in Australia, California, and Tunisia. In Jordan, the volume of recycled water is expected to increase by more than three- or fourfold by the year 2010. A more than tenfold increase in recycled water volume is expected in Egypt by the year 2025.

Many countries have included water reuse as an important dimension of water resource planning (e.g., Australia, Jordan, Israel, Saudi Arabia, Tunisia, the United States). Over 1.7 Mm^3 of recycled water are reused each day in California and Florida, mainly for irrigation of agricultural crops and landscaping. Millions of hectares of crop land are irrigated with sewage effluent in China, India, and Mexico, in many cases without adequate treatment. It is worth noting that irrigation with untreated wastewater leads to bacterial and viral diseases and helminth infections. For this reason, the choice of appropriate treatment of wastewater and the implementation of sound irrigation practices are the two major actions necessary to protect the public health and prevent nuisance conditions and damage to crops, soils, and groundwater.

Besides the well-recognized benefits of water reuse, the use of recycled water for irrigation may have adverse impacts on public health and the environment, depending on treatment level and irrigation practices. Nevertheless, the existing scientific knowledge and practical experience can be used to lower the risks associated with water reuse by the implementation of sound planning and effective management of irrigation with recycled water.

1.2 OBJECTIVES AND CONTENTS OF THIS VOLUME

The purpose of this book is to provide guidelines for the use and management of recycled water for both landscape and agricultural irrigation. The target audience includes water and wastewater engineers, administrators and planners, operators, and recycled water users. Multidisciplinary knowledge in this field is summarized and can be easily used by students and researchers. To bridge the gaps between fundamental science (biology, agronomy, environmental engineering, etc.) and the relatively uncharted areas of economic, institutional, and liability issues and facilitate the successful planning and operation of water reuse projects for various purposes, these guidelines are intended to provide the needed information, analysis of the existing knowledge, and recommendations for agricultural and landscape irrigation. Emerging issues and concerns are also covered, such as adverse effects on plants, groundwater, environment, and public health.

The information is presented in comprehensive tables, charts, figures, photographs, and syntheses of best practices. The reader will find the most important information on water reuse for irrigation and learn how to avoid the health and environmental impacts of irrigation with recycled water, improving economic viability and operational reliability, as well as crop production and landscape plant quality.

The guidelines are intended to be a key reference for the use of recycled water for irrigation in both developed and emerging countries. This volume differs from existing publications on water reuse in the synthetic presentation of the key topics, including recent advances in research and existing operation and management experience. In summary, this publication brings together most of the available information on good engineering and agronomic practices that make possible the use of recycled water for irrigation with minimal negative impact on health, vegetation, soils, and groundwater.

The guidelines for good irrigation practices cover the following topics:

Chapter 2—Presentation of water quality parameters of health and agronomic significance
Chapter 3—Summary and analysis of international guidelines and regulations
Chapter 4—Guidelines for health protection giving the main aspects of the choice of appropriate treatment, as well as other restrictive measures such as the control of crops and public access
Chapter 5—Guidelines for good agronomic practices giving the main restrictions on water quality and recommendations for the selection of crops, irrigation technique, and management practices
Chapter 6—Specific aspects of landscape and golf course irrigation
Chapter 7—The main treatment processes for production of recycled water for irrigation and recommendations for design and troubleshooting
Chapter 8—Potential adverse effects of irrigation with recycled water and their management
Chapter 9—Economic and financial aspects of water reuse
Chapter 10—Societal aspects of irrigation with recycled water
Chapter 11—Legal and institutional issues

Chapter 12—Three representative case studies
Chapter 13—Conclusions and summary of practices

Information provided in the guidelines is applicable for any quality of recycled water from primary to tertiary treated wastewater and thus must be interpreted according to the degree of treatment considered by water-recycling planners.

The guidelines illustrate the complexity of developing water reuse projects, the need for multidisciplinary knowledge, and the importance of prudent decision making regarding public health, environmental impact, economic and financial concerns, social and legal aspects, and design and planning. It is important to recognize that the economic viability and overall benefits of water reuse vary from country to country based on their special needs and local conditions.

Because of the rapid development of the technical aspects of water reuse in recent years, this book should not be considered as a mandatory doctrine. Rather, it represents a general overview of the various aspects that should be included in water reuse projects for irrigation and provides examples and recommendations valuable at the present point of time.

No discussion can address every facet of water reuse. Similarly, the guidelines cannot replace the comprehensive planning, design, and operational programs necessary for any water reuse project. Although essential elements of irrigation with recycled water for various purposes are identified and addressed, there is no substitute for good professional experience and judgment.

1.3 ROLE OF WATER REUSE FOR IRRIGATION

The majority of the water reuse projects developed in the world are for agricultural irrigation. In some arid and semi-arid countries, such as Israel and Jordan, the reuse of treated municipal wastewater provides a large share of irrigation water. In this particular sector, water reuse becomes a vital resource to enhance agricultural production, providing a number of additional benefits such as increased crop yields, improved health safety, and decreased reliance on chemical fertilizers.

Recycled water can be used for both landscape and agricultural irrigation purposes. The list presented in Table 1.1 provides an overview of the major uses. It is important to stress that all irrigation uses for crops, landscapes, and lawns can be satisfied with recycled water if the appropriate management practices are applied.

Recycled water has successfully irrigated a wide array of crops with a reported increase in crop yields from 10 to 30%. It is worth noting, however, that the suitability of recycled water for a given type of reuse depends on water quality and the specific use requirements. Indeed, water reuse for irrigation conveys some risks for health and environment, depending on recycled water

Table 1.1 Possible Uses of Recycled Water for Irrigation

Landscape irrigation	Agricultural irrigation
Community parks and playgrounds	Farms:
Schoolyards and athletic fields	Pastures
Golf courses and golf-related facilities	Fodder, fiber, and seed crops
Other turfgrass areas for sport fields	Crops that grow above the ground such as fruits, nuts, and grapes (orchards and vineyards)
Cemeteries and churches' green areas	Crops that are processed so that pathogenic organisms are destroyed prior to human consumption
Freeway landscaping and street median strips	Vegetable crops
Common area landscaping	Nurseries
Commercial building landscaping	Greenhouses
Industrial landscaping	Community vegetable gardens
Residential landscaping	Commercial woodlands for timber
Open areas	Commercial vegetation for energy (burning)
Woodlands	Plant barriers against wind
River and dry-river banks	Overirrigation for groundwater recharge
Overirrigation for groundwater recharge	

quality, recycled water application, soil characteristics, climate conditions, and agronomic practices.

The main water quality factors that determine the suitability of recycled water for irrigation are pathogen content, salinity, sodicity (levels of sodium that affect soil stability), specific ion toxicity, trace elements, and nutrients. All modes of irrigation may be applied depending on the specific situation. If applicable, drip irrigation provides the highest level of health protection, as well as water conservation potential.

Two separate initiatives are being considered to enhance public health protection. The first is the setting up of a legislative framework through the adoption of stringent standards and regulations that are adapted to the characteristics of different crops. In general, standards are most stringent for vegetables for direct human consumption. For example, South Africa requires potable water quality for this use, and the Californian Water Recycling Criteria[7] calls for almost total removal of total coliforms (2.2 TC/100 mL). The second initiative combines the wastewater treatment process, the irrigation system, the exposed group, and the crops to be irrigated. Localized irrigation such as drip irrigation, which is effective, is emphasized. This type of irrigation avoids direct contact between humans and recycled water and limits contact with cultivated crops.

In addition to public health risk, insufficiently treated effluents can also have detrimental effects on the environment. For example, high salinity in the effluent can lead to a decrease in productivity for certain crops, destabilizing the soil structure. Another important adverse effect is groundwater pollution.

This is the case, for example, of the aquifers in Egypt due to the irrigation of Nile Delta desert fringes with about 800,000 m^3/d of untreated wastewater from the Greater Cairo.

1.4 BENEFITS AND CONSTRAINTS OF IRRIGATION WITH RECYCLED WATER

Planning water reuse projects for irrigation purposes can be very valuable for planners and local authorities, as it has numerous associated benefits.[8–10]

Table 1.2 summarizes some of the most important benefits of water reuse, as well as the major constraints for implementation of water reuse projects. It is important to stress, however, that not all water reuse practices generate immediately detectable benefits. Moreover, a number of constraints should be taken into account. For the successful implementation of water reuse, the main advantages should be balanced against negative impacts or other constraints.[9,11]

According to some water reuse specialists, benefits to be gained from the retrofit of landscape irrigation to recycled water use are numerous and may be greater than the benefits of agricultural irrigation. Many water reuse project planners would prefer landscape irrigation as an outlet for the recycled water they will produce, rather than agricultural irrigation and especially irrigation of food crops. There are several reasons for this[12]:

Most expanses of irrigated turf are located within or adjacent to cities where effluent water is produced, so transportation costs are lower.

Recycled water is produced continuously, and, depending on climate, the turfgrass "crop" may be continuous (i.e., uninterrupted by cultivation, seeding, or harvest, all of which mean stopping irrigation for considerable periods).

Turfgrasses absorb relatively large amounts of nitrogen and other nutrients often found in higher quantities in recycled water than in freshwater. This characteristic may greatly decrease the potential for groundwater contamination by use of recycled water.

Depending on recycled water quality, potential health problems arising from the use of recycled water would appear to be less common when water is applied to turf than when it is applied to food crops.

Soil-related problems that might develop due to the use of recycled water would have less social and economic impact if they develop where turf is cultivated than if they develop where food crops are grown.

The main issues and tools to address disadvantages and constraints related to irrigation with recycled water are presented in detail in the following chapters.

Table 1.2 Advantages and Disadvantages of Water Reuse for Irrigation

Main advantages and benefits	Main challenges and constraints
Alternative resource	**Health and regulatory concerns**
Displace the need for other sources of water	Health problems related to pathogens or chemicals in improperly treated wastewater
Reliable, secure, and drought-proof water source	Lack of regulations and incentives for reuse
Fast and easier implementation than new freshwater supply (occasionally)	Water rights: Who owns and recovers the water reuse revenue?
Independence from the current freshwater purveyor (e.g., for political reasons)	Inadvertent exposure or unreliable operation
Water conservation	**Social and legal concerns**
Closing water cycle	Water reuse acceptability
Saving of high-quality freshwater water for potable water supply	Change of the socioeconomic and cropping patterns of farmers
More efficient water use after the retrofit of distribution systems and repair of leaks and breaks	Marketability of crops might be reduced
Health and regulatory concerns	**Economic concerns**
Improved public health (farmers, downstream users)	Cost of recycled water infrastructure (additional treatment, dual distribution) and O&M, including cross-connection control
Enhanced policy awareness, compatibility with water/wastewater treatment policies and regulations	Difficult revenue and cost recovery (uncertain water reuse patterns)
Economic value	Seasonal variations in demand and need for large storage capacity
Avoided costs for new freshwater resource development, transfer, and pumping	Inadequate water pricing: e.g., low price of water for farmers
Lower water treatment costs for downstream users	Change in market (in particular agriculture) can affect water reuse programs
Avoided costs for advanced wastewater treatment and discharge	Liability for potential loss of potable water revenue
Reduced or eliminated application of commercial fertilizers	Need for well-adapted economic approach
Additional revenue from sale of recycled water and agricultural products	**Environmental and agronomic concerns**
Secondary economic benefits for customers and industries in case of continuous supply during drought	Recycled water quality, especially salts and boron, can have negative effect on crops and soils
Improvement of tourism activity in dry regions	Surface and groundwater may be polluted by several chemical and biological components if irrigation is not properly managed (leaching)
Increase in land and property values	

(*continued*)

Table 1.2 Continued

Main advantages and benefits	Main challenges and constraints
Environmental value	**Technical concerns**
Reduced pollutant discharge into receiving bodies	Reliability of operation
Improved recreational value of waterways	Appropriate choice and design of treatment technologies
Avoided impact of developing new freshwater resources (dams, reservoirs, etc.)	
Alternative water supply for environment enhancement	
Alternative to restrictions in wastewater discharge permits (volumes, nutrients)	
Effective use of nutrients contained in wastewater for irrigation leading to higher crop production and low fertilizer application	
Additional treatment of wastewater through irrigation before dilution with groundwater	
Provide a link between rural and urban areas with joint benefits	
Sustainable development	
Source of additional water that contributes to the sustainable development of dry regions (irrigation, industries, tourism)	
Increased food production	
Improved aquatic life and fish production	

1.5 SPECIFICS OF WATER REUSE PLANNING

Numerous state-of-the-art technologies enable wastewater to become a sustainable water resource for a number of reuse purposes and thus allow high-quality freshwater to be reserved for domestic uses. The development and implementation of water reuse projects, however, remain problematic. The main constraints are the economic viability, availability of funding, sensitive health and environmental issues, and, in some cases, public acceptance. Therefore, planning a water reuse project should be undertaken carefully and along with a well-established methodology. The key components of successful water reuse planning include not only the technical know-how and good engineering design, but also a rigorous market analysis and economic, environmental, and social considerations.

In order to meet the goals of resource substitution, environmental protection, and cost recovery, water reuse projects must be able to distribute sufficient volumes of recycled water and sell it at an adequate price.[9] Failure to attain either planned distribution or sales goals may be a reason to reassess the long-term viability of the system, while handicapping further reuse development in the community.

Whether water reuse will be appropriate in a given situation depends on careful economic considerations, potential types of water reuse, stringency of wastewater discharge requirements, and public policy and acceptance. The desire to conserve rather than develop available water resources may override economic and public health considerations.

A feasible water reuse system needs to provide an acceptable balance between the following main considerations:[9]

1. Economic, social, and environmental benefits
2. Costs
3. Project risks

The main issue to be considered during water reuse planning is a good definition of project objectives and its ability to resolve existing problems and expected benefits (Figure 1.3). On the basis of analysis of local conditions favoring water reuse (water resources and needs, water demand projections, wastewater treatment and disposal, new sewer systems, ocean outfalls) and water reuse market assessment, alternative scenarios will be formulated and the best solutions will be chosen by means of a multicriteria analysis.

It is wise to adopt a systematic and holistic approach when planning a water reuse project from the very beginning. Planning usually evolves through three main phases (Figure 1.4):

1. Phase I: Conceptual planning
2. Phase II: Feasibility investigations
3. Phase III: Facility planning

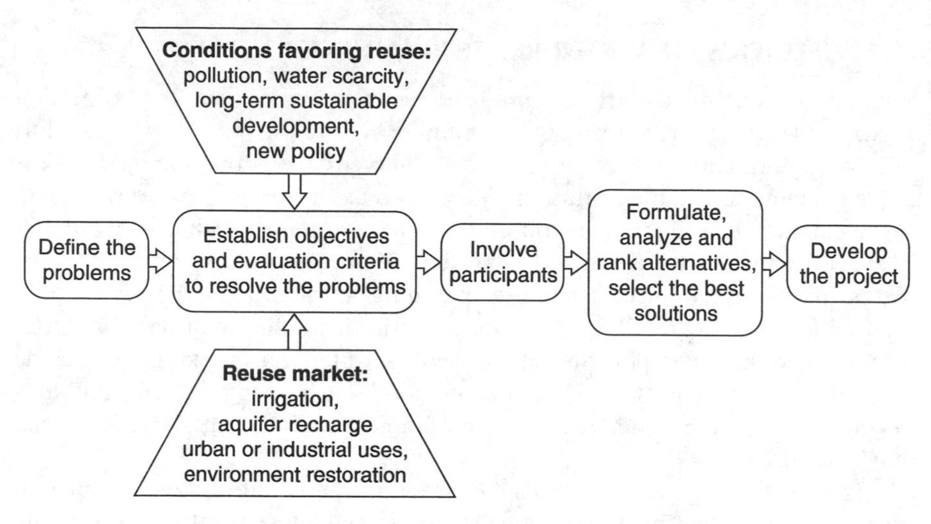

Figure 1.3 Strategy for definition of water reuse project objectives.

The multicriteria screening and evaluation of alternative reuse and nonreuse options involve the following feasibility criteria:

Engineering feasibility: possibility for implementation of wastewater treatment, storage, and distribution
Economic feasibility: reasonable investment and O&M costs
Financial feasibility: available funding and subsidies
Environmental impact: potential negative effects on soils, groundwater, crops, or ecosystems, as well as environmental benefits
Institutional feasibility: water policy, water rights, regulations, enforcement
Social impact and public acceptance: support by stakeholders
Market feasibility: who will use recycled water and under what conditions

The first five criteria in the list are common to all water resource projects. The crucial elements for water reuse are the public acceptance and the market analysis and identification of potential users of recycled water. In this respect, the factors of public health, water quality, and public acceptance create additional complexity in identifying and securing water reuse market, which is generally not the case with freshwater supply. For this reason, the water reuse planning strategy should include a rough market assessment early in the planning process and early involvement of project participants (stakeholders) to better formulate project objectives and the possible alternative scenarios that correspond to local conditions and specific problems to be resolved. As a rule, water reuse should be a cost-competitive solution for long-term water planning and have multiobjective purposes (either complementary water supply, part of water conservation programs, or valorization of existing wastewater treatment and environmental protection).

The conceptual planning, phase I (Figure 1.4), is not an in-depth study; however, it is a very important step. It consists of a preliminary evaluation

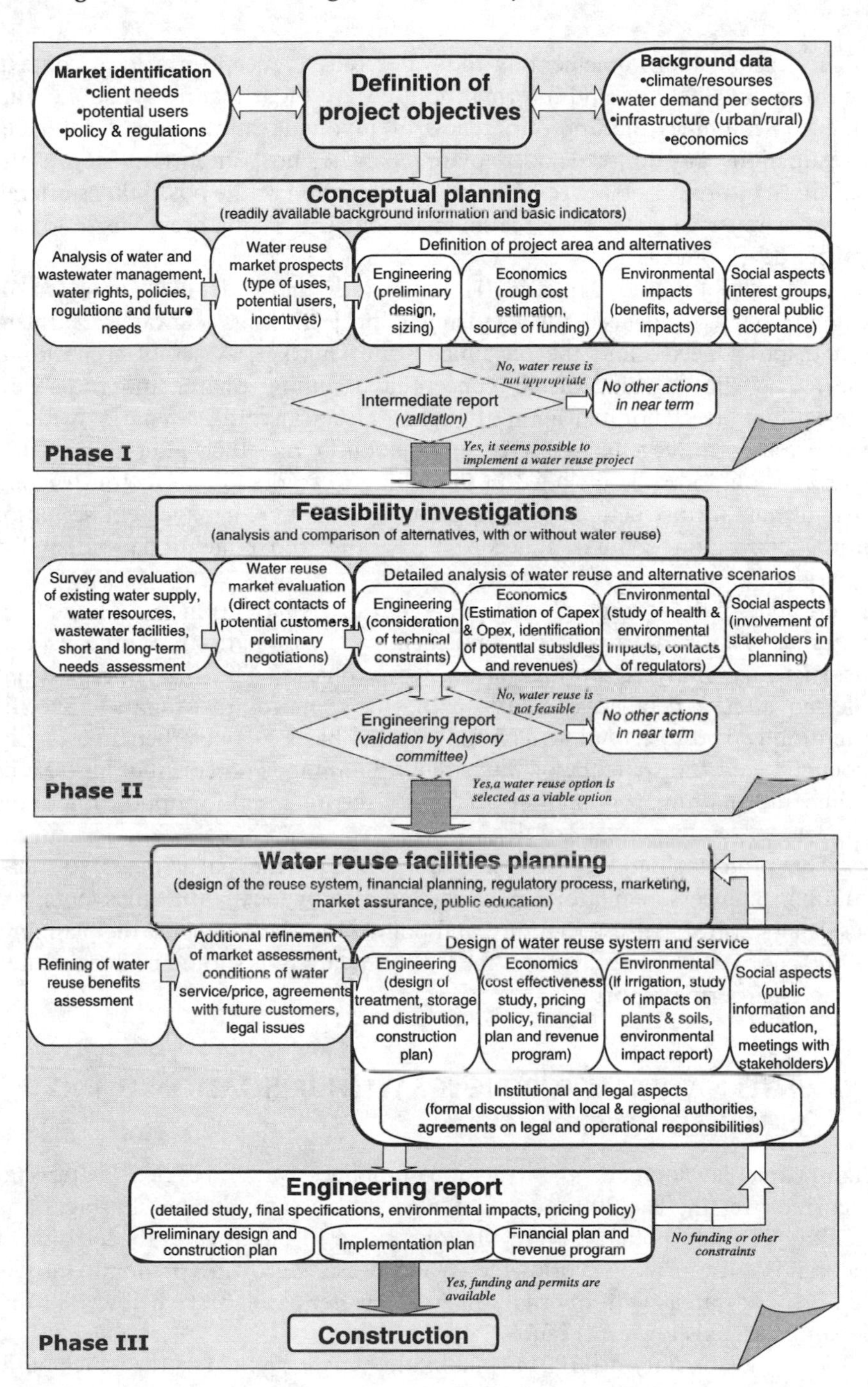

Figure 1.4 Planning phases of water reuse projects. (Adapted from Refs. 9 and 10.)

of the feasibility of implementing the water reuse concept in a local context. At this point, the data and information used are those readily available. This first step examines the appropriateness of involving more funds for further investigations and more detailed planning. The most important step is the definition of project objectives. Having a clear vision of the possibilities offered by water reuse to water resource management is a major prerequisite to any project development.

A feasibility study (phase II) comprises more detailed analyses of water resources and needs: water demand projections, wastewater treatment and disposal needs, and the reclaimed water market. A set of scenarios is selected as the outcome of the conceptual planning phase, and some new alternatives are elaborated upon after the market analysis. Scenarios without water reuse or recycling with the mobilization of other alternative water sources should also be included in this phase (e.g., desalination, construction of hydraulic infrastructures such as dams). The water management scenarios are assessed and compared according to the above-mentioned feasibility criteria.

Fewer solutions are selected for the third phase (phase III) facilities planning. Phase III involves complementary investigations on the aspects insufficiently analyzed during the previous phases. The best alternative is selected after a detailed comparison of the scenarios investigated. Specific attention is paid to economic feasibility on the basis of a cost/benefit analysis. Openness and transparency of water reuse planning is an essential part of the public information program that will reduce the potential for opposition to the project.

The project should be reviewed not only by the participants (owners, funding, engineers, regulatory agencies), but also by local authorities, potential customers, water user associations, and politicians. At the end of the planning, all the basic data and results from the feasibility analysis should be documented in the engineering report.

1.6 INTERNATIONAL EXPERIENCE WITH IRRIGATION USING RECYCLED WATER

Significant development of irrigation practices with recycled water has occurred over the last 20 years, stimulated by increasing water shortages and facilitated by new policies and regulations. Figure 1.5 illustrates the location of the major reuse projects worldwide. Water reuse is growing predominantly in relatively dry areas with internal renewable water availability below 1700 m^3/person/year, as shown in Figure 1.5.

Water availability of 1700 m^3/inhabitant/year is defined as the benchmark, or water stress index, below which most countries are likely to experience water stress on a scale sufficient to impede development and harm human health.[13] At levels less than 1000 m^3/inhabitant/year of water availability, countries experience chronic water scarcity. The World Bank[14] has accepted this value

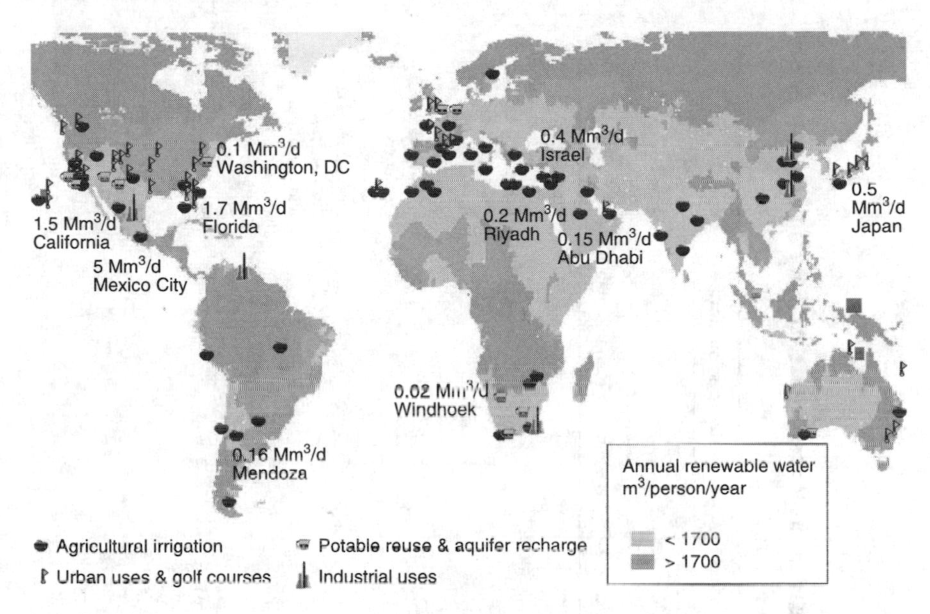

Figure 1.5 Major water reuse projects worldwide: geographic location, type of reuse, and capacity of some leading projects.

as a general indicator of water scarcity. Absolute water scarcity occurs below 500 m^3/inhabitant/year, while the minimum survival level for domestic and commercial use, as mentioned previously, is defined at 100 m^3/inhabitant/year as a rough requirement for basic household needs to maintain good health.[13] These levels should be considered rough benchmarks and not precise thresholds. The exact level of water stress varies from region to region, depending on the climatic conditions, level of economic development, and other factors. Comprehensive programs of water conservation and more efficient technologies can also ease water stress.

Countries with water reuse experience are presented in Table 1.3. Their respective water reuse activity and applications are rated, followed by some specific observations. In most cases, water reuse has been developed in countries with water stress. In this case, another approach is used to evaluate water scarcity: the exploitation rate of renewable water resources, i.e., the ratio between the volume of available renewable water and annual use. When the exploitation rate exceeds 20% of existing reserves, water management becomes a vital element in country's economy.[15,16] This is the case in a number of Mediterranean countries, while in the Middle East the water supply is ensured by the exploitation of nonrenewable deep aquifers and desalination (shown by an exploitation rate of the renewable resources of over 100% in Egypt, Jordan, Oman, Saudi Arabia, and United Arab Emirates). With few exceptions (Canada, France, Sweden, and the United Kingdom), the agricultural sector is characterized by the highest percentage of water consumption, from 60 up to 93%.

Table 1.3 Overview of Water Reuse Worldwide, 1980–2000: Countries Leading in Water Reuse Activity and Countries Still Using Raw Sewage for Irrigation

				Annual water withdrawal[15]		Percent water consumption by sector[15]			Wastewater reuse applications[a]			
Country/ State	Reuse activity[a]	Reuse guidelines or regulations	GNP$/inh (1995)[16]	% of renewable water resources	m^3/inh	Urban	Industry	Agriculture	Irrigation	Urban	Potable	Comments
Argentina	Medium	Yes	8030	4	822 (1995)	16	9	75	++			Large project in Mendoza (irrigation)
Australia	High	Yes	18,720	4	839 (1995)	12	6	70	++	+		Virus regulation for urban use, reuse of 11% of wastewater
Belgium	Medium	No	24,710	75	917 (1980)						+	Increasing issue
Canada	Medium	Yes	19,380	2	1623 (1990)	11	68	7		+		Increasing issue
Chile	Low		4160	2	1629 (1987)	5	11	84	+			Increasing reuse concern
China[b]	Medium	No	620	19	439 (1993)	5	18	77	++			Industrial reuse and irrigation
Egypt[b]	Medium	No	790	3061	920 (1993)	6	8	86	+			Increasing issue
France	Medium	Yes	24,990	23	704 (1994)	15	73	12	++	+		Mostly irrigation
India[b]	High	No	340	40	588 (1990)	5	3	92	+++			Untreated wastewater
Israel	Intensive	Yes	15,920	88	292 (1997)	29	7	64	+++	++		70% of wastewater is recycled
Italy	Medium	Yes	19,020	30	840 (1996)	14	33	53	++			Increasing issue
Japan	High	Yes	39,640	21	735 (1992)	19	17	64	++	+++		Mostly urban uses, toilet flushing
Jordan	High	Yes	1510	145	187 (1993)	22	3	75	++			Agricultural irrigation
Kuwait	Intensive	Yes	17,390		307 (1994)	37	2	60	+	++		High water needs, desalination

Namibia	Medium	No	2000	4	185 (1990)	[illegible]	3	68	+		+	Windhoek direct potable reuse
Mexico[b]	Medium	No	3320	19	812 (1998)	[illegible]	5	78	+++			Reuse of raw effluent, Mexico City
Morocco[b]	Medium	No	1110	37	446 (1992)	[illegible]	3	92	++	+		Reuse of raw effluent
Oman	High	Yes	4820	124	658 (1991)	[illegible]	2	93	++	+		High water needs, desalination
South Africa	High	Yes	3160	30	391 (1990)	[illegible]	11	72	++	+		Potable reuse prohibited, industrial uses
Saudi Arabia	Intensive	Yes	7040	708	1002 (1992)	[illegible]	1	90	+++	+−		High water needs, desalination, reuse of 10% of wastewater
Spain	High	Yes	13,580	32	897 (1997)	[illegible]	18	68	++	+		Regional regulations
Sweden	Low		23,750	2	310 (1995)	[illegible]	30	4	+			Environmental protection
Tunisia	Medium	Yes	1820	76	295 (1995)	[illegible]	3	83	+−			Reuse of 20–30% of collected wastewater
United Arab Emirates	Intensive	Yes	17,400	1405	954 (1995)	[illegible]	10	67	+	++		High water needs, desalination
United Kingdom	Medium	Yes	18,700	6	160 (1995)	[illegible]	8	2	+	++		Greywater recycling
United States	High	Yes	26,980	19	1844 (1990)	1	44	40	++	+	+	State guidelines and regulations
Arizona	High	Yes							++	+	+	Mostly irrigation
California	Intensive	Yes							+++	+	++	All types of reuse
Florida	High	Yes							++	++	+	Mostly residential reuse

[a]Subjective evaluation based on literature,

[b]Countries using raw sewage for irrigation,

GNP, Growth National Product; +++, numerous projects; ++, occasional projects; +, isolated projects.

1.6.1 Water Reuse in Europe, the Mediterranean Region, and the Middle East

In Europe the development of water reuse is being driven by the need for alternative water resources together, in most cases, with the need to protect receiving water bodies from treated effluent discharge, which must meet increasingly stringent water quality regulations. In the Mediterranean region, an imbalance between water resource availability and needs is the main reason for the considerable increase in water reuse. Consequently, water reuse is growing steadily in densely populated northern European countries like Belgium, England, and Germany and more rapidly in tourist coastal areas and on islands in western and southern Europe.

Figure 1.6 illustrates the location of the major reuse projects in Europe. As a rule, water reuse for irrigation has been practiced in most Mediterranean region where water is short. Even if the average water availability of the southern European countries is over 3000 m^3/inhabitant/year, many regions experience chronic water shortages due to recurring droughts and high water demand during the summer tourist season.

Agricultural irrigation with municipal wastewater had been practiced in Europe (Paris, Reims, Milan) and the Mediterranean region for centuries.[6] In recent years, Israel, Jordan, and Tunisia are the leaders in the Mediterranean region in agricultural water reuse, satisfying 20, 10, and 1.3% of the total water demand, respectively. Cyprus has also developed a sound water reuse strategy, which will satisfy 11% of the total water demand with recycled water in the

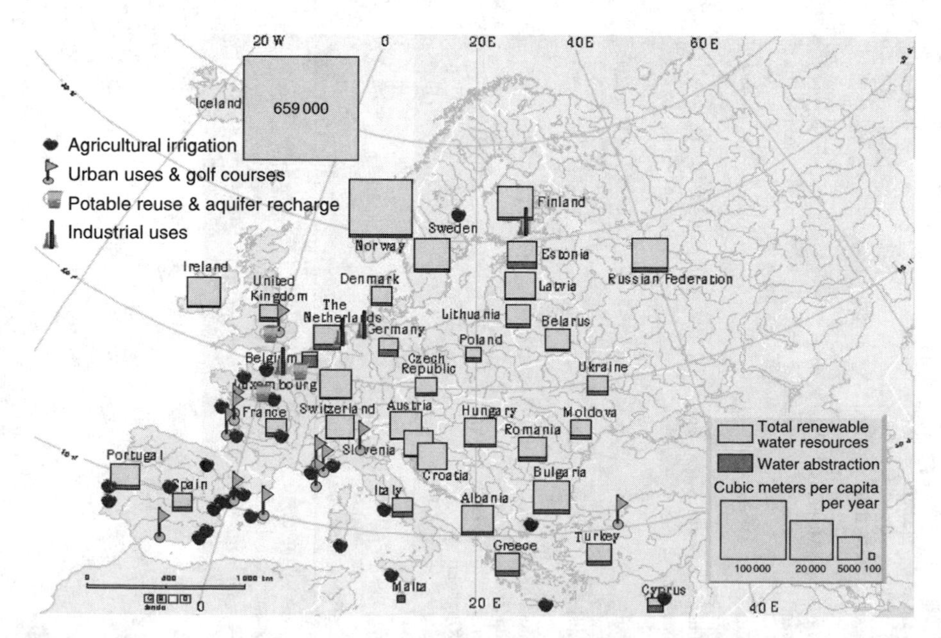

Figure 1.6 Location of major water reuse projects in Europe.

next few years, mainly for irrigation purposes. Another emerging reuse leader in this region is Spain, where new water policies and legislation promote water reuse. In fact, 22% of collected wastewater there is reused in agriculture, with an overall volume of 340 Mm^3/year.

Numerous small projects have been implemented in France and Spain. One of the largest and more recent projects in Europe was started in 1997 in Clermont-Ferrand, France, where more than 10,000 m^3/d of tertiary treated urban wastewater are reused for irrigation of 700 ha of maize. Another large reuse project (250,000 p.e., 20 Mm^3/year of recycled water) was implemented in 1995 in the city of Vitoria in northern Spain. From the total volume of recycled water, 8 Mm^3/year are used for unrestricted irrigation after advanced wastewater treatment and disinfection.

In Italy, more than 4000 ha of various crops are irrigated with recycled water. However, the controlled reuse of municipal wastewater in agriculture is not yet developed in most Italian regions due to stringent standards, including very high levels of disinfection. One of the largest projects in Italy was implemented in Emilia Romagna, where over 450,000 m^3/year of treated effluents are used for irrigation of more than 250 ha.

Israel is the world's leader in water reuse on a percentage basis, with over 70% of collected wastewater recycled in agriculture. The two largest reclamation systems are located in the Dan Region of the Greater Tel Aviv area (95 Mm^3/year, 220 km^2) and in the Kishon Complex near Haifa (32 Mm^3/year). The wastewater in the Dan Region (1.5 million inhabitants) is treated with activated sludge. The secondary effluent is spread over infiltration basins to the regional groundwater aquifer, which serves as multiyear storage system. This soil-aquifer treatment produces high quality effluent, which may be used to irrigate all crops without restriction, including vegetables eaten uncooked.

The reuse in Greater Haifa started in 1983. The reclamation system consists of biological treatment by activated sludge and trickling filters. The recycled effluent is stored in a large seasonal reservoir before use in summer for irrigation.

In some Mediterranean countries, such as Tunisia and Jordan, water reuse has been made an integral part of the water management strategy. In Tunisia, the National Sewage and Sanitation Agency (ONAS), under the Ministry of Agriculture, is responsible for wastewater collection, treatment, and disposal. Currently, 61 wastewater treatment plants that treat approximately 156 Mm^3/year are actively implementing water reuse for agricultural and golf course irrigation.

In Tunisia the use of recycled water for recreational purposes and more particularly for golf course irrigation is an important component of the development of tourism. Since the beginning of the 1970s and with the development of tourism, a policy was set up for water reuse for golf course irrigation. Golf course irrigation requires a high rate of water reuse and a water demand that lasts all year long through varying climatic conditions. Eight existing golf courses are irrigated with secondary treated effluent. Some are irrigated with recycled water blended with conventional water (surface or groundwater) or with desalinated water (Jerba). Secondary treated wastewater, conveyed and stored in ponds on the golf course during different periods depending on the operational regime, is used for irrigation. Precautions for use, such as night irrigation with low-range sprinklers, are applied. Irrigation water is in compliance with the WHO[17] guidelines for water reuse in recreational areas with free access to the public (2.3 log units of fecal coliform/100 mL) during winter and part of spring.[18] Polishing secondary effluents through lagooning or seasonal storage would lower health hazards and help increase the demand for recycled water.

Agricultural irrigation is also a growing practice in the Middle East: 20,000, 26,000, 55,000, 67,000, and 115,000 m^3/d are reused in Taif, Dubai, Jubail, Doha, and Abu Dhabi, respectively. In these applications, chlorinated tertiary effluent is used for irrigation. The required water quality of the disinfected tertiary effluent varies from 100 FC/100 mL in Dubai to 2.2 FC/100 mL in Taif and Jubail. An important step in the extension of reuse practices has been taken in Riyadh, the capital of Saudi Arabia, where 200,000 m^3/d of treated wastewater is used to irrigate 175 farms with an area of 1200 ha.

The construction of the world's largest water reuse project started in 2003 at Sulaibiya, near Kuwait City in Kuwait.There 375,000 m^3/d of high-quality recycled water will be produced by conventional biological treatment followed by prefiltration by disk filters, ultrafiltration and reverse osmosis (RO).[19] RO will be used to desalinate municipal effluents, which has an average TDS (total dissolved solids) of about 1280 mg/L with maximum values up to 3000 mg/L. The recycled water will be used mostly for agricultural irrigation after blending with existing brackish water.

Analysis of the existing practices of irrigation with recycled water in the Mediterranean region shows that water recycling has a vital part to play as a reliable alternative source for irrigation. Moreover, water recycling is a good preventive measure against degradation of collecting water bodies and environment. Water reuse projects have been successful not only in arid and semi-arid regions, but also in regions with temperate climates to protect sensitive areas, to expand recreational activities and water-intensive economic sectors, and to cope with water crises caused by repeated droughts. In several Mediterranean countries, recycled water is considered the only significant low-cost alternative water resource for irrigation.

1.6.2 Water Reuse in the United States

Numerous factors, including chronic and temporary water shortages, fast-growing water demand in urbanized areas, more stringent standards for wastewater discharge, and increased costs for mobilization of new water resources and environmental constraints, are giving impetus to explore the use of recycled water in the United States to augment the existing water resources and provide environmentally acceptable wastewater disposal.

Irrigation and thermoelectric power are the largest consumers of freshwater in the United States. Irrigation accounts for 42% of the total use, whereas thermoelectric power accounts for 52% of the return flow. Residential water use is 40% for potable uses and 60% for toilet flushing and outdoor use. In Florida and California, 60 and 44%, respectively, of all potable water produced is used for outside purposes, principally irrigation of lawns and gardens.

Table 1.4 illustrates the main reuse applications and volumes of recycled water in six U.S. states. Four states included water reuse in their official water policies: California, Florida, Hawaii, and Washington. Florida and California reused the greatest volumes of treated wastewater, 1.67 and 1.52 Mm^3/d, respectively, in 1997, while Nevada, Florida, and Arizona have the highest percentage of treated wastewater that is reused for different purposes, 80, 40, and 35%, respectively.[20–23]

The state of Washington intends to significantly accelerate its water reuse program as a result of a drought in 2001. This state has regulations (Reclaimed Water Act 1992) with four classes of recycled water (stringent water quality similar to those of the California Water Recycling Criteria[7]). New state legislation (House Bill 1852) promotes water conservation and reuse. The total capacity of potential reuse projects is estimated at 95,000 m^3/d, with the greatest potential in King County and the city of Tacoma.

The state of Nevada is also expecting the implementation of new water reuse projects, with higher requirements for tertiary treatment and new applications, including industrial reuse. More than 150,000 m^3/d will be recycled in Las Vegas alone.

Agricultural irrigation is the most common current water reuse practice in the United States (Table 1.5). The project capacities vary from 1000 to 190,000 m^3/d. Agricultural irrigation is often coupled with other uses such as golf course and landscape irrigation. The main irrigated crops are food crops, pastures, orchards, fodder and fiber crops, sugar cane, and flowers. It is worth noting that this practice is often associated with the need for storage systems and alternative disposal.

Agricultural irrigation with recycled water is practiced on a large scale in Arizona, California, Florida, and Texas. In 1995, 34% (340,000 m^3/d) and 63% (570,000 m^3/d) of the total volume of recycled water in California and Florida, respectively, were used for various agricultural purposes. Two important irrigation facilities in Florida are those of Water Conserv II near Orlando (> 6000 ha, 130,000 m^3/d) and Tallahassee (1750 ha, 68,000 m^3/d). It is important to mention Hawaii, where some wastewater treatment plants reached zero effluent discharge through water recycling.

Unrestricted landscape irrigation and irrigation of food crops eaten raw require extensive disinfection (2.2 TC/100 mL, California Water Recycling Criteria[7]). To achieve this objective, the most common treatment process includes pretreatment, primary settling, activated sludge, clarification, coagulation/flocculation, filtration, and disinfection (chlorine with residual chlorine concentration multiplied by contact time $CT \geq 450$ min.L/mg). In some cases, additional advanced treatment is used, as in the Goleta project (California), where reverse osmosis is implemented after storage for public health protection.

Table 1.4 Main Characteristics of Water Reuse in Most Proactive U.S. States

State	% treated wastewater that is reused	Total volume (m^3/d)		Main uses of recycled water	Ref.
		1997	2020		
Arizona	35			Agricultural irrigation Landscape irrigation (Tucson, 37 mgd) Groundwater recharge (Phoenix 200 mgd)	20
California	13	1,520,000 200 utilities	5,100,000	48% agricultural irrigation 21% groundwater recharge 12% landscape irrigation	21
Florida	40	1,670,000 451 utilities	9,600,000	40% urban uses 21% agricultural irrigation 21% groundwater recharge 13% industrial uses	22
Hawaii	Kauai, 91 Hawaii, 23 Maui, 12 *Objective 2000: 20*	50,000		Agricultural irrigation Urban uses Industrial applications	23
Nevada	80			Agricultural irrigation Urban uses Industrial applications	20
Texas	8.5	600,000 190 utilities	2,000,000 (in 2010)	41% agricultural and landscape irrigation 38% industrial applications 21% other uses	21

Table 1.5 Examples of U.S. Projects for Agricultural Irrigation

Location	Name of project	Capacity (m^3/d)	Irrigated crops	Start-up	Treatment
California	Antelope Valley (LA)	35,000	Fodder (plus park irrigation, wetlands)	1988	Secondary and tertiary effluents
	Bakersfield	64,000	Cotton, cereals, sugar beet	1912	Secondary effluent
	Castroville, Salinas Valley	68,000	Market gardening (vegetables)	1980	Tertiary disinfected effluent
	Eastern Water District (Perris valley)	111,000 (total)	Agricultural irrigation (golf course, environmental enhancement)		Tertiary disinfected effluent
	Goleta	11,300 (RO 1135)	Agricultural irrigation (golf irrigation)	1991	Tertiary disinfected effluent
	Irvine Ranch (Los Angeles)	13,500	Avocado, maize, orange, cabbage, asparagus (plus landscape irrigation, toilet flushing, industry)	1967	Tertiary disinfected effluent
	Laguna (Santa Rosa)	75,000 (total)	Agricultural irrigation (golf course, environmental enhancement)		Tertiary disinfected effluent
	Monterey County	47,300	4800 ha agricultural crops		Tertiary disinfected effluent + ASR
	South Bay (San Jose)	473,000 (total)	Agricultural irrigation (landscape irrigation, industry)		Tertiary disinfected effluent
	Tuolumne County (Sonora)		Ranch land, 530 ha		Secondary effluent
Florida	Leesbourg	11,400	Meadow, 134 ha	1981	Secondary effluent
	Orlando (Water Cons.II)	190,000 (70% irrig)	Citrus fruits, 6000 ha	1986	
	Sarasota	29,000	Ranch land, fodders, citrus fruits	1987	
	Tallahassee	68,000	Cereals, fodders	1966	Secondary effluent
	Venice (Project Wave)	7600	Agricultural irrigation (golf course, residential)		
Hawaii	Haleakala Ranch (Maui)		Cereals, meadow	1970	
	Honolulu, Hawaii Reserves	1100	Tropical fruits and flowers	1995	Tertiary disinfected effluent
	Kikiaola (Kauai)	950	Sugarcane, cereals	1973	Chlorinated secondary effluent
	Walalua (Hawaii)	17,000	Sugarcane	1928	Chlorinated secondary effluent
Texas	Amarillo	27,600	Agricultural irrigation (golf course, environmental)		
Arizona	Chandler	61,600	Agricultural irrigation (golf course, landscape)		

RO: reverse osmosis; ASR: aquifer storage and recovery

In California, the Irvine Ranch Water District has been using reclaimed water for irrigation of orange, avocado, and row crops since 1967. In Florida, the first authorized projects for food crop irrigation began in 1986.

The main wastewater reclamation and reuse studies carried out on agricultural irrigation in the United States are as follows:

1. Pomona Virus Study, 1976, California. This study served as the basis for a change in California's Water Recycling Criteria to allow direct filtration in lieu of the full coagulation/clarification step prior to filtration.
2. Monterey Wastewater Reclamation Study for Agriculture (MWRSA), 1980–1985, California. The main conclusions from MWRSA are: (1) virtually pathogen-free effluents could be produced from municipal wastewater via tertiary treatment and extended disinfection with chlorine; (2) there is an equivalence between direct filtration as a treatment and long-term safety of field application of reclaimed water in vegetable crops; and (3) food crops that are consumed uncooked could be successfully irrigated with reclaimed water without adverse environmental or health effects (Figure 1.7).

Landscape and golf course irrigation is a rapidly growing reuse application because it is easy to implement, especially wherever potable water is used in urban areas (Figure 1.8 and Figure 1.9). Several hundred small and large projects (e.g., $150\,m^3/d$ for park irrigation in Hawaii and $76{,}000\,m^3/d$ in St. Petersburg, Florida) have been implemented in states where water reuse

Figure 1.7 Lettuce seedlings irrigated with recycled water in northern Monterey County, California.

Figure 1.8 View of a golf course in southern California irrigated with recycled water.

Figure 1.9 Landscape irrigation in Irvine Ranch Water District, California.

is practiced. As a rule, tertiary disinfected effluent is used for all urban reuse purposes. The treatment process allowed in California is coagulation of the secondary effluent followed by filtration and disinfection (chlorine, UV). UV disinfection is becoming one of the most popular and cost-effective disinfection alternatives (e.g., in California, Washington, and Arizona).

1.6.3 Water Reuse in Central and South America

One major example of agricultural reuse of water is Mexico City, Mexico. Almost all collected raw wastewater (45–300 m^3/s dry and wet flows, respectively), is reused for irrigation of over 85,000 ha of various crops. Of the total wastewater generated, 4.25 m^3/s (367,000 m^3/d) will be reused for urban uses (filling recreational lakes, irrigating green areas, and washing cars); 3.2 m^3/s will be used for filling a part of a dry lake called Texcoco and for other uses in the neighborhood; 45 m^3/s of wastewater transported to the Mezquital Valley (at a distance of 6.5 km) has been allocated for irrigation. The reuse of this wastewater represents an opportunity for development of one of the most productive irrigation districts in the country. However, health problems have resulted from this practice because of the use of untreated wastewater.

Driven by water scarcity, the largest water reuse system in South America is located in the arid region of Mendoza, in the western part of Argentina near the Andes. Over 160,000 m^3/d of urban wastewater (1 million inhabitants, 100 Mm^3/year) is treated by one of the largest lagooning system in the world at the Campo Espejo wastewater treatment plant, which has a total area of 290 ha to meet WHO[17] for unrestricted irrigation. Reused effluents in this region are a vital water resource, enabling the irrigation of over 3640 ha of forest, vineyards, olives, alfalfa, fruit trees, and other crops. To avoid contamination of aquifers, best reuse practices are under development, including establishment of special areas for restricted crops and restrictions in the choice of irrigation technologies. An extension-of-water-reuse scheme is planned in the north of the Mendoza City Basin, where treated effluent from the Paramillo wastewater treatment plant (100,000 m^3/d, series of stabilization ponds) is used for irrigation of an oasis of 20,000 ha after dilution with the flow of the Mendoza River.

1.6.4 Water Reuse in Asia and Oceania

Japan has well-organized advanced water-recycling programs, as the country needs to intensively develop regional and on-site systems to face high urbanization and fluctuations in rainfall throughout the year.[24] In 1991, 228,000 m^3/d of recycled water was produced by over 1369 treatment plants, with in-building water-recycling installations accounting for 56% of this production. Sixty-one percent of domestic water-recycling was used for toilet flushing, 23% for irrigation, 15% for air conditioning, and 1% for cleaning purposes.

In Australia, water reuse has received increasing attention since the late 1990s, when new water policy and resource protection legislations were adopted. Recently, growing support has been observed for a number of demonstration reuse projects as a result of drought and need to maintain or increase sustainable yields. The Virginia Pipeline project (start-up 1999) is considered the largest water reuse project in Australia; it includes recycled water use of 120,000 m^3/d, 150 km of distribution network, irrigation of 20,000 ha of vegetables, and advanced tertiary treatment by dissolved air flotation, filtration, and disinfection. To overcome the high economic risks related to the use of recycled water, over US$1.65 million (AU$2.5 million) have been invested in contract agreements and commercial and technical investigations.

The largest urban reuse project, the Rouse Hill project in Sydney, was initiated in 1994[25,26] and recycles tertiary treated municipal wastewater from Rouse Hill Wastewater Treatment Plant for toilet flushing and landscape irrigation. The first phase of this project planned to provide recycled water to 17,000 individual houses with approximately 50,000 inhabitants. Recently, the reuse scheme was extended to serve 35,000 houses. This project was developed with subsidiaries, and recycled water cost was set to 30% of the cost of potable water. Recycled water is being supplied to consumers at a price of US$0.18/$m^3$ (AU$0.27/$m^3$), with a quarterly connection charge of US$3.8 (AU$5.8). Sydney Water has undertaken a detailed risk-management assessment of the entire system to identify actions needed to fine-tune the operation of the recycled water system.

In the Sydney Olympic Park, the Water Reclamation and Management Scheme at Homebush Bay will save up to 850,000 m^3/year by using recycled greywater and rainwater for landscape irrigation and toilet flushing. Overall, it is predicted that the use of freshwater at Sydney Olympic Park and Newington will be halved. The Sydney Olympics athletes' village at Newington has been redeveloped as a permanent suburb with about 6000 residents. Recycled water is supplied for residential garden watering and toilet flushing in the Newington Village from the recycled water system at Sydney Olympic Park at a cost to the consumer of US$0.1/$m^3$ (AU$0.15/$m^3$), less expensive than drinking water at US$0.6/$m^3$ (AU$0.9/$m^3$). The recycled water meets the requirements of the NSW Guidelines for Urban and Residential Use of Recycled Water.

In recent years, more than 55 cities in China, in particular in the northern part of the country, such as Beijing, Dalian, Handan, Qingdao, Shenyang, Shenzhen, Shijiazhuang, Taiyan, Tianjin, and Zhaozhuang, have been practicing water reuse for various purposes, including agriculture. Water shortage and pollution concerns in the Beijing-Tianjin region have driven the implementation of one of the largest reuse projects in China. The Gaobeidian sewage treatment plant (2.4 million p.e., 1.5 Mm^3/d), in operation since 1994, provides over 0.5 Mm^3/d of secondary effluent for reuse in industry and agriculture (50/50%). During the nonirrigation season, half of this volume is discharged into the Tonghui River to supplement its flow.

In India, over 73,000 ha of land were irrigated with wastewater in 1985 on at least 200 sewage farms. Only a small portion of the collected wastewater is treated (24% in 1997). In numerous locations, raw sewage is discharged into rivers and used downstream for agricultural irrigation. Consequently, enteric diseases, anemia, and gastrointestinal illnesses are high among sewage farm workers. The use of raw sewage for crop irrigation is also common in Pakistan, where it is practiced in 80% of urban settlements. The irrigated crops include vegetables, fodder, wheat, cotton, etc.

1.7 MANAGEMENT ACTIONS FOR IMPROVEMENT OF IRRIGATION WITH RECYCLED WATER

The success of water reuse projects and, in particular, irrigation with recycled water depends greatly on the implementation of proper management practices. The main management actions could be structured in three major groups (Figure 1.10):

1. Policy and institutional measures (see Chapters 3 and 11)
2. Engineering initiatives
3. Agronomic practices

As a rule, policy decisions need to be made before the implementation of engineering or agronomic practices. On the other hand, for a number of irrigation practices, clear distinctions between engineering and agronomic actions do not exist.

For each group of measures, management actions can be categorized in three levels, depending on the final objective (see Figure 1.10):

Health protection measures (see Chapter 4), including improved design and operation of wastewater treatment and reuse facilities (see Chapter 7)
Good practices to improve food production (see Chapter 5) and quality of turf grass and landscape ornamentals (see Chapter 6), as well as recommended actions to prevent degradation of soils and water bodies (see Chapter 8)

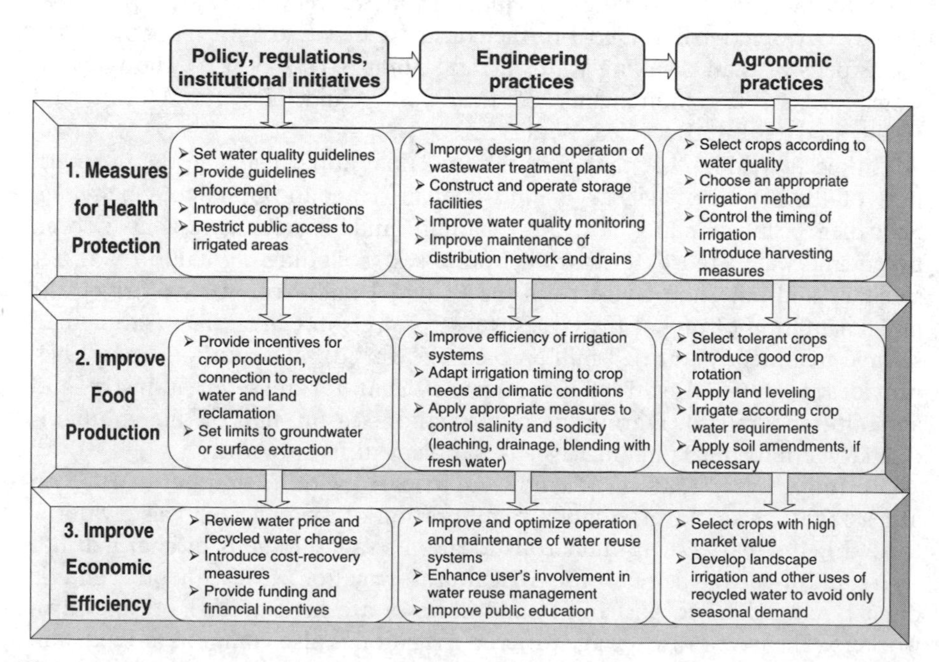

Figure 1.10 Major management activity for improvement of efficiency and competitiveness of irrigation with recycled water.

Management practices aiming to improve economic competitiveness (see Chapter 9) and consequently to enhance public acceptance (see Chapter 10)

The main objective of engineering actions is to enhance water quality, improve water use efficiency, and remove polluted water (drainage, for example). Various decision-support systems are available to improve the efficiency of irrigation. These measures are associated with high capital and operation costs (treatment, storage, distribution) and require strong governmental support for legislative and financial incentives or sanctions to help their implementation. The benefits are not immediate, but in the long term lead to improvement of water resource management and sustainable development.

Agronomic actions have great importance for the mitigation of soil or aquifer degradation and crop production. The key option is crop diversification. The appropriate choice of crops and type of turf grass can help to manage water scarcity and salinity. Of course, this option is feasible only in the presence of a market for such crops. Soil fertilization, leveling, or amendments could further improve crop production.

In most countries, establishing or adopting a regulatory framework is an essential step for the development and social acceptance of water reuse. Decision makers and politicians need clear, sound, reliable standards to endorse reuse projects. Moreover, regulations have a major influence on the choice of treatment technologies and, hence, on the cost of water reuse projects. Other policy actions to change current irrigation practices are water pricing and penalties for exceeding extraction of groundwater. However, in many countries, water is heavily subsidized. In such cases, recycled water is often not cost-competitive with subsidized water. Finally, the involvement of end users (farmers, irrigation communities) in the management of the water reuse system is an important factor for the project success and efficiency.

REFERENCES

1. IWMI, Global Water Scarcity Study, http://www.iwmi.cgiar.org/home/wsmap.htm, 2000.
2. Winpenny, J.T., Managing water scarcity for water security, Paper prepared for FAO, http://www.fao.org/ag/agl/aglw/webpub/scarcity.htm, 2003.
3. Merrette, S., *Water for Agriculture: Irrigation Economics in International Perspective*, ed. Spon Press, London, chapter 1, 2002.
4. Earth Policy Institute, Eco-economy Indicators: world irrigated area, http://www.earth-policy.org/Indicators/indicator7_data1.htm, 2001.
5. Asano, T., Water from wastewater the dependable water resource, *Wat. Sci. Techn.*, 45, 8, 23, 2002.
6. Lazarova, V., Cirelli, G., Jeffrey, P., Salgot, M., Icekson, N. and Brissaud, F., Enhancement of integrated water management and water reuse in Europe and the Middle East, *Wat. Sci. Techn.*, 42, 1/2, 193, 2000.

7. State of California, *Water Recycling Criteria.* California Code of Regulations, Title 22, Division 4, Chapter 3. California Department of Health Services, Sacramento, CA, 2000.
8. Anderson, J., The environmental benefits of water recycling and reuse, *Wat. Sci. Tech.: Water Supply*, 4, 4, 1–10, 2003.
9. Asano, T. (ed.), *Wastewater Reclamation and Reuse*, Water Quality Management Library, vol. 10, CRC Press, Boca Raton, FL, 1998.
10. Lazarova, V. (ed.), Role of water reuse in enhancing integrated water resource management, Final Report of the EU project CatchWater, EU Commission, 2001.
11. Sheikh, B., Rosenblum, E., Kasower, S., and Hartling, E., Accounting for the benefits of water reuse, *AWWA/WEF Water Reuse Conf. Proc.*, February 1–4, Lake Buena Vista, FL, 211, 1998.
12. Harivandi, A., Effluent water for turfgrass irrigation, Cooperative extension—University of California, Division of Agriculture and Natural Resources, 11p., 1989.
13. Falkenmark, M., and Widstrand, C., Population and water resources: a delicate balance. *Population Bulletin.* Population Reference Bureau, 1992.
14. World Bank, *World Development Report 1992*, New York: Oxford University Press, 1992.
15. World Resources Institute, http://earthtrends.wri.org/datatables/, 2000–2001.
16. INED, Population & Sociétés, ISSN 0184 77 83, 1997.
17. WHO, World Health Organization, *Health Guidelines for the Use of Wastewater in Agriculture and Aquaculture*, Report of a WHO Scientific Group, Technical Report Series 778, World Health Organization, Geneva, Switzerland, 1989.
18. Bahri, A., Basset, C., Oueslati, F., and Brissaud, F., Reuse of reclaimed water for golf course irrigation in Tunisia, *Proc. IWA 1st World Water Congress*, CD-rom, Paris, July 3–7, 2000.
19. Gottberg, A., and Vaccaro, G., Kuwait's giant membrane plant starts to take shape, *Desalin. Water Reuse*, 13, 2, 30, 2003.
20. Crook, J., Okun, D.A., and Pincince, A.B., Water reuse, *WERF Project 92-WRE-1*, 1994.
21. WateReuse Association, *Recycling Water to meet the World's Needs*, WateReuse Association, Sacramento, CA, 1999.
22. York, D.W. and Wadsworth, L., Reuse in Florida: moving toward the 21st century, in: *Proceedings of Water Reuse 98*; Feb. 1–4; Lake Buena Vista, FL, 1998.
23. Parabicoli, S., *Water Reuse in Hawaii: An Overview—County of Maui,* Wastewater Reclamation Division, Hawaii, 1997.
24. Asano, T., Maeda, M., and Takaki, M., Wastewater reclamation and reuse in Japan: overview and implementation examples, *Wat. Sci. Tech.*, 34, 11, 219, 1996.
25. Anderson, J., Current water recycling initiatives in Australia: scenarios for the 21st century, *Wat. Sci. Tech.*, 33, 10–11, 37, 1996.
26. Law, I.B., Rouse Hills - Australia's first full-scale domestic non-potable reuse application, *Wat. Sci. Tech.*, 33, 10–11, 71, 1996.

2

Water Quality Considerations

Valentina Lazarova, Herman Bouwer, and Akiça Bahri

CONTENTS

2.1 Introduction 31

2.2 Parameters with health significance 36
  2.2.1 Chemicals 36
  2.2.2 Pathogens 40

2.3 Parameters with agronomic significance 45
  2.3.1 Salinity 45
  2.3.2 Toxic ions 46
  2.3.3 Sodium adsorption ratio 48
  2.3.4 Trace elements 49
  2.3.5 pH 53
  2.3.6 Bicarbonate and carbonate 53
  2.3.7 Nutrients 54
  2.3.8 Free chlorine 55

2.4 Sampling and monitoring strategies 55

References 58

2.1 INTRODUCTION

Water quality is the most important issue in water reuse systems that determines the acceptability and safety of the use of recycled water for a given reuse application. For each category of water reuse, the definition of appropriate water quality is driven by a number of health, safety, socio-psychological, and technical-economic criteria (Figure 2.1).

As a rule, water quality objectives are set by guidelines and regulations, which in turn determine the treatment technology to be used. Table 2.1 shows

1-56670-649-1/05/$0.00 + $1.50

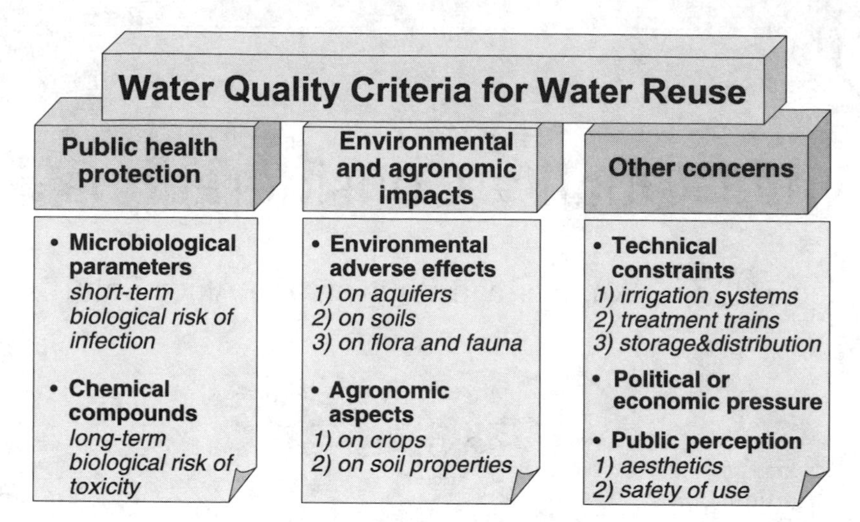

Figure 2.1 Main criteria influencing the choice of water quality in reuse systems for irrigation.

the list of parameters used in the evaluation of water quality for irrigation (the most important parameters are given in bold). The typical concentrations in raw municipal wastewater and the main characteristics and impacts of these parameters for water reuse are given in Table 2.2.[1–3]

Of the four categories, microbiological parameters have received the most attention. Since monitoring for all pathogens is not realistic, specific target organisms such as fecal or total coliforms are being used as indicators of potential health risk.

In addition, other parameters are chosen and used for regulatory purposes and to monitor the treatment efficiency of a process or before reuse, depending on the type of reuse or regional specificities. For example, water salinity is of great concern in agricultural reuse, while trace organics are of lesser concern for agricultural reuse but are an important issue for potable reuse.

Wastewater quality data routinely measured and reported are mostly in terms of general parameters (e.g., biochemical oxygen demand [BOD_5], suspended solids [SS], chemical oxygen demand [COD], which are of interest in water pollution control in receiving water bodies. While monitoring of suspended solids can be useful to predict clogging problems in irrigation systems, COD and BOD_5 usually are not directly used in irrigation project planning, although organic constituents can be problematic if present in high concentrations. The evaluation of nutrient content of wastewater (N and P) is becoming increasingly important to avoid eutrophication, as well as to assess the fertilizing value of these waters.

The main factors that affect recycled water quality include source control, type of sewage system, wastewater treatment and operation, as well as storage

Table 2.1 Parameters Used for Evaluation of Water Quality for Irrigation

General parameters	Organics	Microorganisms	Anions	Cations	Trace elements		Specific parameters
Acidity/ Alkalinity: pH	BOD, mgO_2/L	Bacteria	Bicarbonate (HCO_3^-)	Ammonia ($N{-}NH_4^+$)	Aluminum (Al)	Arsenic (As)	Boron (B)
Color/Turbidity	COD, mgO_2/L	Total coliforms	Carbonate (CO_3^{2-})	Calcium (Ca^+)	Beryllium (Be)	Cadmium (Cd)	Sodium adsorption ratio (SAR)
Electrical conductivity, (EC_w), dS/m	Organic nitrogen	Fecal coliforms	Chloride (Cl^-)	Magnesium (Mg^{2+})	Cobalt (Co)	Chromium (Cr)	
Hardness, $mEqCaCO_3/L$	Organic phosphorus	*Salmonella*	Fluoride (F^-)	Potassium (K^+)	Iron (Fe)	Copper (Co)	
Total suspended solids (TSS), mg/L	Target compounds: pesticides and other trace organics	Viruses	Nitrate ($N{-}NO_3^-$)	Sodium (Na^+)	Lithium (Li)	Lead (Pb)	
Temperature		Helminths	Phosphate ($P{-}PO_4^{-3}$)		Manganese (Mn)	Mercury (Hg)	
Total dissolved solids (TDS), mg/L		Ascaris	Sulfate (SO_4^{2-})		Molybdenum (Mo)	Zinc (Zn)	
		Taenia eggs			Selenium (Se)		
		Protozoa			Tin (Sn)		
		Giardia, Cryptosporidium			Tungsten (W)		
		Pathogens			Vanadium (V)		
		Salmonella, Legionella					

Table 2.2 Characteristics and Impacts of Important Water Quality Parameters in Wastewater

Category	Constituents	Parameters of interest	Typical range in wastewater	Characteristics and impacts
Pathogenic organisms	Protozoa	*Entamoeba histolytica*	10^{-1}–10^{3}	Infectious dose between 1–20
		Giardia lamblia		High survival time in water 10–30 days
		Cryptosporidium		
		Cyclospora		
	Helminths eggs	*Ascaris lumbricoides*	10^{1}–10^{3}	Infectious dose 1–10 helminth eggs
		Taenia	10^{-2}–10^{1}	Highly persistent in the environment (many months)
		Hookworm		Considered the main pathogenic risk in irrigation with recycled water
		Clonorchis		
	Bacteria	Total coliforms, cfu/100 mL	10^{5}–10^{9}	Infectious dose highly variable (10–10^{7})
		Fecal coliforms, cfu/100 mL	10^{4}–10^{9}	High survival time in water 10–60 days
		Streptococci	10^{3}–10^{7}	Total and fecal coliforms commonly used as indicator
		E. coli	10^{4}–10^{9}	*E. coli* is proposed as a more specific indicator of fecal contamination
		Shigella	10–10^{3}	
		Salmonella	10^{2}–10^{4}	
	Viruses	Enterovirus	10^{1}–10^{4}	Detection methods not sensitive
		Hepatitis		Variable incubation times, very high survival time in water 50–120 days
		Rotavirus		Person-to-person contamination main mode of transmission
		Adenovirus		Infectious dose 1–10
		Norwalk virus		
General parameters	Suspended solids	Total suspended solids (TSS), mg/L	100–350	Sorb organic pollutants trace elements, and heavy metals
				Shield microorganisms
				Plug irrigation systems and soil
	Nutrients	Nitrogen, mgN_{tot}/L	20–85	Induce eutrophication when combined with high concentrations
				Nitrogen can lead to nitrate build-up in groundwater after after leaching
		Phosphorus, mgP_{tot}/L	4–15	Induce eutrophication
	Hydrogen ion concentration	pH	6.5–8.5	Impact on coagulation, disinfection, metal solubility, and soils

Inorganics	Dissolved inorganics	Total dissolved solids (TDS), mg/L	250–3000	High salinity may damage crops
		Electrical conductivity (EC), dS/m	0–5	Salt can contaminate ground water
		Boron, mg/L	0–2	Sodium can be toxic to plants and can cause soil structure deterioration
		Sodium adsorption ratio	0–15	Affects soil's permeability and structure
		Bicarbonate, mg/L	0–600	Sprinkler irrigation leads to foliar deposits (> 90 mg/L)
		Sodium, mg/L	20–170	Essential for plant growth but toxic effects are possible if > 1 mg/L
	Trace elements	Specific elements (Cd, Zn, Hg, Ni, Pb, Hg), mg/L	0.001–0.6	Accumulate in certain plants and animals Limit suitability of the reclaimed water
Organics	Biodegradable organics	Biochemical oxygen demand (BOD_5), mg/L	110–400	Aesthetic and nuisance problems
		Chemical oxygen demand (COD), mg/L	250–1000	Contribute to chlorine demand
		Total organic carbon (TOC), mg/L	80–290	
	Stable organics	Specific compounds (pesticides, PAH, chlorinated hydrocarbons), μg/L	100–400	Toxic to the environment and public health Limit suitability of recycled water

Source: Adapted from Refs. 1–3.

and distribution. Assurance of treatment reliability and the good operation of water reuse systems are the major water quality control measures. In addition, the control of industrial wastewater discharge lowers the risks related to toxic inorganic and organic compounds.

Thorough knowledge and appropriate monitoring of water quality is needed to protect public health and minimize the negative impact of recycled water on irrigated crops.

2.2 PARAMETERS WITH HEALTH SIGNIFICANCE

Biological risks related to water reuse have been recognized since the very beginning of irrigation with wastewater. On the other hand, the considerations related to chemical risks have been developed recently following improvements in analytical capabilities. Additionally, biological risks have a relatively immediate outcome (illnesses develop in a short period of time), while chemical risks are translated into time-delayed illnesses (carcinogens, long-term toxicity, etc.).

2.2.1 Chemicals

Municipal wastewater that has limited industrial wastewater input generally contains concentrations of organic and inorganic compounds that do not present health concerns when the recycled water is used for irrigation. Moreover, up to 90% of the added chemicals might be removed from wastewaters and sometimes are concentrated in biological sludge. Human health-related issues involving toxic chemicals have been reported only for irrigation with wastewater heavily polluted by industrial waste discharge.

The majority of irrigation water quality criteria give numerical levels only for some potentially toxic elements such as As, Cu, Cr, Cd, Pb, and Hg (see §2.3.4), which usually have concentrations below the guideline level in raw sewage. In some countries (e.g., China, Hungary), a few trace organic compounds are also included as a measure for groundwater protection (benzene, petroleum hydrocarbon, trichloroacetylaldehyde). Currently, the World Health Organization (WHO)[4] is working on new health-related chemical guidelines on the reuse of municipal wastewater and sludge.

The principal health hazards associated with chemical constituents of recycled water arise from the contamination of crops or groundwater by the following compounds:

- Cumulative poisons, principally trace elements (heavy metals)
- Carcinogens, mainly organic chemicals
- Pharmaceuticals (antibiotics, synthetic drugs) and personal care products
- Other compounds suspected to exert endocrine disruption properties (hormones or other chemicals such as PCBs, octilphenol, nonilphenol, etc.)

Existing drinking water regulations[5–7] include maximum values for several organic and toxic substances based on acceptable daily intakes (ADI) (Table 2.3 and Table 2.4). These limits can be adopted directly for groundwater protection purposes or in view of the possible accumulation of certain toxic

Table 2.3 Drinking Water Standards for Inorganic and Organometallic Substances

	MCL[a] (μg/L)		
Contaminant	**USEPA 2001**	**EU 98/83/CE**	**WHO 1998**
Inorganic compounds			
Antimony	6 (6)	5	5
Arsenic[e]	50 (NA)	10	10
Asbestos	7 MFL[b]		
Barium	2000		700
Beryllium	4 (4)		
Boron		1000	500
Bromate	10 (0)	10	25
Cadmium[e]	5 (5)	5	3
Chlorite	1000 (800)		
Chromium	100 (100)[c]	50	50
Copper	(1300)	2000	2000
Cyanide	200 (200)	50	70
Fluoride	4000 (4000)	1500	1500
Lead[e]	(0)	10	10
Manganese			500
Mercury[e]	2 (2)	1	1
Molybdenum			70
Nickel[e]	100 (100)	20	20
Nitrate	10,000[d]	50,000	50,000
Nitrite	1000[d]	500	3000
Selenium	50 (50)	10	10
Thallium	2 (0.5)		
Uranium	30 (0)		20
Organometallic compounds			
Tributyltin compounds[e]:			
– *Dialkyltin*			
– *Tributyltin oxide*			2

[a]Maximum contaminant level, MCL (maximum contaminant level goal, MCLG).
[b]MFL, million fibers per liter.
[c]Total chromium.
[d]As N.
[e]Priority substances in water policy included in European Directive 2000/60/EC, Decision 2455/2001/EC.

Table 2.4 Drinking Water Standards for Organic Substances

	MCL[a] (μg/L)				MCL[a] (μg/L)		
Contaminant	**USEPA 2001**	**EU 98/83/CE**	**WHO 1998**	**Contaminant**	**USEPA 2001**	**EU 98/83/CE**	**WHO 1998**
Acrylamide	(0)	0.1	0.5	Fenoprop			9
Alachlor*	2 (0)		20	Fluoranthene*			
Aldicarb	3 (1)		10	Glyphosate	700 (700)		
Aldrin/dieldrin			0.03	Heptachlor and heptachlor epoxide	0.2 (0)		0.03
Anthracene*				Hexachlorobenzene*	1 (0)		1
Atrazine*	3 (3)		2	Hexachlorobutadiene*			0.6
Bentazone			30	Hexachlorocyclohexane*			
Benzene*	5 (0)	1	10	Hexachlorocyclopentadiene	50 (50)		
Benzo-a-pyrene*	0.2 (0)	0.01	0.7[b]	Isoproturon*			9
Brominated diphenylethers*				Lindane*	0.2 (0.2)		2
Bromodichloromethane	(0)		60	Methoxychlor	40 (40)		20
Bromoform	(0)		100	Molinate			6
Carbofuran	40 (40)		5	Monochlorobenzene	100 (100)		300
Carbon tetrachloride	5 (0)		2	MCPA			2
Chlordane	2 (0)		0.2	Naphthalene*			
Chlorfenvinphos*				Nitrilotriacetic acid			200
C_{10-13}-chloroalkanes*				Nonylphenols*			
Chloroform*	NA	100	200	Octylphenols*			
				Oxamyl (vydate)	200 (200)		
Chlorpyrifos*				Pendimethalin			20
Chlorotoluron	–		30	Pentachlorobenzene*			
Cyanazine			0.6	Pentachlorophenol*	1 (0)		9
2,4-Dichlorophenoxyacetic acid (2,4-D)	70 (70)		30	Permethrin			20
DDT			2	Pesticides/total/		0.1/(0.5)	

Dalapon	200 (200)		–	Polycyclic aromatic hydrocarbons*		0.1	
Dibromochloromethane	NA (60)		100	Benzofluoranthene*			
1,2-Dibromo-3-chloropropane	0.2 (0)		1	Benzoperylene*			
1,2-Dibromoethane			0.4–15	Indenopyrene			
Dichloroacetic acid	NA (0)		50	Polychlorinated biphenyls	0.5 (0)		
1,4- or *p*-Dichlorobenzene	75 (75)		300	Propanil			20
1,2- or *o*-Dichlorobenzene	600 (600)		1000	Pyridate			100
1,2-Dichloroethane*	5 (0)	3	30	Simazine*	4 (4)		2
1,1-Dichlorethylene	70 (70)		30	Styrene	100 (100)		20 (*4–2600*)
1,2-Dichlorethylene	5 (0)		50	2,3,7,8-TCDD (dioxin)	0.00005 (0)		
Dichloromethane*	5 (0)		20	Tetrachlorethylene	5 (0)	10	40
1,2-Dichlopropane (1,2-DCP)			20	Toluene	1000 (1000)		700 (*24–170*)
1,3-Dichlopropene			20	Toxaphene	5 (0)		
Di(2-ethylhexyl)adipate			80	2,4,5-TP (silvex)	50 (50)		
Di(2-ethylhexyl)phthalate*			8	1,1,1-Trichloroethane	200 (200)		2000
Diuron*				1,1,2-Trichloroethane	5 (3)		2
Edetic acid (EDTA)			200 (*600*)[d]	2,4,6 Trichloroethylene	5 (0)	10	70
Endosulfan*				Trichlorobenzene*	70 (70)[c]		20
Endothall	100 (100)			Trihalomethanes (total)	80 (NA)	100	
Endrin	2 (2)			2,4,6-Trichlorophenol			200
Epichlorohydrin		0.1	0.4	Trifuraline*			20
Ethylbenzene	70 (70)		300 (*2–200*)	Vinyl chloride	2 (0)	0.5	5
Ethylene dibromide	0.05 (0)			Xylene	10,000		500 (*20–1800*)

*Priority substances in water policy included in European Directive 2000/60/EC, Decision 2455/2001/EC.

[a]Maximum contaminant level, MCL (maximum contaminant level goal, MCLG).

[b]Benzo(3,4)pyrene.

[c]1,2,4-Trichlorobenzene.

[d]WHO 2000 values (in italics).

NA, not applicable.

elements in plants (e.g., cadmium and selenium) ingested by humans. Some adaptation of these criteria should be envisaged for establishing limits for recycled water.

It is worth noting, however, that the intake of most toxic elements or compounds by consumption of crops irrigated with contaminated recycled water is very low. Occasionally some amount of water can be ingested (10–20 mL), but the content of toxic compounds in such volume is extremely low (maximum $6 \pm 2E\text{-}5$ mg/kg/d for trace organics and heavy metals).

During the last 10 years, an impressive improvement in analytical capacity has led to the discovery in natural waters of a huge amount of substances capable of exerting negative effects on humans. Health-related concerns (endocrine disruption, feminization of fishes, antibiotic resistance of pathogens, etc.) pertaining to endocrine disruptors, pharmaceuticals, and personal care products in raw wastewater, recycled water, and other waters are receiving increased attention, as is the removal of these constituents during wastewater treatment and soil percolation. This is currently a fertile field for research, because such compounds enter the water cycle through wastewater disposal. This knowledge creates an additional motivation to find treatments to eliminate such substances, which are usually present in very low concentrations.

As indicated in Chapter 8, irrigation with adequately treated recycled water does not seem to present significant or unacceptable chemical risks. Nevertheless, irrigation with recycled water must be applied with proper precautions to protect human health and the environment.

2.2.2 Pathogens

The greatest health concern when using recycled water for irrigation is related to pathogens that could be present. It is widely known that it is not practical to establish the presence or absence of all pathogenic organisms in wastewater or recycled water in a timely fashion. For this reason, the indicator organism concept was established many years ago to allow monitoring of a limited number of microbiological constituents. Table 2.5 enumerates the microbiological organisms (pathogens and indicator organisms) that are usually analyzed for to establish the presence or absence of health hazards. Table 2.6 provides the survival time of some common pathogens under different conditions in fresh water, sludge, soil, and crops.[8,9]

Epidemiological studies conducted to date have not established definitive adverse health impacts attributable to the use of appropriately treated recycled water for irrigation. Some experts have concluded that the annual risk of enteric virus and bacterial ingestion from eating lettuce irrigated with water meeting WHO guideline levels ranges from 10^{-5} to 10^{-9}.[10] Compared to the accepted risk of infection by enteric disease from drinking water of 10^{-4} in the United States, the risk arising from irrigation with recycled water would appear

Table 2.5 Waterborne Pathogens and Microbiological Indicators Used in Water Reuse Systems

Types of waterborne pathogens	Indicators	Observations
1. Bacteria 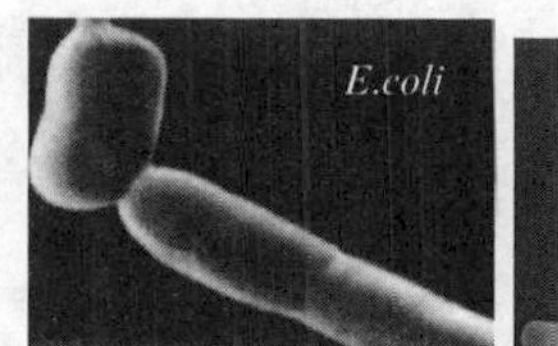 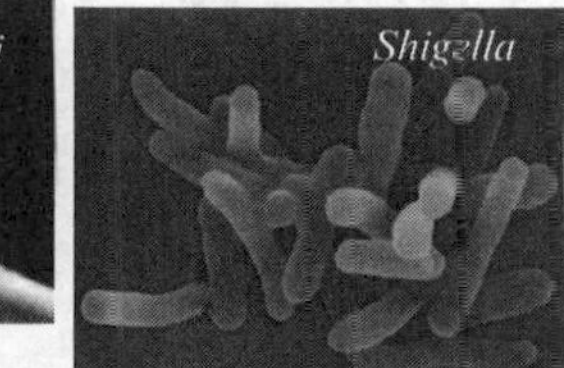  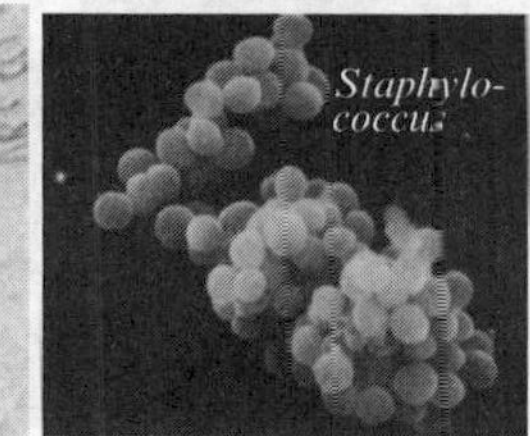  Salmonella typhi	Fecal coliform Total coliform *Streptococcus* *E. coli* *Clostridium perfringens* 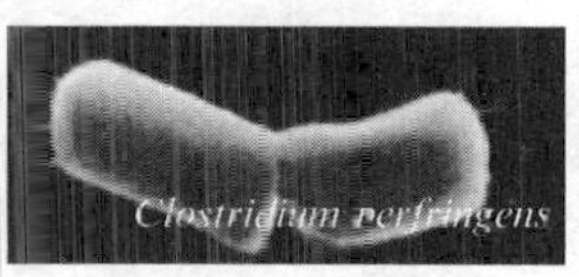	Fecal coliform is the most common indicator; *E. coli* determination is slowly substituting it (more representative of human fecal contamination) *Salmonella* is used in some new revisions of water reuse or bathing zone regulations *Clostridium perfringens* indicates the presence of spore-forming bacteria
2. Viruses	Enteroviruses Bacteriophages  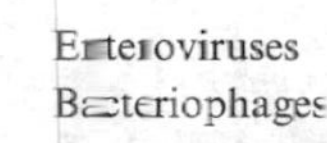	An accepted indicator still does not exist; bacteriophages (F+, somatic) are being studied Monitoring of enteroviruses is required by bathing zone regulations and by some reuse standards for irrigation

(continued)

Table 2.5 Continued

Types of waterborne pathogens	Indicators	Observations
3. Helminths and Nematodes		
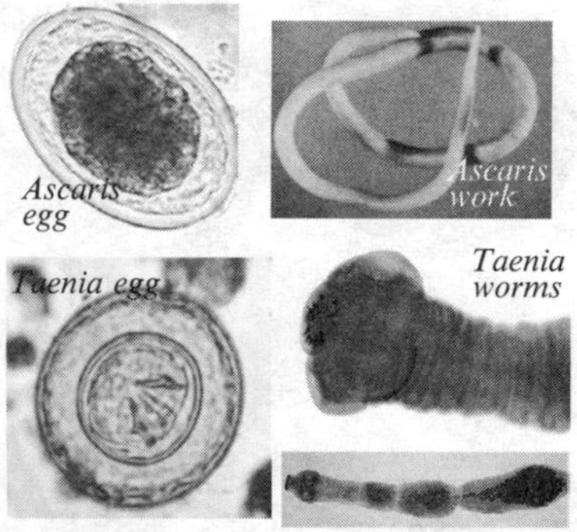 *Ascaris egg*; *Ascaris work*	Nematode eggs *Ascaris* *Trichuris* *Ancylostoma* Hookworm *Clonorchis* Others *Taenia eggs*	Discouraging negative results found in wastewater in many countries High importance only for emerging countries In some cases other helminths are important for risk related to animal health, in particular *Taenia* eggs
Taenia egg; *Taenia worms*	Indicators not known	
4. Protozoa and cysts 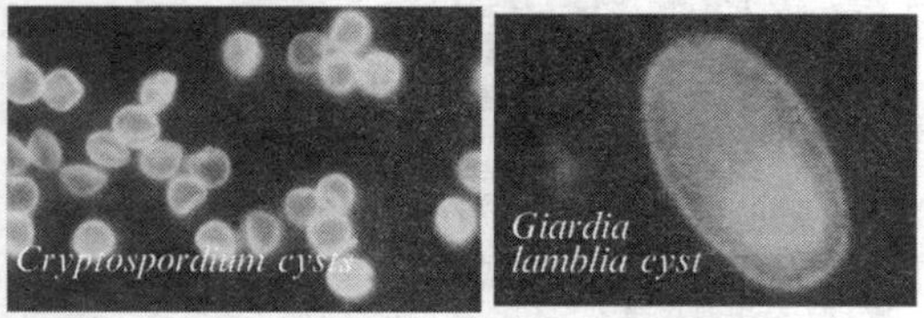 *Cryptosporidium cysts*; *Giardia lamblia cyst*	Indicators not known	Analytical tools not well developed
5. Fungi, algal toxins 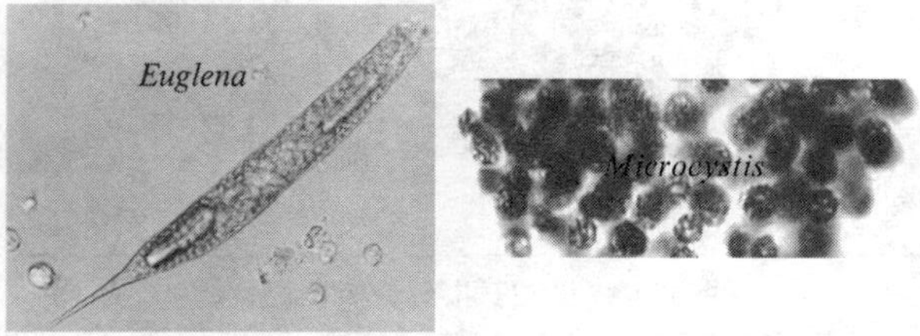 *Euglena*; *Microcystis*	Indicators not known	Few cases detected, mostly in reservoirs

Table 2.6 Survival of Excreted Pathogens at 20–30°C

	Survival time (days)			
Type of pathogen	In feces, nightsoil, and sludge	In freshwater and sewage	In soil	On crops
Viruses				
Enteroviruses	< 100 (< 20)[a]	< 120 (< 50)	< 100 (< 20)	< 60 (< 15)
Bacteria				
Fecal coliforms	< 90 (< 50)	< 60 (< 30)	< 70 (< 20)	< 30 (< 15)
Salmonella spp.	< 60 (< 30)	< 60 (< 30)	< 70 (< 20)	< 30 (< 15)
Shigella spp.	< 30 (< 10)	< 30 (< 10)	—	< 10 (< 5)
Vibrio cholerae	< 30 (< 5)	< 30 (< 10)	< 20 (< 10)	< 5 (< 2)
Protozoa	< 30 (< 15)	< 30 (< 15)	< 20 (< 10)	< 10 (< 2)
Entamoeba histolytica cysts	< 30 (< 15)	< 30 (< 15)	< 20 (< 10)	< 10 (< 2)
Helminths	Many	Many	Many	< 60 (< 30)
Ascaris lumbricoides eggs	months	months	months	

Source: Adapted from Refs. 8, 9.
[a]The usual survival time.

to offer a similar level of protection. In emerging countries the greatest health risk is associated with spray irrigation of recycled water when concentrations of nematode eggs are over 1 egg/L, particularly for children who eat vegetables irrigated with such water. Nevertheless, there is clear epidemiological evidence of health problems when raw or improperly treated wastewater is used for irrigation in areas where such infections are endemic.[10–14] Table 2.7 summarizes the findings of some of these studies.

No strong evidence has been found to suggest that population groups residing near wastewater treatment plants or recycled water irrigation sites are subject to increased risk from pathogens resulting from aeration processes or sprinkler irrigation.[10,12,15] Adverse health effects have been detected only in association with the use of raw or poorly settled wastewater, while inconclusive evidence suggested that appropriate wastewater treatment could provide a high level of health protection.

Several studies have been undertaken to test the hazards related to food-crop irrigation with tertiary treated reclaimed municipal wastewater. One is the 5-year field pilot study in Monterey, California, the Monterey Wastewater Reclamation Study for Agriculture (MWRSA). Raw-eaten food crops, including lettuce, broccoli, and celery, were irrigated with recycled water having received tertiary treatment plus disinfection. This research project included 96 randomized field plots, each receiving a different combination of water type and fertilizer application rate. Four replicates of each combination were provided to ensure reliable statistical analysis of data. Water types included two tertiary recycled waters and a control (local well water). The MWRSA study indicated that there was an absence of microorganisms of

Table 2.7 Summary of Major Epidemiological Studies Related to Agricultural Irrigation with Recycled Water

Pathogens of concern	Provenance	Level of treatment	Transmission route	Infected population	Ref.
Helminths (*Ascaris, Trichuris* spp.)	Diseases endemic in the population	Raw untreated sewage	Irrigation of vegetables eaten uncooked;	High infection of children (< 15 years); population consuming such crops; sewage and farm workers	10, 11, 13, 14
Protozoa (*Amoeba, Giardia, Cryptosporidium*)			Oral through aerosols or spray and surface irrigation		
Beef tapeworm (*Taenia saginata*)			Irrigation of pastures or presence of raw wastewater canals or ponds, where cattle can drink	Cattle grazing on such fields; possibility of infection of population consuming meat of infected cattle[a]	14
Vibrio cholerae	Cholera endemic or not in the population		Oral through aerosols or spray and surface irrigation; irrigation of vegetables eaten uncooked	Sewage farm workers, population consuming irrigated crops	8, 14
Bacteria and viruses	Diseases endemic in the population		Idem	Very few outbreaks associated with irrigation; evidence in inexperienced sewage workers	10
Helminths	Diseases endemic in the population	Treated wastewater	Direct contact; irrigation of vegetables eaten uncooked; Oral through aerosols	No infection of farm workers or surrounding population	12
Bacteria and viruses			Idem	No conclusive evidence for all risk groups	10

[a]There is strong evidence of infection of cattle grazing on fields irrigated with raw wastewater, but limited evidence for the contamination of human beings.

concern for food safety in the water and on the edible and residual plant tissues.[16]

Natural barriers also reduce the threat of crop contamination by pathogens. Cell walls of plant roots and leaves filter the irrigation water, and microorganisms cannot readily pass through and into the edible tissues of the crops unless the cell walls are injured. Moreover, drying and solar radiation further prevent any organisms remaining in irrigation water from continuing to be viable on plant surfaces as long as there is an adequate drying period after the last irrigation and before harvest. These mechanisms normally provide a high level of natural protection against contamination of food crops from many pathogens that might be present in recycled water.

Potential risks induced by the presence of pathogenic microorganisms in wastewater or on crops may become actual risks if the following four criteria occur:

1. The pathogen must reach the plant or be able to multiply to the number required for an infective dose.
2. A human host must come into contact with the infective dose of the pathogen.
3. The host must become infected.
4. Disease results from the infection or leads to further transmission.

2.3 PARAMETERS WITH AGRONOMIC SIGNIFICANCE

Important agricultural water quality parameters include a number of specific properties of water that are relevant in relation to the yield and quality of crops, maintenance of soil productivity, and protection of the environment. The quality of irrigation water is of particular importance in arid zones where extremes of temperature and low relative humidity result in high rates of evaporation with consequent deposition of salt, which tends to accumulate in the soil profile.

The physical and mechanical properties of the soil, such as soil structure (stability of aggregates) and permeability, are very sensitive to the type of exchangeable ions present in irrigation water. Thus, when water reuse is being planned, several factors related to soil properties must be taken into consideration.

2.3.1 Salinity

Compared to many other irrigation waters, recycled water generally has a low to medium salinity with electrical conductivity of 0.6 to 1.7 dS/m. Some dissolved mineral salts are identified as nutrients and are beneficial for plant growth, while others may be phytotoxic or may become so at high concentrations (see Chapter 5, §5.5).

The major salinity sources in recycled water are drinking water (especially hardness and naturally occurring salts), salts added by urban or industrial water use, infiltration of brackish water into sewers, and agricultural irrigation (impact on groundwater salinity). As a rule, residential use of water typically adds about 300 ± 100 mg/L of dissolved salts. Consequently, if the drinking water used by a given municipality is of acceptable quality for irrigation, the treated municipal water will also be of acceptable quality. The main exceptions would be the coastal areas, where infiltration of saline water in sewers is a concern, or where industrial wastes with unacceptable contaminants are discharged into urban sewers (e.g., brines).

Salinity in the soil is related to, and often determined by, the salinity of irrigation water. The rate at which salts accumulate to undesirable levels in soils depends on the following factors:

Their concentration in the irrigation water
The amount of water applied annually
Annual precipitation
Evapotranspiration
Soil characteristics, both physical and chemical

Dissolved salts increase the osmotic pressure of soil water and consequently lead to an increase in the energy plants must expend to take up water from the soil. As a result, respiration is increased and the growth and yield of most plants decline progressively as osmotic pressure increases.

Water salinity can be reported either as total dissolved solids (TDS, mg/L) or as electrical conductivity (EC_w), measured in mmhos/cm or most correctly in dS/m. The relationship between EC_w and TDS is approximately EC_w (dS/m) $\times 640 =$ TDS (mg/L). The symbol EC_e is used to designate the electrical conductivity of the soil saturation extract.

Recently, the classification of saline water has been reconsidered (Table 2.8) on the basis of research and practical observations.[17,18] This classification must be used only as a guideline to determine the level of salinity of irrigation waters. It is important to stress that Table 2.8 cannot be used to assess the suitability of saline water for irrigation, because a number of other conditions must be taken into account, including crop, climate, soil, irrigation method, and management practices (see Chapter 5, §5.5.1). Generally, nonsaline water is characterized by TDS < 500 mg/L and $EC_w < 0.7$ dS/m, maximum salt content in slightly saline waters is TDS < 2000 mg/L and $EC_w < 3$ dS/m, and water is considered as brine when TDS $> 30{,}000$ mg/L and $EC_w > 42$ dS/m. As a rule, recycled urban water salinity is below 2 dS/m, with some exceptions in dry countries and coastal areas.

2.3.2 Toxic Ions

Many of the ions that are harmless or even beneficial at relatively low concentrations may become toxic to plants at high concentration (see

Table 2.8 Classification of Irrigation Water According to Salinity

Salinity class	Range of variation	
	Electrical conductivity, EC_W (dS/m)	Total dissolved solids, TDS (mg/L)
Nonsaline water	<0.7	<500
Saline water	0.7–42	500–30,000
Slightly saline water	0.7–3	500–2000
Medium saline water	3–6	2000–4000
Highly saline water	6–14	4000–9000
Very highly saline water	14–42	9000–30,000
Brine	>42	>30,000

Source: Adapted from Refs. 17, 18.

Chapter 5, §5.5). This effect could result either from direct interference with the metabolic processes or through indirect effects on other nutrients, which might be rendered unavailable. Toxicity normally results in impaired growth, reduced yield, changes in the morphology of the plant, and even its death. The degree of damage depends on the crop, its stage of growth, concentration of the toxic ion or ions, its relationships, climate, and soil conditions.

The most common phytotoxic ions that may be present in municipal effluents in concentrations high enough to cause toxicity are boron (B), chloride (Cl), and sodium (Na). Each can cause damage individually or in combination.

Sodium and chloride are usually absorbed by the roots but can also enter directly into the plant through the leaves when moistened during sprinkler irrigation. This typically occurs during periods of high temperature and low humidity. Leaf absorption speeds up the rate of accumulation of a toxic ion and may be a primary source of toxicity.[19] The concentration of these ions should be determined on an individual case basis to assess the suitability of wastewater quality for agricultural or landscape irrigation, although concentration changes are usually not relevant for short and medium periods of time.

Boron can become toxic at levels only slightly greater than those required by plants for good growth. The predominant source of anthropogenic boron is domestic effluents, due to the use of perborate as a bleaching agent. As a result, boron can be found in urban wastewater at concentration levels as high as 5 mg/L (dry countries and concentrated sewage), with an average level around 1 mg/L. It should be noted that boron at concentrations of less than 1 mg/L is essential for plant development, but higher levels can cause problems in sensitive plants. Most plants exhibit toxicity problems when the concentration of boron exceeds 2 mg/L.

2.3.3 Sodium Adsorption Ratio

Sodium is a unique cation because of its effect on soil. When present in the soil in exchangeable form, sodium causes adverse physical-chemical changes, particularly to soil structure, which results in dispersion of particles and, consequently, reduced infiltration rates of water and air into the soil. As a rule, recycled water could be a source of excess Na in the soil compared to other cations (Ca, K, Mg), and for this reason it should be monitored.

The most reliable index of the sodium hazard of irrigation water is the sodium adsorption ratio SAR. The sodium adsorption ratio is defined by Equation (2.1), where the ion concentrations are expressed in mEq/L:

$$SAR = \frac{Na}{\sqrt{Ca + Mg/2}} \tag{2.1}$$

It should also be noted that the SAR from Equation (2.1) does not take into account changes in calcium ion concentration in the soil water due to changes in solubility of calcium resulting from precipitation or dissolution during or following irrigation. However, this calculated SAR is considered an acceptable evaluation procedure for most irrigation waters.

If significant precipitation or dissolution of calcium due to the effect of carbon dioxide (CO_2), bicarbonate (HCO_3^-) and total salinity (EC_w) is suspected, an alternative procedure for calculating an adjusted sodium adsorption ratio (SAR_{adj}) can be used. This method to calculate SAR has not been widely accepted.

When the water in the soil is concentrated by transpiration and evaporation, which causes a tendency for calcium and possibly magnesium to precipitate as carbonates, and the proportion of sodium dissolved in water increases, another indicator of the sodium hazard can be used: the residual sodium carbonate (RSC). The RSC expressed in mEq/L is given by Equation (2.2):

$$RSC = (CO_3^- + HCO_3^-) - (Ca^{2+} + Mg^{2+}) \tag{2.2}$$

This concept is controversial, and the above relation is not widely used and few data can be found in the literature. When the RSC is below 1.25, the water is considered safe, whereas if it exceeds 2.5 the water is considered unsuitable for irrigation. In most countries, RSC is no longer used and has been replaced by SAR.

The threshold value of SAR of less than 3 indicates no restriction on the use of recycled water for irrigation, while severe damage could be observed when SAR is over 9, in particular for surface irrigation. At a given SAR, the infiltration rate increases as salinity increases or decreases when salinity decreases. Therefore, SAR and EC_w should be used in combination to evaluate the potential problem. Recycled water is often high in sodium, and the resulting high SAR is a major concern in planning water reuse projects.

2.3.4 Trace Elements

In addition to sodium, chloride, and boron, many trace elements are toxic to plants at low concentrations (see Chapter 5, §5.5.3). Fortunately, most irrigation supplies and sewage effluents contain low concentrations (usually less than a few mg/L) of these compounds, and trace elements are generally not a problem for irrigation with recycled water.

Urban wastewater may contain trace elements at concentrations that will give rise to high levels of such elements in the soil and cause undesirable accumulations in plant tissues and crop growth reduction. Trace elements are readily fixed and accumulate in soils with repeated irrigation with such recycled waters and may render them nonproductive or the product unusable. Surveys of irrigation with recycled water have shown that more than 85% of the applied trace elements are likely to accumulate in the soil, most at or near the surface, and may be leached to groundwater.

Trace elements are not normally included in the routine analysis of regular irrigation water, but attention should be paid to them when using treated municipal effluents, particularly if contamination with industrial wastewater discharge is suspected. These include (see Table 2.1) aluminum (Al), beryllium (Be), cobalt (Co), fluoride (F), iron (Fe), lithium (Li), manganese (Mn), molybdenum (Mo), selenium (Se), tin (Sn), titanium (Ti), tungsten (W), and vanadium (V). Heavy metals include a special group of trace elements that have been shown to create definite health hazards when taken up by plants. In this group are included arsenic (As), cadmium (Cd), chromium (Cr), copper (Cu), lead (Pb), mercury (Hg), and zinc (Zn).

Table 2.9 presents phytotoxic threshold levels of some selected trace elements.[illegible] According to the recommendations of the National Academy of Sciences (NAS[22]), a distinction is made between permanent irrigation of all soils (low maximum contaminant levels) and up to 20 years of irrigation of fine-textured neutral to alkaline soil, where higher concentrations of trace elements can be tolerated. These concentrations are set because of concern for long-term build-up of trace elements in the soil and for protection of agricultural soils from irreversible damage. Under normal irrigation practices, these suggested levels should prevent a build-up that might limit future crop production or utilization of the final product.

The concentration limits given in Table 2.9 reflect the current information available, but as they are supported by only limited, long-term field experience, they are necessarily conservative, which means that if the suggested limit is exceeded, phytotoxicity still may not occur. Compliance with the suggested limits will help ensure that the site can be used for all future crops. It is recommended that the given values be considered as the maximum long-term average concentration based upon normal irrigation application rates. When more reliable data become available, the levels may be adjusted. If water above or close to the given levels is considered for use, an up-to-date review of more recent information is suggested to prevent possible future problems.

Table 2.9 Recommended Maximum Concentrations of Trace Elements in Irrigation water

Element	MCL[a] (mg/L)		Comments
	Permanent irrigation[b]	< 20 years irrigation[c]	
Al Aluminum	5.0	20	Can cause nonproductivity in acid soils (pH < 5.5), but more alkaline soils at pH > 7.0 will precipitate the ion and eliminate any toxicity.
As Arsenic	0.10	2.0	Toxicity to plants varies widely, ranging from 12 mg/L for Sudan grass to < 0.05 mg/L for rice.
Be Beryllium	0.10	0.50	Toxicity to plants varies widely, ranging from 5 mg/L for kale to 0.5 mg/L for bush beans.
Cd Cadmium	0.01	0.05	Toxic to beans, beets, and turnips at concentrations as low as 0.1 mg/L in nutrient solutions. Conservative limits recommended due to its potential for accumulation in plants and soils to concentrations that may be harmful to humans.
Cr Chromium	0.10	1.0	Not generally recognized as an essential growth element. Conservative limits recommended due to lack of knowledge on its toxicity to plants.
Co Cobalt	0.05	5.0	Toxic to tomato plants at 0.1 mg/L in nutrient solution. Tends to be inactivated by neutral and alkaline soils.
Cu Copper	0.20	5.0	Toxic to a number of plants at 0.1–1.0 mg/L in nutrient solutions.
F Fluoride	1.0	15	Inactivated by neutral and alkaline soils.
Fe Iron	5.0	20	Not toxic to plants in aerated soils, but can contribute to soil acidification and loss of availability of essential phosphorus and molybdenum. Overhead sprinkling may result in unsightly deposits on plants, equipment, and buildings.
Li Lithium	2.5		Tolerated by most crops up to 5 mg/L; mobile in soil. Toxic to citrus at low concentrations (< 0.075 mg/L). Acts similarly to boron.
Mn Manganese	0.20	10	Toxic to a number of crops at a few tenths to a few mg/L, but usually only in acid soils.
Mo Molybdenum	0.01	0.05	Not toxic to plants at normal concentrations in soil and water. Can be toxic to livestock if forage is grown in soils with high concentrations of available molybdenum.
Ni Nickel	0.20	2.0[d]	Toxic to a number of plants at 0.5–1.0 mg/L; reduced toxicity at neutral or alkaline pH.
Pd Lead	5.0	10	Can inhibit plant cell growth at very high concentrations.

(continued)

Table 2.9 Continued

Element		MCL[a] (mg/L) Permanent irrigation[b]	< 20 years irrigation[c]	Comments
Se	Selenium	0.02	0.02	Toxic to plants at concentrations as low as 0.025 mg/L and toxic to livestock if forage is grown in soils with relatively high levels of added selenium. An essential element to animals but in very low concentrations.
Sn	Tin	—		Effectively excluded by plants; specific tolerance unknown.
Ti	Titanium			Idem
W	Tungsten			Idem
V	Vanadium	0.10	1.0	Toxic to many plants at relatively low concentrations.
Zn	Zinc	2.0	10	Toxic to many plants at widely varying concentrations; reduced toxicity at pH > 6.0 and in fine-textured or organic soils.

Source: Adapted from Refs. 20–22.
[a]Maximum concentration level (MCL) is based on water application rate consistent with good irrigation practices (10,000 m^3/ha/year). If water application rate greatly exceeds this, maximum concentrations should be adjusted downward accordingly. No adjustment should be made for application rates < 10,000 m^3/ha/year. Values given are for water used on a continuous basis at one site.
[b]Irrigation of all soils.
[c]Irrigation of fine textured neutral to alkaline soils (pH 6–8.5).
[d]For acid soils only.

Several long-term field experiments have been conducted in different countries on the impact of land application of recycled water on soils, microorganisms, and plants. Long-term environmental impact from irrigation with recycled water was reported to be minimal.[4] It was demonstrated that heavy metals such as Cu, Cr, Ni, and Zn accumulated at the top of the soil (1–2 m) after 20 years of irrigation with recycled water in the Dan region of Israel.[3]

Some trace elements are essential at low concentrations (Table 2.10) but toxic at elevated concentrations (e.g., Cu, Cr, Mo, Ni, Se, and Zn). As, Cr^{6+}, Fl, Pb, Hg, Mo, and Se are considered to be of environmental concern because they are taken up by plants in amounts potentially harmful to animals and humans. B, Cd, Cu, Cr^{6+}, Ni, Zn, and Se are of concern because of their phytotoxicity. These elements can be transferred to animals or humans through different pathways and, depending on their concentration, may cause human health effects.

Table 2.10 Concentration of Selected Trace Elements Normally Found in Soil and Plant Tissue and Their Impact on Plant Growth

Element	Soil concentration (μg/g)		Typical concentration (μg/g) in plant tissue (range)	Impact on plant growth[a]
	Range	Typical		
As	0.1–40	6	0.1–5	Not required
B	2–200	10	5–30	Required, wide species differences
Be	1–40	6	–	Not required: toxic
Bi	–	–	–	Not required: toxic
Cd	0.01–7	0.06	0.2–0.8	Not required: toxic
Cr	5–3000	100	0.2–1.0	Not required: low toxicity
Co	1–40	8	0.05–0.15	Required by legume at <0.2 ppm
Cu	2–100	20	2–15	Required at 2–4 ppm: toxic at >20 ppm
Pb	2–200	10	0.1–10	Not required: low toxicity
Mn	100–400	850	15–100	Required: toxicity depends on Fe/Mn ratio
Mo	0.2–5	2	1–100	Required at <0.1 ppm: low toxicity
Ni	10–1000	40	1–10	Not required: toxic at >50 ppm
Se	0.1–0.2	0.5	0.02–2.0	Not required: toxic at >50 ppm
V	20–500	100	0.1–10	Required by some algae; toxic at >10 ppm
Zn	10–300	50	15–200	Required: toxic at >200 ppm

Source: Adapted from Refs. 23–28.

[a]Concentrations on a dry-weight (70°C) basis.

Trace element accumulation in soils in relation to uptake by plants depends on the chemical forms of the elements, which can be in exchangeable, sorbed, organic-bound, carbonate, and sulfide forms. Their accumulation by plants depends on the soil supplying these elements to plant roots, on the rhizosphere environment, and on the characteristics of the plant root system. Soil pH has been shown to have a significant effect on plant uptake of trace elements in biosolids, much more consistently than other soil variables such as organic matter content, cation exchange capacity, and soil texture. Trace element toxicities to plants are more common in acid soils. Other soil components such as clay, organic matter, hydrous iron and hydrous manganese oxides, organic acids, amino acids, and humic and fulvic acids can also react to prevent trace element movement.

Persistent synthetic organic compounds and some organochlorine pesticides are potential hazards to the environment and public health. Knowledge about the health effects of these chemicals and the technology for their monitoring and removal from municipal water supply will always lag behind the development of new chemicals, which are introduced into commerce and industry at the approximate rate of 1000 per year and ultimately find their way into the watercourses that drain urban and industrial areas. In China, it is reported that toxicity to crops occurred often when trichloro-acetaldehyde was present in sewage. The chemical reduced yield markedly, affecting about 1.5% of the total sewage-irrigated area. Hence, the concentration of these compounds should be minimized by eliminating or reducing the discharge of contaminants into wastewaters or removing the contaminants via wastewater treatment (e.g., lime coagulation).

As a general rule, the order of magnitude of the acceptable limits of organic trace elements can be considered as being identical to those used in the potable water supply (see Table 2.4). The transformation of trace organic substances can occur in the soil by adsorption, volatilization, and biodegradation at different rates depending on the compound. Since the most dramatic and severe impact of pollution generally occurs from wastewater discharged by industry, source treatment of pollutants should be carried out and made the legal responsibility of the industry concerned.

2.3.5 pH

pH is an indicator of the acidity or alkalinity of water but is seldom a problem by itself. The normal pH range for irrigation water is from 6.5 to 8.4. pH values outside this range provide an indication that the water is abnormal in quality. In this case, irrigation water may cause a nutritional imbalance affecting plant growth and health. Moreover, abnormal pH can be very corrosive to such appurtenances as pipelines, sprinklers, and control valves.

Normally, pH is a routine measurement in irrigation water quality assessment as it may be an indication of the presence of toxic ions.

2.3.6 Bicarbonate and Carbonate

Substantial bicarbonate levels (>3–4 mEq/L or >180–240 mg/L) can increase soil pH and, in combination with carbonate, may affect soil permeability. Bicarbonate ion may combine with calcium or magnesium and precipitate as calcium carbonate or magnesium carbonate, increasing the SAR in the soil solution due to a lowering of the dissolved calcium concentration.

Water containing excess bicarbonate and carbonate can leave white lime deposits on leaves of plants irrigated with overhead sprinklers during hot periods. These white formations reduce the aesthetic quality of the plants and certainly their marketability. In addition, these deposits can accumulate to cause clogging of small openings in irrigation equipment such as drip emitters and spray nozzles.

The water quality limit for bicarbonate (HCO_3) to avoid foliar deposits in the case of sprinkler irrigation is 90 mg/L (1.5 mEq/L).[19–20] Severe plant damage could occur when bicarbonate concentration is over 500 mg/L (8.5 mEq/L).

2.3.7 Nutrients

The most important nutrients for crops are nitrogen, phosphorus, potassium, zinc, boron, and sulfur. Usually, recycled water contains enough of these elements to supply a large portion of a crop's needs.

The most beneficial nutrient for plants is nitrogen. Both the concentration and forms of nitrogen (nitrate and ammonium) need to be considered in irrigation water. The relative proportion of each form varies with the origin and treatment of the wastewater, but most commonly ammonium is the principal form, usually present in a concentration range of 5 to 40 mgN–NH_4/L. The organic fraction, which may be either soluble or fine particulates, consists of a complex mixture including amino acids, amino sugars, and proteins. All of these fractions are readily convertible to ammonium through the action of microorganisms in the wastewater or in soil to which the wastewater is applied. During aerobic wastewater treatment, some ammonium could be oxidized to nitrates through the action of nitrifying bacteria. Common nitrate concentrations in urban wastewater range from 0 to 30 mg N–NO_3/L.

Nitrogen is a macronutrient for plants that is applied on a regular basis. Nevertheless, at very high concentrations (over 30 mgN_{tot}/L) it can overstimulate plant growth, causing problems such as lodging and excessive foliar growth and also delay maturity or result in poor crop quality. Nitrogen sensitivity varies with the development stage of the crops. It may be beneficial during growth stages, but causes yield losses during flowering/fruiting stages. The long-term effects of excess nitrogen include weak stalks, stems, and/or branches unable to support the weight of the vegetation under windy or rainy conditions.

Pollution of groundwater from the percolation of nitrogen presents a health concern. This usually results from excessive application of nutrients in areas having permeable soils. When nitrogen is washed from soils and reaches streams, lakes, canals, and drainage ditches, it stimulates algae growth, which can result in plugged filters, valves, pipelines, and sprinklers. In addition, excessive nitrogen application to pastures may be hazardous to livestock that consume the vegetation.

Potassium in recycled water has little effect on crops. The phosphorus content in recycled water is too low to meet a crop's needs. Over time, phosphorus can build up in the soil and reduce the need for supplementation. Although excessive phosphorus does not appear to cause serious immediate problems to crops, it may affect future land use because some plants species are sensitive to high phosphorus concentrations.

Phosphorus can also be a problem in surface water runoff as a limiting factor in eutrophication.

2.3.8 Free Chlorine

For sprinkler irrigation, excessive residual chlorine in recycled water causes plant damage if high residual chlorine exists at the time of irrigation. As free chlorine (Cl_2) is highly reactive and unstable in water, a high level of residual chlorine rapidly dissipates if the treated water is stored in reservoirs for more than few hours.

Residual free chlorine concentrations below 1 mg/L are not likely to affect plant foliage. Some damage may occur on very sensitive species at relatively low levels of about 0.5 mg/L. Severe plant damage of a burning nature can occur in the presence of excessive free chlorine. Most reuse strategies will not face this problem if an intermediate storage facility is used, but care is needed during any period where the storage facility is bypassed for direct irrigation from the treatment plant.

2.4 SAMPLING AND MONITORING STRATEGIES

The development and implementation of an appropriate sampling procedure is a crucial step that influences the precision and reliability of water quality data. There are no strict rules for sampling location, timing, and handling in water reuse irrigation. However, depending on the type of monitored parameters, some basic sampling rules are described in the standard methods for water analysis or defined by the laboratory.

The main requirements for sampling recycled water are:

The type of samples can be either grab or composite samples to be used for water quality monitoring depending on the final objectives.

All samples should be well labeled, indicating the type of water, site location, date, time, and other pertinent data.

The water reuse permit usually defines sampling frequency. For better planning and management of the irrigation process, it is recommended to take seasonal samples in spring, summer, autumn, and winter in order to obtain representative data on the variation in water quality, in particular, nitrogen levels and salinity. The most important period for the sampling of trace elements is the crop's germination period.

The sample location should be, as closely as possible, representative of the recycled water quality at the point of use. It is recommended to take samples before and after the treatment plant, at different steps of the treatment process, as well as after the storage reservoir and at the point of use, if possible.

Table 2.11 Recommendations for Sample Preparation and Conservation

Parameter	Type of bottle	Addition of chemicals	Conservation	Comments
Anions and cations (chloride, sulfate, etc.), all forms of nitrogen and phosphorus, as well as general physicochemical parameters (pH, suspended solids, conductivity, etc.)	1 L plastic, with or without air	No additive	Dark, 4°C	Temperature and dissolved oxygen should be measured on site
COD	100 mL, plastic, no air	Sulfuric acid	Dark, 4°C	No additive is needed if the samples are analyzed within 48 h
BOD	500 mL, plastic, no air	No additive	Dark, 4°C	
Trace elements	250 mL, plastic, with or without air	Nitric acid	Dark, 4°C	A special bottle and additive is needed for the analysis of mercury (Hg)
Trace organics and pesticides	1 L, dark glass bottle, no air	No additive	Dark, 4°C	
Microbiological parameters (total and fecal coliforms, helminths, viruses, etc.)	1–5 L, sterile plastic bottle, with air	No additive	Dark, 4°C	Additive should be added only to disinfected effluent (sodium thiosulfate in presence of residual chlorine)

Table 2.12 Typical Minimum Monitoring Requirements and Sampling Frequency in Water Reuse Systems for Irrigation

Monitored parameters	Raw wastewater and recycled water	Receiving soils	Groundwater	
			Shallow aquifers	Deep aquifers
Coliforms[a]	Weekly to monthly	—	Bi-annual	Annual
Turbidity	On-line for unrestricted irrigation	—	—	—
Chlorine residual	On-line for unrestricted irrigation	—	—	—
Volume	Monthly	—	—	—
Water level	—	—	Bi-annual	—
pH	Monthly	Annual	Bi-annual	Annual
Suspended solids	Monthly	—	—	—
Total dissolved solids	Monthly	—	Bi-annual	Annual
Conductivity (EC_w)	Monthly	Bi-annual (EC_e)	Bi-annual	Annual
BOD	Monthly	—	—	—
Ammonia	Monthly	—	Bi-annual	Annual
Nitrites	Monthly	—	Bi-annual	Annual
Nitrates	Monthly	Annual (exchangeable NO_3)	Bi-annual	Annual
Total nitrogen	Monthly	Bi-annual	Bi-annual	Annual
Total phosphorus	Monthly	Bi-annual (extractable P)	Bi-annual	Annual
Phosphates (soluble)	Monthly	Bi-annual	Bi-annual	Annual
Major solutes (Na, Ca, Mg, K, Cl, SO_4, HCO_3, CO_3)	Quarterly	Bi-annual	Bi-annual	Bi-annual
Exchangeable cations (Na, Ca, Mg, K, Al)	—	Annual	—	—
Trace elements	Annual	—	—	—

[a]Unrestricted landscape irrigation and irrigation of food crops may require higher sampling frequency and additional monitoring parameters.

Sampling bottles should be clean. Plastic bottles are preferred because certain types of glass bottles yield boron to the samples. The sample quantity depends on the type of analysis to be performed. For the analysis of basic water characteristics and the main anions and cations, 1 L of sample is usually sufficient.

Sampling and handling should be done safely with suitable precautions to avoid disease transmission (i.e., plastic gloves or other protection).

Table 2.11 gives some basic recommendations for the sampling and handling of raw wastewater and treated recycled effluents. Minimum monitoring requirements and sampling frequency for water reuse projects for irrigation are summarized in Table 2.12.

It is important to stress that monitoring requirements and sampling frequency differ from one country to another and depend on the type of irrigation. As a rule, reuse criteria require sampling at the outlet of the treatment facility. In some cases the groundwater monitoring is required at the agricultural site, depending on the behavior of aquifers. Soil and plant monitoring are recommended to assess the influence of recycled water on soil characteristics and plant composition.

REFERENCES

1. U.S. Environmental Protection Agency, *Guidelines for Water Reuse*, Manual, EPA & USAID (United States Agency for International Development), 1992.
2. Metcalf & Eddy, *Wastewater Engineering: Treatment and Reuse*, 4th ed., McGraw-Hill Inc., New York, 2003.
3. Lazarova, V. (ed.), Role of water reuse in enhancing integrated water resource management, Final Report of the EU project CatchWater, EU Commission, 2001.
4. Chang, A.C., Page, A.L., and Asano, T., Developing human health-related chemical guidelines for reclaimed wastewater and sewage sludge applications in agriculture, Report submitted to Community Water Supply and Sanitation Unit, Division of Environmental Health, World Health Organization, Geneva, Switzerland, 1993.
5. U.S. Environmental Protection Agency, *National Interim Primary Drinking Water Regulations*, EPA 570/9-76-003, Washington, DC, 1976.
6. World Health Organization (WHO), *Directives de Qualité pour L'eau de Boisson*, 10th ed., Geneva, 2000.
7. Commission of the European Communities (CEC), Directive 2000/60/EC of the European Parliament and of the Council establishing a framework for the Community action in the field of water policy. *Published at Official Journal (OJ L 327)*, December 22, 2000.
8. World Bank, Wastewater irrigation in developing countries: health effects and technical solutions. *Technical Paper No. 51. World Bank*, by Shuval H.I., Adin A., Fattal B., Rawitz E. and Yekutiel P., Washington, DC, 1986.

9. Cifuentes, E., The epidemiology of enteric infections in agricultural communities exposed to wastewater irrigation: perspectives for risk control, *Int. J. Environ. Health Res.*, 8, 203, 1998.
10. Blumenthal, U.J., Peasey, A., Ruiz-Palacios, G., and Mara, D.D., Guidelines for wastewater reuse in agriculture and aquaculture: recommended revisions based on new research evidence, *WELL Study, Task n°68, Part 1. pp.67*, http:/ www.lboro.ac.uk/well/, 2000.
11. Bouhoum, K., Habbari, K., and Schwartzbrod, J., Epidemiological study of helminthic infections in Marrakech wastewater spreading zone, in *Proc. IAWQ Second International Symposium on Wastewater Reclamation and Reuse,* Iraklio, Greece, Oct. 1995, 2, 679, 1995.
12. Devaux, I., Intérêt et limites de la mise en place d'un suivi sanitaire dans le cadre de la réutilisation agricole des eaux usées traitées de l'agglomération Clermontoise, PhD thesis, Université Grenoble I, Grenoble, France, 1999.
13. Peasey, A.E., Human exposure to *Ascaris* infection through wastewater reuse in irrigation and its public health significance. PhD thesis, London, University of London, 2000.
14. Shuval, H.I., Yekutiel, P., and Fattal, B., Epidemiological evidence for helminth and cholera transmission by vegetables irrigated with wastewater. Jerusalem—case study, *Wat. Sci. Techn.*, 1, 4/5, 433, 1985.
15. U.S. Environmental Protection Agency, Indoor Air Facts N°4 (revised). April 7 1998, Sick Building Syndrome (SBS). [Online] http://www.epa.gov/iaq/pubs/sbs. Html, 1998.
16. Sheikh, B., Cooper, R., and Israel, K., Hygienic evaluation of recycled water used to irrigate food crops—a case study, *Wat. Sci. Techn.*, 40, 4–5, 261, 1999.
17. Kandiah, A., Environmental impacts of irrigation development with special reference to saline water use, in *Water, Soil and Crop Management Relating to Use of Saline Water*, AGL/MISC/16, FAO, Rome, 1990, 152.
18. FAO, Food and Agriculture Organization of the United Nations, The use of saline waters for crop production, Irrigation and Drainage Paper n°48, by Rhoades, J.D., Kandiah, A. and Mashall, A.M., Rome, 1992.
19. FAO, Food and Agriculture Organization of the United Nations, Water quality for irrigation, Irrigation and Drainage Paper n°29, by Ayers, R. S. and Wescot, D. W., Rome, Italy, 1985
20. Pettygrove, G.S. and Asano, T., *Irrigation with Reclaimed Municipal Wastewater—A Guidance Manual*, Lewis Publishers Inc., Chelsea, 1985.
21. FAO, Food and Agriculture Organization of the United Nations, Irrigation and Drainage Paper 47, Wastewater treatment and use in agriculture, by Pescod, M., Food and Agriculture Organization, Rome, 1992.
22. National Academy of Sciences and National Academy of Engineering, Water quality criteria, Report of the Committee on Water Quality Criteria, Ecological Research Series, EPA-R3-73-033, 1973.
23. Allaway, W.H., Agronomic controls over the environmental cycling of trace elements, *Adv. Agron.*, 20, 236, 1994.
24. Bowen, H. (ed.), Environmental Chemistry of the Elements, Academic Press, New York, 1979.
25. Chapman, H.D. (ed.), *Diagnostic Criteria for Plants and Soils*, Quality Printing Co. Inc., Abilene, TX, 1965.
26. Lisk, D.J., Trace metals in soils, plants and animals, *Adv. Agron.*, 24, 267, 1972.

27. Page, A.L., Fate and effects of trace elements in sewage sludge when applied to agricultural soils. A literature review study, EPA-670/2-74-005, USEPA, Cincinnati, OH, 1974.
28. Chang, A.C. and Page, A.L., The role of biochemical data in assessing the ecological and health effects of trace elements, in *Proc. 15th World Congr. Soil Sci.*, Riverside, CA, 1994.

3

International Health Guidelines and Regulations

James Crook and Valentina Lazarova

CONTENTS

3.1 WHO guidelines for irrigation 64
3.2 USEPA guidelines for water reuse 66
3.3 California water recycling criteria 66
3.4 Other water reuse regulations 73
3.5 Standards for urban uses of recycled water and landscape irrigation 76
3.6 Standard enforcement and perspectives 79
References 80

All water reuse standards and guidelines are directed principally at health protection. Contact, inhalation, or ingestion of reclaimed water containing pathogenic microorganisms or toxic chemicals creates the potential for adverse health effects. The most common health concern associated with nonpotable reuse of treated municipal wastewater is the potential transmission of infectious disease by microbial pathogens. Waterborne disease outbreaks of epidemic proportions have, to a great extent, been controlled, but the potential for disease transmission through the water route has not been eliminated. For example, the irrigation of market crops with poorly treated wastewater in developing countries has been associated with enteric disease.[1] The occurrence and concentration of microbial pathogens in raw municipal wastewater depend on a number of factors, and it is not possible to predict with any degree

1-56670-649-1/05/$0.00+$1.50

of assurance what the general characteristics of a particular wastewater will be with respect to infectious agents.

Effects of physical parameters, e.g., pH, color, temperature, particulate matter, and chemical constituents, e.g., chlorides, sodium, and heavy metals, in reclaimed water used for irrigation are well known, and recommended limits have been established for many constituents (see Chapters 2 and 5). With a few exceptions, minimal health concerns are associated with chemical constituents where reclaimed water is not intended to be consumed. While there has been some concern regarding irrigation of food crops with reclaimed water, available data indicate that potentially toxic organic pollutants do not readily enter edible portions of plants that are irrigated with treated municipal wastewater.[2] However, use of poorly treated wastewater or wastewater containing a significant fraction of industrial wastes for irrigation may present hazards to crops or consumers of the crops. Both organic and inorganic constituents need to be considered where reclaimed water utilized for food crop irrigation reaches potable groundwater supplies.

Water reuse standards or guidelines vary with type of application, regional context, and overall risk perception. In practice, these factors are expressed through different water quality requirements as well as treatment process requirements and criteria for operation and reliability. The safe and beneficial implementation of water reuse schemes could be better guaranteed by the development of appropriate codes of good practices that are as important for farmers and operators as the quality requirements for water reuse (see Chapters 4 and 5).

Table 3.1 provides a summary of water quality parameters of concern in water reuse guidelines and regulations with respect to their significance for reclaimed water used for irrigation as well as approximate ranges of the selected parameters in secondary effluent and reclaimed water.[3] The treatment of municipal wastewater is typically designed to meet water quality objectives based on particulate matter (TSS or turbidity), organic content (BOD), biological indicators (e.g., total or fecal coliforms, *Escherichia coli*, helminth eggs, enteroviruses), nutrient levels (N and P), and, in some cases, chlorine residues.

Historically, agricultural water reuse was the first application for which water reuse standards were developed. Different countries have developed different approaches to protect the public health and the environment. A major factor in the choice of regulatory strategy in many countries is economics, i.e., costs of treatment, monitoring, and distribution of the recycled water. Some developed countries have tended to develop conservative low-risk guidelines or standards based on relatively costly high technology. California's Water Recycling Criteria[4] is an example of conservatively based regulations. A number of other countries advocate another strategy of controlled health risk based on the World Health Organization (WHO), Health Guidelines for the Use of Wastewater in Agriculture and Aquaculture.[5]

A key element in water reuse regulations is enforcement and the effectiveness of applied treatment processes. Extensive monitoring data are

Table 3.1. Summary of Water Quality Parameters of Concern in Water Reuse Regulations and Guidelines for Irrigation

Parameter	Significance for irrigation with recycled water	Range in secondary and tertiary effluents	Treatment goal in recycled water
Total suspended solids	Measures of particles; can be related to microbial contamination; can interfere with disinfection; clogging of irrigation systems; deposition	5–50 mg/L	< 5–35 mgTSS/L
Turbidity		1–30 NTU	< 0.2–35 NTU
BOD_5	Organic substrate for microbial growth; can favor bacterial regrowth in distribution systems and microbial fouling	10–30 mg/L	< 5–45 mgBOD/L
COD		50–150 mg/L	< 20–200 mgCOD/L
Total coliforms	Measure of risk of infection due to potential presence of pathogens; can favor biofouling of sprinklers and nozzles in irrigation systems	< 10–10^7 cfu/100 mL	< 1–200 cfu/10 mL
Fecal coliforms		< 1–10^6 cfu/100 mL	< 1–10^4 cfu/100 mL
Helminth eggs		< 1–10/L	< 0.1–5/ L
Viruses		< 1–100/L	< 1/40 L to < 1/50 L
Heavy metals	Specific elements (Cd, Ni, Hg, Zn, etc) are toxic to plants, and maximum concentration limits exist for irrigation	—	< 0.001 mgHg/L < 0.01 mgCd/L < 0.02–0.1 mgNi/L
Inorganics	High salinity and boron are harmful for irrigation of some sensitive crops	—	< 450–4000 mgTDS/L < 1 mgB/L
Chlorine residual	Recommended to prevent bacterial regrowth; excessive amount of free chlorine (> 0.05 mg/L) can damage some sensitive crops	—	0.5– > 5 mgCl/L
Nitrogen[a]	Fertilizer for irrigation; can contribute to algal growth and eutrophication in storage reservoirs, corrosion (N-NH_4), or scale formation (P)	10–30 mgN/L	< 10–15 mgN/L;
Phosphorus[a]		0.1–30 mgP/L	< 0.1–2 mgP/L

Source: Adapted from Ref. 3.

[a]Only in specific cases for groundwater protection or to limit eutrophication in storage reservoirs.

available for reclaimed water that complies with the California Water Recycling Criteria. An analysis[6] of a 10-year compilation of data on six tertiary treatment plants concluded that the California criteria for reclaimed water applications that require a high level treatment (coagulation/flocculation, filtration, chlorination) produced an essentially virus-free effluent. Surveys of agricultural reuse systems conducted in California have indicated no deterioration in quality or quantity of the irrigated crops.[7] Information regarding the treatment effectiveness associated with the WHO recommendations is not readily available at this time. Information on treatment effectiveness for pathogen removal is necessary to validate the safety of the WHO guidelines. Some experts are of the opinion that the WHO recommendations are too lenient and do not provide sufficient public health safety,[8] while others[9] suggest that the WHO recommendations are too conservative and, therefore, overregulate water reuse.

3.1 WHO GUIDELINES FOR IRRIGATION

The first WHO water reuse guidelines were published in 1973 and included recommended criteria for several uses, including crop irrigation and potable reuse.[10] In 1985 WHO and other international organizations sponsored a meeting of experts to review the use of reclaimed water for agriculture and aquaculture.[11] The experts at the meeting concluded that the health risks for those uses were minimal, the then-current guidelines were overly restrictive, and WHO should initiate revision of the 1973 guidelines. WHO subsequently developed revised guidelines, which were published in 1989. These guidelines, entitled Health Guidelines for the Use of Wastewater for Agriculture and Aquaculture[5], are summarized in Table 3.2. The technology recommended by WHO for water reuse is stabilization ponds or any equivalent treatment processes. Some countries have used the WHO guidelines as the basis for their agricultural reuse standards. In the absence of recommendations for suspended solids in the WHO guidelines, these standards typically use TSS concentrations varying between 10 and 30 mg/L.

It is noteworthy that the original WHO guidelines of 1973 were more stringent for food crops eaten raw than the 1989 guidelines with respect to fecal coliforms, i.e., the recommended limit increased from 100 to 1000 FC/100 mL, and that the 1989 WHO guidelines recommend more stringent standards for public lawns (200 FC/100 mL) than for crops eaten raw (1000 FC/100 mL).

The 1989 WHO guidelines are currently under revision. WHO is reviewing epidemiological evidence related to disease transmission resulting from irrigation with reclaimed water and is updating its approach to microbiological risk assessment. The new guidelines will include complementary options for health protection, such as treatment of wastewater, crop restrictions, application controls, and control of human exposures. Water quality requirements may be tightened where there is a high rate of infection from certain pathogens. The multibarrier approach throughout the water cycle is also considered to be an important element. In addition to guidelines for irrigation uses of reclaimed

Table 3.2 WHO Guidelines for Safe Use of Wastewater in Agriculture, 1989[a]

Category	Reuse conditions	Exposed groups	Intestinal nematodes[b] (arithmetic mean no. of eggs/L)[c]	Faecal coliforms (geometric mean no./100 mL)[c]	Wastewater treatment expected to achieve required microbiological quality
A	Irrigation of crops likely to be eaten uncooked, sports fields, public parks[d]	Workers, consumers, public	<1	<1000[d]	A series of stabilization ponds designed to achieve the microbiological quality indicated, or equivalent treatment
B	Irrigation of cereal crops, industrial crops, fodder crops, pasture, and trees[e]	Workers	<1	No standard recommended	Retention in stabilization ponds for 8–10 days or equivalent helminth and fecal coliform removal
C	Localized irrigation of crops in category B if exposure of workers and the public does not occur	None	Not applicable	Not applicable	Pretreatment as required by the irrigation technology, but not less than primary sedimentation

Source: Ref. 5.

[a] In specific cases, local epidemiological, sociocultural, and environmental factors should be taken into account and the guidelines modified accordingly.

[b] *Ascaris* and *Trichuris* species and hookworms.

[c] During the irrigation period.

[d] A more stringent guideline (<200 fecal coliforms/100 mL) is appropriate for public lawns, such as hotel lawns, with which the public may come into direct contact.

[e] In the case of fruit trees, irrigation should cease 2 weeks before fruit is picked, and no fruit should be picked off the ground. Sprinkler irrigation should not be used.

water, WHO is developing guidelines for aquaculture (shellfisheries), artificial recharge exclusively for potable supply, and urban settings.

3.2 USEPA GUIDELINES FOR WATER REUSE

Regulations in the United States are developed at the state level, although the U.S. Environmental Protection Agency (USEPA), in conjunction with the U.S. Agency for International Development, published Guidelines for Water Reuse[12] in 1992, which are currently under revision. The recommendations in the 1992 USEPA guidelines for reclaimed water used for irrigation applications are provided in Table 3.3. The guidelines are not intended to be used as definitive water reuse criteria. They are intended to provide reasonable guidance for states that have not developed their own criteria or guidelines.

3.3 CALIFORNIA WATER RECYCLING CRITERIA

The State of California has been a leader in the development of comprehensive water reuse regulations, and the California Department of Health Services last revised its criteria in 2000. The Water Recycling Criteria[4] provide a very comprehensive set of water quality and other requirements and have served as the basis for similar criteria in other states and countries. Concerning water quality for the irrigation of food crops, both USEPA[12] recommendations and almost all U.S. state-specific regulations require a high level of disinfection with inactivation of total or fecal coliforms (e.g., $\leq$2.2 total coli/100 mL or no detectable fecal coli/100 mL), with total coliforms being the more conservative indicator. The California criteria include conservative requirements for both water quality monitoring, treatment train design, and operation. However, the criteria do not include some factors such as irrigation rates or storage requirements.

Table 3.4 summarizes the California Water Recycling Criteria. Besides the sections that are identical to those in previous regulations adopted in 1978 for design and reliability of operation, engineering report, personnel, maintenance, bypassing, and other general requirements, several significant changes and additions have been included:[13]

1. Change in terminology: the words “reclaimed” and “reuse” have been replaced with “recycled” and “recycling” in all regulations.
2. Changes in quality and treatment requirements for some recycled water applications: primary effluent is no longer acceptable for irrigation of industrial crops and is replaced by oxidized wastewater, i.e., undisinfected secondary-treated effluent. For high-level recycled water uses, additional requirements for the turbidity are included where membranes are used in lieu of media filtration. If membranes are used, the turbidity cannot exceed 0.2 NTU more than 5% of the time within a 24-hour period and cannot exceed 0.5 NTU at any time.

3. Pathogen monitoring in reclaimed water used for nonrestricted recreational impoundments: the use of reclaimed water for nonrestricted recreational impoundments is the only application for which so-called conventional treatment is required. Conventional treatment means a treatment chain that includes a chemical clarification process between the coagulation and filtration processes and produces an effluent that meets the definition for disinfected tertiary recycled water. In consideration of the likelihood of ingesting reclaimed water while swimming in nonrestricted recreational impoundments and the paucity of information regarding pathogen removal where a discrete chemical clarification process is omitted, the regulations are more restrictive for this use than other nonpotable uses. Disinfected tertiary reclaimed water may be used in lieu of water that has received conventional treatment if the reclaimed water is monitored for enteric viruses, *Giardia lamblia*, and *Cryptosporidium parvum*. Monthly monitoring for these pathogens is required during the first year of operation and quarterly during the second year of operation. Monitoring may be discontinued after the first 2 years upon approval by California Department of Health Services (DHS). It should be noted that, without exception, chemical coagulation prior to filtration is a required process for reclaimed water used for nonrestricted recreational impoundments.
4. Requirements were included for several additional recycled water applications. Table 3.5 illustrates the different uses of recycled water allowed in California.[14]
5. Addition of more specific and restrictive chlorine disinfection requirements: a residual chlorine concentration times modal contact time (CT) value of at least 450 mg-min/L at all times with a modal contact time of at least 90 min or a disinfection process that, when combined with the filtration process, has been demonstrated to reduce the concentration of MS2 bacteriophage or poliovirus by 5 logs
6. Inclusion of use area requirements that previously were used as guidelines. Reclaimed water use area requirements include the following:

   no irrigation or impoundment of undisinfected reclaimed water within 50 m of any domestic water supply well

   no irrigation of disinfected secondary-treated reclaimed water within 30 m of any domestic water supply well

   no irrigation with tertiary-treated (secondary treatment, filtration, and disinfection) reclaimed water within 15 m of any domestic water supply well unless special conditions are met

   no impoundment of tertiary-treated reclaimed water within 30 m of any domestic water supply well

   only tertiary-treated reclaimed water can be sprayed within 30 m of a residence or places where more than incidental exposure is likely to occur

Table 3.3 USEPA Suggested Guidelines for Reuse of Municipal Wastewater for Irrigation Applications

Types of reuse	Treatment	Reclaimed water quality[a]	Reclaimed water monitoring	Setback distances[b]	Comments
Urban reuse					
All types of landscape irrigation (e.g., golf courses, parks, cemeteries)—also vehicle washing, toilet flushing, use in fire-protection systems and commercial air conditioners, and other uses with similar access or exposure to the water	Secondary[c] Filtration[d] Disinfection[e]	pH 6–9 ≤10 mg/L BOD[f] ≤2 NTU[g] No detectable fecal coli/100 mL[h,i] ≥1 mg/L Cl_2 residual[j]	pH—weekly BOD—weekly Turbidity—continuous Coliform—daily Cl_2 residual—continuous	15 m to potable water supply wells	Consult recommended agricultural (crop) limits for metals. A lower level of treatment, e.g., secondary treatment and disinfection to achieve ≤14 fecal coli/100 mL, may be appropriate at controlled-access irrigation sites where design and operational measures significantly reduce the potential of public contact with reclaimed water. Chemical (coagulant and/or polymer) addition prior to filtration may be necessary to meet water quality recommendations. The reclaimed water should not contain measurable levels of pathogens.[k] Reclaimed water should be clear, odorless, and contain no substances that are toxic upon ingestion. Higher chlorine residual and/or longer contact time may be necessary to assure that viruses and parasites are inactivated or destroyed. Chlorine residual of ≥0.5 mg/L in the distribution system is recommended to reduce odors, slime, and bacterial regrowth. Provide treatment reliability.

Restricted access area irrigation					
Sod farms, silviculture sites, and other areas where public access is prohibited, restricted, or infrequent	Secondary[c] Disinfection[e]	pH 6–9 ≤30 mg/L BOD[f] ≤30 mg/L SS ≤200 fecal coli/100 mL[h,l,m] ≥1 mg/L Cl_2 residual[j]	pH—weekly BOD—weekly SS—daily Coliform—daily Cl_2 residual—continuous	90 m to potable water supply wells 30 m to areas accessible to the public (if spray irrigation)	Consult recommended agricultural (crop) limits for metals. If spray irrigation, SS < 30 mg/L may be necessary to avoid clogging of sprinkler heads. Provide treatment reliability.
Agricultural reuse—food crops not commercially processed[n]					
Surface or spray irrigation of any food crop, including crops eaten raw	Secondary[c] Filtration[d] Disinfection[e]	pH 6–9 ≤ 10 mg/L BOD[f] ≤2 NTU[g] No detectable fecal coli/100 mL[h,l] ≥1 mg/L Cl_2 residual[j]	pH—weekly BOD—weekly Turbidity—continuous Coliform—daily Cl_2 residual—continuous	15 m to potable water supply wells	Consult recommended agricultural (crop) limits for metals. Chemical (coagulant and/or polymer) addition prior to filtration may be necessary to meet water quality recommendations. Reclaimed water should not contain measurable levels of pathogens.[k] Higher chlorine residual and/or a longer contact time may be necessary to assure that viruses and parasites are inactivated or destroyed. High nutrient levels may adversely affect some crops during certain growth stages. Provide treatment reliability
Agricultural reuse—food crops commercially processed[n]; surface irrigation of orchards and vineyards	Secondary[c] Disinfection[e]	pH 6–9 ≤30 mg/L BOD[f] ≤30 mg/L SS ≤200 fecal coli/100 mL[h,l,m] ≥1 mg/L Cl_2 residual[j]	pH—weekly BOD—weekly SS—daily Coliform—daily Cl_2 residual—continuous	90 m to potable water supply wells 30 m to areas accessible to the public	Consult recommended agricultural (crop) limits for metals. If spray irrigation, SS < 30 mg/L may be necessary to avoid clogging of sprinkler heads. High nutrient levels may adversely affect some crops during certain growth stages. Provide treatment reliability.

(*continued*)

Table 3.3 Continued

Types of reuse	Treatment	Reclaimed water quality[a]	Reclaimed water monitoring	Setback distances[b]	Comments
Agricultural reuse—non-food crops					
Pasture for milking animals; fodder, fiber, and seed crops	Secondary[c] Disinfection[e]	pH 6–9 $\le$30 mg/L BOD[f] $\le$30 mg/L SS $\le$200 fecal coli/100 mL[h,I,m] $\ge$1 mg/L Cl_2 residual[j]	pH—weekly BOD—weekly SS—daily Coliform—daily Cl_2 residual—continuous	90 m to potable water supply wells 30 m to areas accessible to the public (if spray irrigation)	Consult recommended agricultural (crop) limits for metals. If spray irrigation, SS < 30 mg/L may be necessary to avoid clogging of sprinkler heads. High nutrient levels may adversely affect some crops during certain growth stages. Milking animals should be prohibited from grazing for 15 days after irrigation ceases. A higher level of disinfection, e.g., to achieve $\le$14 fecal coli/100 mL, should be provided if this waiting period is not adhered to. Provide treatment reliability.

Source: Adapted from Ref. 12.

[a]Unless otherwise noted, recommended quality limits apply to reclaimed water at the point of discharge from the treatment facility.

[b]Setbacks are recommended to protect potable water supply sources from contamination and to protect humans from unreasonable health risks due to exposure to reclaimed water.

[c]Secondary treatment processes include activated sludge processes, trickling filters, rotating biological contactors, and many stabilization pond systems. Secondary treatment should produce effluent in which both the BOD and SS do not exceed 30 mg/L.

[d]Filtration means the passing of wastewater through natural undisturbed soils or filter media such as sand and/or anthracite.

[e]Disinfection means the destruction, inactivation, or removal of pathogenic microorganisms by chemical, physical, or biological means. Disinfection may be accomplished by chlorination, ozonation, other chemical disinfectants, UV radiation, membrane processes, or other processes.

[f]As determined from 5-day BOD test.

[g]The recommended turbidity limit should be met prior to disinfection. The average turbidity should be based on a 24-hour time period. The turbidity should not exceed 5 NTU at any time. If SS is used in lieu of turbidity, the average SS should not exceed 5 mg/L.

[h]Unless otherwise noted, recommended coliform limits are median values determined from the bacteriological results of the last 7 days for which analyses have been completed. Either the membrane filter or fermentation tube technique may be used.

[i]The number of fecal coliform organisms should not exceed 14/100 mL in any sample.

[j]Total chlorine residual after a minimum contact time of 30 minutes.

[k]It is advisable to fully characterize the microbiological quality of the reclaimed water prior to implementation of a reuse program.

[l]The number of fecal coliform organisms should not exceed 800/100 mL in any sample.

[m]Some stabilization pond systems may be able to meet this coliform limit without disinfection.

[n]Commercially processed food crops are those that, prior to sale to the public or others, have undergone chemical or physical processing sufficient to destroy pathogens.

Table 3.4 California Water Recycling Criteria: Treatment and Quality Requirements for Nonpotable Uses of Reclaimed Water

Type of use	Total coliform limits[a]	Treatment required
Irrigation of fodder, fiber, and seed crops, orchards[b] and vineyards,[b] processed foodcrops, non–food-bearing trees, ornamental nursery stock,[c] and sod farms[c]; flushing sanitary sewers	None required	Secondary
Irrigation of pasture for milking animals, landscape areas,[d] ornamental nursery stock, and sod farms where public access is not restricted; landscape impoundments; industrial or commercial cooling water where no mist is created; nonstructural fire fighting; industrial boiler feed; soil compaction; dust control; cleaning roads, sidewalks, and outdoor areas	≤23/100 mL ≤240/100 mL in more than one sample in any 30-day period	Secondary Disinfection
Irrigation of food crops[b]; restricted recreational impoundments; fish hatcheries	≤2.2/100 mL ≤23/100 mL in more than one sample in any 30-day period	Secondary Disinfection
Irrigation of food crops[e] and open access landscape areas[f] toilet and urinal flushing industrial process water; decorative fountains; commercial laundries and car washes; snow-making; structural fire fighting; industrial or commercial cooling where mist is created	≤2.2/100 mL ≤23/100 mL in more than one sample in any 30-day period 240/100 mL (maximum)	Secondary Coagulation[g] Filtration[h] Disinfection
Nonrestricted recreational impoundments	≤2.2/100 mL ≤23/100 mL in more than one sample in any 30-day period 240/100 mL (maximum)	Secondary Coagulation Clarification[i] Filtration[h] Disinfection

Source: Ref. 4

[a]Based on running 7-day median.

[b]No contact between reclaimed water and edible portion of crop.

[c]No irrigation for at least 14 days prior to harvesting, sale, or allowing public access.

[d]Cemeteries, freeway landscaping, restricted access golf courses, and other controlled access areas.

[e]Contact between reclaimed water and edible portion of crop; includes edible root crops.

[f]Parks, playgrounds, schoolyards, residential landscaping, unrestricted access golf courses, and other uncontrolled access irrigation areas.

[g]Not required if the turbidity of the influent to the filters is continuously measured, does not exceed 5 nephelometric turbidity units (NTU) for more than 15 minutes, and never exceeds 10 NTU, and there is capability to automatically activate chemical addition or divert the wastewater if the filter influent turbidity exceeds 5 NTU for more than 15 minutes.

[h]The turbidity after filtration through filter media cannot exceed 2 NTU within any 24-hour period, 5 NTU more than 5% of the time within a 24-hour period, and 10 NTU at any time. The turbidity after filtration through a membrane process cannot exceed 0.2 NTU more than 5% of the time within any 24-hour period and 0.5 NTU at any time.

[i]Not required if reclaimed water is monitored for enteric viruses, *Giardia*, and *Cryptosporidium*.

Table 3.5 Reclaimed Water Uses for Irrigation Allowed in California[a]

Type of irrigation	Required treatment level			
	Disinfected tertiary recycled water	Disinfected secondary 2.2 recycled water	Disinfected secondary 23 recycled water	Undisinfected secondary recycled water
Food crops where reclaimed water contacts the edible portion of the crop, including all root crops	Allowed	Not allowed	Not allowed	Not allowed
Parks and playgrounds	Allowed	Not allowed	Not allowed	Not allowed
School yards	Allowed	Not allowed	Not allowed	Not allowed
Residential landscaping	Allowed	Not allowed	Not allowed	Not allowed
Unrestricted-access golf courses	Allowed	Not allowed	Not allowed	Not allowed
Any other uses not prohibited by other provisions of the California Code of Regulations	Allowed	Not allowed	Not allowed	Not allowed
Food crops where edible portion is produced above ground and not contacted by reclaimed water	Allowed	Allowed	Not allowed	Not allowed
Cemeteries	Allowed	Allowed	Allowed	Not allowed
Freeway landscaping	Allowed	Allowed	Allowed	Not allowed
Restricted-access golf courses	Allowed	Allowed	Allowed	Not allowed
Ornamental nursery stock and sod farms	Allowed	Allowed	Allowed	Not allowed
Pasture for milk animals	Allowed	Allowed	Allowed	Not allowed
Any nonedible vegetation with access control to prevent use as a park, playground, or school yard	Allowed	Allowed	Allowed	Not allowed
Orchards with no contact between edible portion and reclaimed water	Allowed	Allowed	Allowed	Allowed
Vineyards with no contact between edible portion and reclaimed water	Allowed	Allowed	Allowed	Allowed
Non–food-bearing trees, including Christmas trees not irrigated < 14 days before harvest	Allowed	Allowed	Allowed	Allowed
Fodder crops (e.g., alfalfa) and fiber crops (e.g., cotton)	Allowed	Allowed	Allowed	Allowed
Seed crops not eaten by humans	Allowed	Allowed	Allowed	Allowed
Food crops that must undergo commercial pathogen-destroying processing before consumption by humans (e.g., sugar beets)	Allowed	Allowed	Allowed	Allowed

Source: Ref. 14.

[a]This table is a summary of the California Water Recycling Criteria; 2.2 and 23 stand for 2.2 and 23 total coliforms/100 mL. The tertiary treatment is chemical coagulation, filtration, and disinfection. Note that chemical addition is not required under certain conditions.

confinement of runoff to the reclaimed water use area unless otherwise authorized by the regulatory agency
protection of drinking water fountains against contact with reclaimed water
signs at sites using reclaimed water that are accessible to the public, although educational programs or other approaches to assure public notification may be acceptable to DHS
prohibition of hose bibbs on reclaimed water piping systems accessible to the public

7. Specific requirements for dual plumbed systems that supply recycled water for residential irrigation or to plumbing outlets within a building.
8. Inclusion of cross-connection control requirements.

In California, laws and regulations exist that mandate water reuse under certain conditions. Section 13550 of the California Water Code[15] states that the use of potable domestic water for nonpotable uses, including, but not limited to, cemeteries, golf courses, highway landscaped areas, and industrial and irrigation uses, is a waste or an unreasonable use of the water if reclaimed water is available that meets certain conditions, i.e., adequate quality, reasonable cost, and no adverse effect on public health and environment. Moreover, in 2000 the California legislature passed the Water Recycling in Landscape Act,[16] which created a state-mandated local program. Some local jurisdictions in the state have taken action to require the use of recycled water in certain situations.

3.4 OTHER WATER REUSE REGULATIONS

Other countries (e.g., Australia, Canada, Jordan, Israel, and South Africa) have developed their own standards. Water quality requirements generally limit coliform organisms such as total and fecal coliforms (TC and FC) or *E. coli*. Other pathogens, such as viruses and protozoa, are seldom determined and are rarely required as control criteria.

The wide range of approaches applied to water reuse regulation is characterized by a number of inconsistencies and differences not only among nations, but also within a given country (e.g., Australia, Italy, Spain, and the United States). These differences pertain mostly to the existing irrigation practices, local soil conditions, desire to protect public health, choice of irrigation or wastewater treatment technologies, and economic feasibility. A comparison of water quality requirements for irrigation with recycled water in some countries (stringent limits for irrigation) is provided in Table 3.6.

Some countries such as Israel and South Africa, and more recently Japan and Australia, have chosen criteria more or less similar to those of California and do not use the WHO guidelines, which are considerably less restrictive. Around the Mediterranean basin however, and particularly in Europe, many of the existing regulations and guidelines are based on the WHO guidelines,

Table 3.6 Water Quality Criteria (Maximum Limits) for Agricultural Irrigation

Parameter	California (2000) *regulations*	Arizona (2000) *regulations*	USEPA (1992) *guidelines*	Israel (1978) *regulations*	WHO (1989) *guidelines*	WHO (2001) *draft guidelines*	France (2001) *draft guidelines*	Spain (2001) *draft regulations*
Total coliforms/100 mL	2.2[a]							
Fecal coliforms/100 mL		Not detected	Not detected	2.2 (50%); 12 (80%)	1000	1000		
E. coli/100 mL							1000	200
Salmonella							Absence	
Viruses								
Helminths, eggs/L					1	0.1	1[d]	1[e]
Total BOD_5, mg/L			10	15				
COD, mg/L							125	
Suspended solids, mg/L				15			35	20
Turbidity, NTU	2	2	2	—				5
pH			6–9					
Conductivity, dS/m								
Dissolved O_2, mg/L				≥0.5				
Residual Cl, mg/L	Dependent on contact time		≥1	≥0.5				
Minimum treatment required	Tertiary + disinfection	Tertiary + disinfection	Tertiary + disinfection	Secondary[b]	Stabilization ponds[c]	Secondary + disinfection, lagooning		

Depending on the analytical method, total and fecal coliforms are measured in terms of MPN (most probable number) or cfu (colony forming unis).
[a]TC based on running 7-day median.
[b]Seasonal storage may constitute an equivalent to tertiary treatment.
[c]Stabilisation ponds in series with proper retention time.
[d]Absence of *Taenia* eggs.
[e]Legionella pneumophila 0/100 mL for greenhouse irrigation and cooling and *Taenia saginata* and *solemn* < 1/L for irrigation if pasture.

even though regulatory authorities recognize their limitations. Additional criteria, such as treatment requirements or use limitations, are often required in order to improve public health safety. France and Spain (Andalusia, Balearic Islands), for example, include additional recommendations in their draft guidelines. Traditional practices and economic considerations have a strong influence on water reuse guidelines/regulations developed in European countries.[17]

While there appears to be a general agreement that the WHO guidelines are insufficient for implementation in developed countries, there is no general consensus to date on the best approach to follow. There is a wide discrepancy between the California Water Recycling Criteria and the WHO guidelines. Some experts favor an approach that lies between the California criteria and the WHO guidelines. Developing a scientifically sound rationale is critical to implementation of safe and acceptable water reuse criteria, particularly for areas where international tourism and export of agricultural products are significant or where water reuse is mainly performed for environmental protection.[17]

The most recent draft water reuse guidelines of Spain provide an example[18] of guidelines that are not as restrictive as the California criteria but more restrictive than the WHO guidelines: *E. coli* is proposed as a microbiological indicator, with concentrations from 0 to 10,000 cfu/mL, depending on the type of irrigation and public access. Israeli standards incorporate a "multiple barrier" approach that allows irrigation with low-quality effluent in the presence of a sufficient number of barriers to prevent any transmission of pathogens. The number of barriers depends on the type of treatment, type of irrigation, type of crop, and distance from the crops.

No water reuse regulations or guidelines are based on strict risk assessment methodology. Treatment and quality limits are based on factors such as pathogen destruction or inactivation, degree of direct or indirect human contact with reclaimed water, operational experience, research results, attainability, and, ultimately, judgment by the regulators. Different countries ascribe different treatment and quality requirements for specific uses based on perceived risk. It is important to stress, however, that use categories in the countries that have criteria are defined differently. As such, direct comparison of risk levels is difficult.

Despite the complexity of the existing use category definitions, a general decision tree can be developed. The first stage of identification is whether the crop is edible or not. For edible crops, the second level is whether the crop is eaten cooked or raw. In this case, cooking is viewed as an additional treatment or barrier favoring public protection. It should be recognized, however, that if low-quality reclaimed water is used to irrigate food crops, selling such uncooked crops to the public, restaurants, etc., presents risks of pathogen transmission from handling the crop or contamination of cooking environments. Because of opportunities for transmission of infectious organisms created by handling crops that may be contaminated, some U.S. states prohibit selling or distributing the crops before processing. This provision assures that the transmission chain is severed and that contaminated raw foods are not brought into food-preparation environments. Many water reuse regulations neglect to take this fact into consideration. For those crops that are not edible, it is important to consider whether the irrigated area is a restricted area or not. In a restricted area, the likelihood for public exposure is less than in a nonrestricted area. The risk of disease transmission is related to the reclaimed water quality and degree of human contact with the reclaimed water. Finally, sprinkler irrigation is associated with higher risk due to the potential for disease transmission from aerosols or windblown spray if a low level of disinfection is provided.

In many countries, crops eaten raw are generally considered to present the greatest potential for disease transmission associated with irrigation using reclaimed water. However, this is not always the case. For example, the WHO guidelines recommend more stringent standards for public lawns than for crops eaten raw.

3.5 STANDARDS FOR URBAN USES OF RECYCLED WATER AND LANDSCAPE IRRIGATION

Nonpotable urban water reuse programs are becoming increasingly important for the development of cities that lack adequate water resources, where recycled water is used for applications that do not require potable water quality (e.g., landscape irrigation, car or street cleaning, and toilet flushing). Landscape irrigation involves the irrigation of golf courses, parks, cemeteries, school grounds, freeway medians, residential lawns, and similar areas. The concern for pathogenic microorganisms is somewhat different than for agricultural irrigation in that landscape irrigation frequently takes place in urban areas where control over the use of the reclaimed water is more critical. Depending on the area being irrigated, its location relative to populated areas, and the extent of public access or use of the grounds, the water quality requirements and operational controls placed on the system may differ. Irrigation of areas not subject to public access have limited potential for creating public health problems, whereas microbiological requirements become more restrictive as the expected level of human contact with reclaimed water increases. Despite the absence of direct consumption, public exposure to recycled water is an important consideration due to inhalation, contact, or accidental ingestion. Therefore, strict regulations are highly recommended. Pubic acceptance of landscape irrigation with reclaimed water is widespread, although health concerns occasionally are raised where reclaimed water is to be used for the irrigation of parks, school grounds, or athletic fields. Except for TDS, the chemical composition of the irrigation water is not usually limiting.

Table 3.7 illustrates examples of water reuse regulations for urban uses of recycled water. The table indicates the trend to impose criteria for urban uses that are at least as stringent as agricultural reuse standards.

Table 3.7 Water Quality Criteria for Unrestricted Landscape Irrigation and Other Urban Uses Compared to Bathing Zone Regulations

Location	Fecal coliforms (cfu/100 mL)	Total coliforms (cfu/100 mL)	*E. coli* (cfu/100 mL)	BOD (mg/L)	Turbidity (NTU)	TSS (mg/L)	Comments
Arizona[a]	ND in 4 of last 7-daily samples; 23 (single sample)	—	—		2 (24 h average) 5 (any time)		Class A reclaimed water; secondary treatment, filtration and disinfection
Australia, EPA Victoria[a]	—	—	10 (median)	10	2 (24 h median value)	5	Tertiary treatment to achieve < 1 helminth egg/L, virus < 2/50L, parasites < 1/50 L,
Australia, NSW[a]	1	10 (95% of samples)	—		2 (geometric mean) 5 (95% of samples)		monitoring at the plant outlet; virus < 2/50 L, parasites < 1/50 L, color < 15 TCU, pH 6.5–8
California[a]	—	2.2 (7-day median); 23 (1 from any 30-day period) 240 (any sample)	—	—	2 (within a 24 h period) 5 (5% in any 24 h) 10 (any time) 0.2 (membranes, 5% in any 24 h) 0.5 (membranes, any time)	—	Disinfected tertiary recycled water; chlorine residual ≥ 5 mg/L, CT ≥ 450 mg-min/L; TC monitoring once a day; continuous monitoring of turbidity
Canada[a] BC*	2.2 (median); 14 (single sample)	—	—	10	5	10	
Florida[a]	Not detected (75% of samples, 30-day period); 25 (single sample)	—	—	20 (annual average)	—	5 (single sample)	Secondary treatment, filtration and high-level disinfection; chlorine residual ≥ [illegible] mg/L with on-line monitoring; storage min volume of 1–3 times the average daily flow

(*continued*)

Table 3.7 Continued

Location	Fecal coliforms (cfu/100 mL)	Total coliforms (cfu/100 mL)	*E. coli* (cfu/100 mL)	BOD (mg/L)	Turbidity (NTU)	TSS (mg/L)	Comments
Germany[b]	100	500	—	20	1–2	30	Oxygen saturation 80–120%
Greece[c]			0		2	10	80% of samples per month; < 1 helminth egg/10 L
Japan[a]	not detected	—	—	—	—	—	Chlorine residual ≥0.4 mg/L
Texas[a]	20 (geometric mean) 75 (any sample)	—	—	5 (30-day average)	3 (30-day average)	—	Type 1 reclaimed water; monitoring twice/week
Spain[c]	—	—	0	—	2	10	< 1 helm.egg/10 L
WHO[a], lawn irrigation	200	—	—	—	—	—	
UK[c], in-building recycling	14 for any sample; 0 for 90% samples	—	—	—	—	—	
US EPA[b]	Not detected (7-day median); 14 (any sample)	—	—	10	2	—	Chlorine residual ≥1 mg/L; CT ≥ 30 mg-min/L
EC bathing water	100[b] 2000[a]	500[b] 10,000[a]			2[b] 1[a]		Oxygen saturation 80–120%

Source: Adapted from Refs. 12, 13, 18–22, 24, 25.

[a]Guidelines.

[b]Mandatory regulations.

[c]Proposal of new regulations.

*British Columbia.

The potential transmission of infectious disease by pathogenic agents is the most common concern associated with nonpotable reuse of treated municipal wastewater. The infectious agents that may be present in untreated municipal wastewater can be classified into three broad groups: bacteria, parasites (protozoa and helminths), and viruses. Most pathogenic bacteria are readily destroyed by common disinfection practices, but parasites and viruses may require greater levels of treatment and disinfection to assure their removal or inactivation. Many viruses and parasites are infectious at low doses and can survive for extended time periods in the environment. Further, sampling and analysis for these organisms is both time-consuming and costly.

In most U.S. states that have water reuse criteria, both treatment unit processes and water quality limits (e.g., turbidity and coliform organisms) are specified to reduce or eliminate pathogenic organisms from reclaimed water.

In addition to water quality requirements, safety measures for urban reuse applications often include the following measures.[13,26–28]

1. Conformance to storage and distribution system requirements
2. Use of color-coding to distinguish potable and nonpotable systems
3. Cross-connection control via backflow-prevention devices to protect potable water supplies
4. Periodic tracer studies to detect cross connections between potable and nonpotable systems
5. Off-hour usage to further minimize potential human contact
6. Information signs at sites using reclaimed water

3.6 STANDARD ENFORCEMENT AND PERSPECTIVES

Presently, there are significant differences among different countries' water reuse criteria. Moreover, water reuse regulations are often inconsistent or incompatible with standards applied to other types of water. No existing water reuse criteria are based on strict risk assessment methodology, and the rationale supporting some criteria lack a sufficiently sound scientific basis. Some experts are of the opinion that economic viability should drive the standard-setting process and indicate that restrictive standards would impede implementation of water reuse projects. While this is undoubtedly true for developing countries, it is questionable that such is the case in technically advanced countries. Since the overarching objective of water reuse criteria is to increase available water resources in a safe and reliable manner, public health considerations should be the overriding factor in determining acceptable standards. In the United States and several other countries, the philosophy has always been that any use of reclaimed water should not present undue health risks, and conservative criteria have been adopted to achieve that goal.

Developing a single European framework for water reuse would have numerous benefits. A single set of European regulations would improve public understanding and favor the development of water reuse projects. Moreover, new research programs and elaboration of appropriate treatment, distribution, and use practices could be better targeted. One of the basic criteria in setting new European guidelines would likely be the balance of risk and affordability. The most rational approach may be to lower risks by implementing sound water reuse practices, new irrigation techniques, better control of public access, and public education.

REFERENCES

1. Shuval, H.I., Adin, A., Fattal, B., Rawitz, E., and Yekutiel, P., Wastewater irrigation in developing countries—health effects and technical solutions, *World Bank Technical Paper Number 51*, The World Bank, Washington, DC, 1986.
2. National Research Council, Use of Reclaimed Water and Sludge in Food Crop Irrigation. National Academy Press, Washington, DC, 1996.
3. Lazarova, V. (ed.), Role of water reuse in enhancing integrated water resource management, Final Report of the EU project CatchWater, EU Commission, 2001.
4. State of California, Water recycling criteria, California Code of Regulations, Title 22, Division 4, Chapter 3. California Department of Health Services, Sacramento, CA, 2000.
5. World Health Organization (WHO), Health guidelines for the use of wastewater in agriculture and aquaculture. Report of a WHO Scientific Group, Technical Report Series 778, World Health Organization, Geneva, Switzerland, 1989.
6. Yanko, W.A., Analysis of 10 years of virus monitoring data from Los Angeles County treatment plants meeting California reclamation criteria, *Wat. Env. Res.*, 65, 3, 221, 1993.
7. Wright, R.R. and Missimer, T.M., Reuse: the US experience and trend direction, *Desalin. Water Reuse*, 5/3, 28, 1996.
8. Shelef, G. and Azov, Y., The coming era of intensive wastewater reuse in the Mediterranean region, *Wat. Sci. Tech.*, 33, 10–11, 115, 1996.
9. Murcott, S.E., Dunn, A., and Harleman, D.R.F., Chemically enhanced wastewater treatment for agricultural irrigation in Mexico, in *Proc. Metropolitain areas and Rivers IAWQ/IWSA symposium*, Rome, May 27–31, 1996.
10. World Health Organization (WHO), *Reuse of effluents: methods of wastewater treatment and health safeguards*, Report of a WHO Meeting of Experts, Technical Report Series No. 17, World Health Organization, Geneva, Switzerland, 1973.
11. International Reference Centre for Waste Disposal, Health aspects of wastewater and excreta use in agriculture and aquaculture: The Engelberg Report, *IRCWD News*, No. 23, Dubendorf, Switzerland, 1985.
12. U.S. Environmental Protection Agency. *Guidelines for Water Reuse*, EPA/625/R-92/004, U.S. Environmental Protection Agency, Center for Environmental Research Information, Cincinnati, OH, 1992.
13. Crook, J., Johnson, L., and Thompson K., California's new water recycling criteria and their effect on operating agencies, in *Proc. AWWA*, Las Vegas, 2001.
14. WateReuse Association, *Recycling Water to meet the World's Needs*, WateReuse Association, Sacramento, CA, 1999.
15. State of California, *Porter-Cologne Water Quality Control Act*, California Water Code, Division 7, Compiled by the State Water Resources Control Board, Sacramento, CA, 1998.

16. State of California, Water Recycling in Landscaping Act, Government Code, Title 7, Division 1, Chapter 3, Sacramento, CA, 2000.
17. Angelakis, A.N., Marecos do Monte, M.H., Bontoux, L., and Asano, T., The status of wastewater reuse practice in the Mediterranean basin, *Wat. Res.*, 33, 10, 2201, 1999.
18. CEDEX, Centro de Estudios y Experimentación de Obreas Publicas Propuesta de calidades mínimas exigidas para la reutilización directa de efluentes depurados según los distintos usos posibles, así como de aspectos relativos a la metodología, frecuencia de muestreo y criterios de cumplimiento de los análisis establecidos, para incluir en una normativa de carácter estatal, Madrid, Spain (in Spanish), 1999.
19. Angelakis, A.N., Tsagarakis, K.P., Kotselidou, O.N., and Vardakou, E., The Necessity for Establishment of Greek Regulations on Wastewater Reclamation and Reuse, Report for the Ministry of Public Works and Environment and Hellenic Union of Municipal Enter. for Water Supply and Sewage, Larissa, Greece (in Greek), 2000.
20. British Columbia, Ministry of Environment, Lands and Parks, 1999.
21. BSRIA, Building Services Research and Information Association, Rainwater and greywater in buildings: project report and case studies, Report 13285, BSRIA, Bracknell, UK, 2001.
22. EPA Victoria, Guidelines for environmental management: use of reclaimed water, EPA Victoria, Publication 464.2, ISBN 0 7306 7622 6, 2003.
23. NSW Guidelines for urban and residential use of reclaimed water, New South Wales, Australia, 1993.
24. Surendran, S. and Wheatley, A.D., Greywater reclamation for non-potable reuse, *J. CIWEM*, 12, 406, 1998.
25. Lazarova, V., Hills, S., and Birks, R., Using recycled water for non potable urban uses: a review with particular emphasis on toilet flushing, *Wat. Sci. Tech.: Water Supply*, 3, 4, 69, 2002.
26. Asano, T., Irrigation with treated sewage effluent, in *Adv. Series in Agricultural Sciences*, vol. 22, Tanji, K.K., and Yaron, B., Eds, Springer, Berlin, 1994, 199.
27. Crook, J. and Surampalli, R.Y., Water reclamation and reuse criteria in the USA, *Wat. Sci. Techn.*, 33, 10–11, 475, 1997.
28. Sakaji, R. H. and Funamizu, N., Microbial risk assessment and its role in the development of wastewater reclamation policy, in *Wastewater Reclamation and Reuse*, Asano, T., ed., Technomic Publ. Co., Lancaster, PA, 1998, 705.

4 Code of Practices for Health Protection

Valentina Lazarova and Akiça Bahri

CONTENTS

4.1 Introduction 83
4.2 Specific wastewater treatment for reuse purposes 86
4.2.1 Typical schemes used for production of recycled water for irrigation 87
4.2.2 Main disinfection processes used in water reuse systems 89
4.2.3 Requirements for recycled water storage and distribution 92
4.2.4 Requirements for reliability of operation of water reuse systems 93
4.3 Control of recycled water application 94
4.4 Restrictions on crops and public access 96
4.5 Human exposure control 98
References 101

4.1 INTRODUCTION

The major health risk associated with irrigation with recycled water is infection from pathogenic microorganisms, including viruses, bacteria, helminths, and protozoa. As underlined in the World Health Organization (WHO) Health Guidelines for the Use of Wastewater in Agriculture and Aquaculture,[1] a risk

1-56670-649-1/05/$0.00 + $1.50

of disease transmission exists when the following conditions are met:

Humans or animals are exposed to pathogens in recycled water by direct or indirect contact in irrigated areas or by eating contaminated crops
The number of pathogens reaching a human or animal host is higher than the infective dose
The host becomes infected
The infection results in disease and/or further transmission of pathogens to other persons or animals

Consequently, the main objective of health-protection measures in water reuse projects is to prevent the first two conditions from occurring. This means that appropriate practices must be applied to reduce the number of pathogens in recycled water as well as to implement several barriers and other measures to reduce the probability of contact with potentially infectious microorganisms. The choice of such good practices for health protection depends on specific practical conditions and economic concerns.

The principal health-protection measures that can be applied to irrigation with recycled water include the following four groups of practices (Figure 4.1)[2]:

1. Wastewater treatment and storage
2. Control of wastewater application by the choice of appropriate irrigation methods and cultivation practices
3. Crop restriction for agricultural irrigation and access restriction for landscape irrigation
4. Human exposure control, harvesting measures, education, and promotion of hygienic practices

Figure 4.2 illustrates some possible combinations of these health-control measures that can provide a suitable level of safety for irrigation with recycled water.[3] Each specific measure is characterized by a given number of safety credit units. The highest level of safety is guaranteed by a high level of wastewater treatment with almost total disinfection (4 safety credit units). Other safety measures give an additional 1 to 3 safety credit units, with the highest value given to localized irrigation (subsurface, trickle).

On the basis of the existing experience, it can be stated that good health protection can be achieved by a combination of control measures ensuring at least 6 safety credit units, as shown in the Figure 4.2. The arrows illustrate some combinations of good practices depending on the wastewater treatment level. For low-quality recycled water, a combination of different control measures should be applied, such as appropriate choice of irrigation method, access or crop-restriction measures, and human exposure control. In all cases, public education and information remains a critical issue for the successful operation of water reuse projects. For example, in the case of localized irrigation, the use of secondary treated effluent may not require additional restriction measures, while other irrigation methods will need

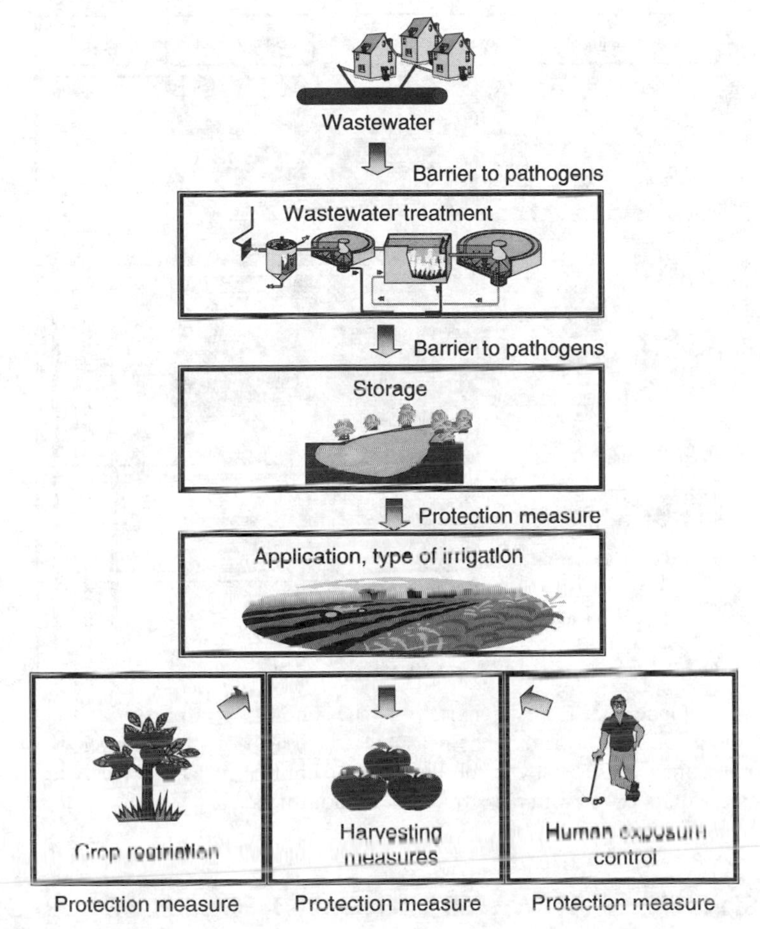

Figure 4.1. Flow chart of irrigation with recycled water showing different health-protection measures that can be implemented to prevent transmission of pathogens.

additional restrictions on cultivation practices, crop selection, and public access. After extended treatment and disinfection (tertiary treatment, such as that required by California Water Recycling Criteria), irrigation with recycled water does not require other specific restriction measures. In the case of urban uses of such effluents, cross-connection control provisions are generally required.

Figure 4.2 provides possible combinations of different groups of safety measures and should be used only as guidelines for the selection of appropriate health-protection actions. Evaluation of the most appropriated health-protection practices must be made on the basis of a case-by-case study with good professional judgment and practical experience, taking into account specific water reuse project objectives and local conditions.

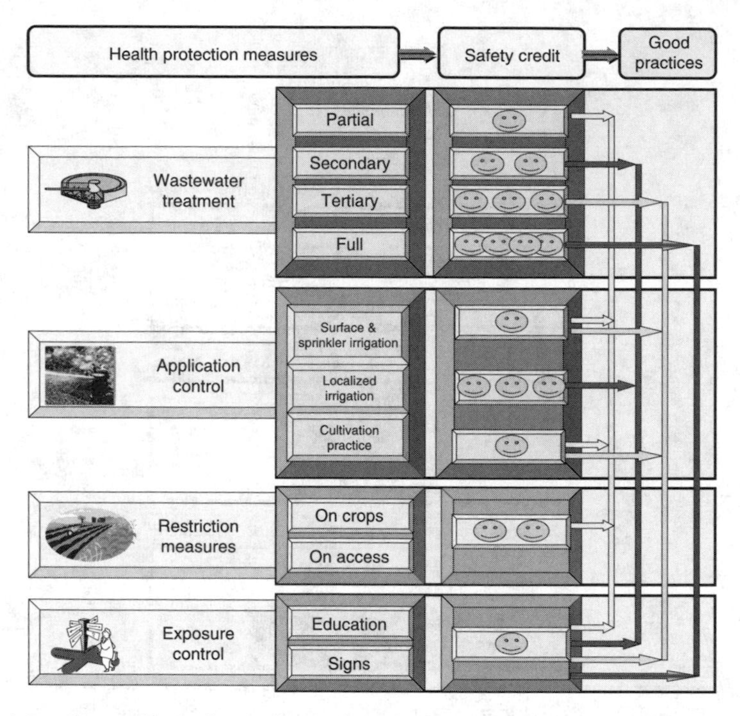

Figure 4.2 Generalized flow chart, for the choice of appropriate actions and combinations of health-protection measures for irrigation with recycled water: low health risk can be ensured by a combination of control measures receiving at least 6 safety credit units (wastewater treatment is mandatory).

4.2 SPECIFIC WASTEWATER TREATMENT FOR REUSE PURPOSES

Wastewater treatment is probably the most effective measure to reduce the health risks associated with the use of recycled water for irrigation. It must be appropriate for the intended use, with special attention to impacts on soils and crops and nuisance conditions during storage and application. Advanced treatment to remove wastewater constituents that may be phytotoxic or harmful to certain crops is technically possible, but may not always be justified economically, especially in the case of agricultural irrigation. To use water containing such constituents, farmers or gardeners should apply appropriate agronomic practices (see Chapter 5).

The design of wastewater treatment plants has traditionally been based on the need to reduce suspended solids, pathogens, and organic and nutrient loads to limit pollution of the environment. For wastewater reuse in irrigation, pathogen removal is of primary concern and treatment processes should be selected and designed accordingly. In many countries, the recycled water quality criteria for restricted irrigation are at a minimum equivalent to the quality of secondary treatment, while additional tertiary treatment is often required for unrestricted irrigation (see Chapter 3).

The necessary recycled water quality, and consequently the level of the treatment needed, depend on the following factors:

Water quality requirements and regulations
Degree of worker and public exposure
Type of distribution and irrigation systems
Soil characteristics
Irrigated crops

4.2.1 Typical Schemes Used for Production of Recycled Water for Irrigation

Some typical treatment schemes for the production of recycled water for irrigation according to different water quality criteria are summarized in Figure 4.3 and Figure 4.4.[3,4]

The main technical challenge for the production of recycled water for restricted agricultural irrigation is the removal of microbial pathogens. The main technical challenge in developing countries is the removal of helminth eggs that represent the most important health risk, in particular for children.[5] In this case, there may be limited restrictions on BOD removal, and physicochemical primary treatment appears to be a cost-competitive solution well adapted to the requirements of restricted irrigation.[6] The main advantage of this treatment is the conservation of the fertilization capacity of wastewater.

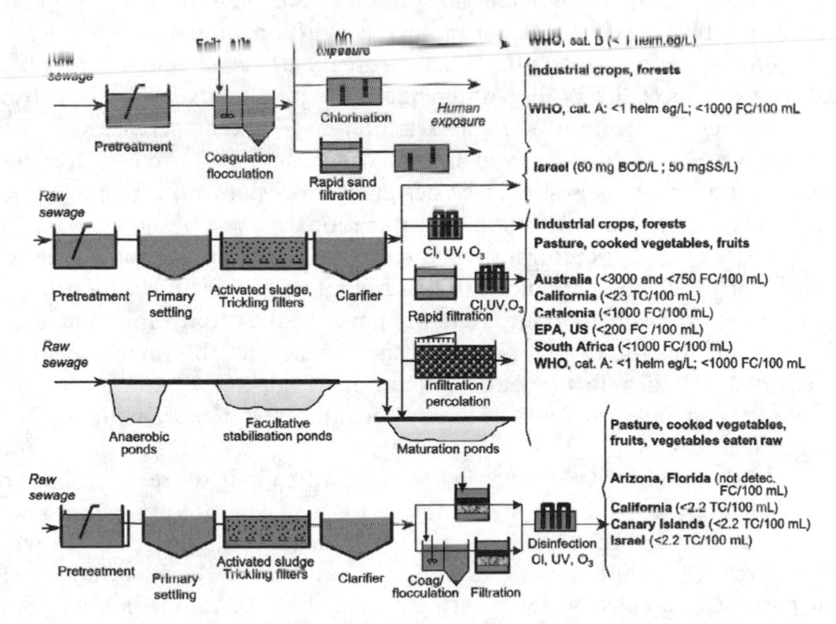

Figure 4.3 Typical treatment schemes for production of recycled water according to different requirements for agricultural irrigation.

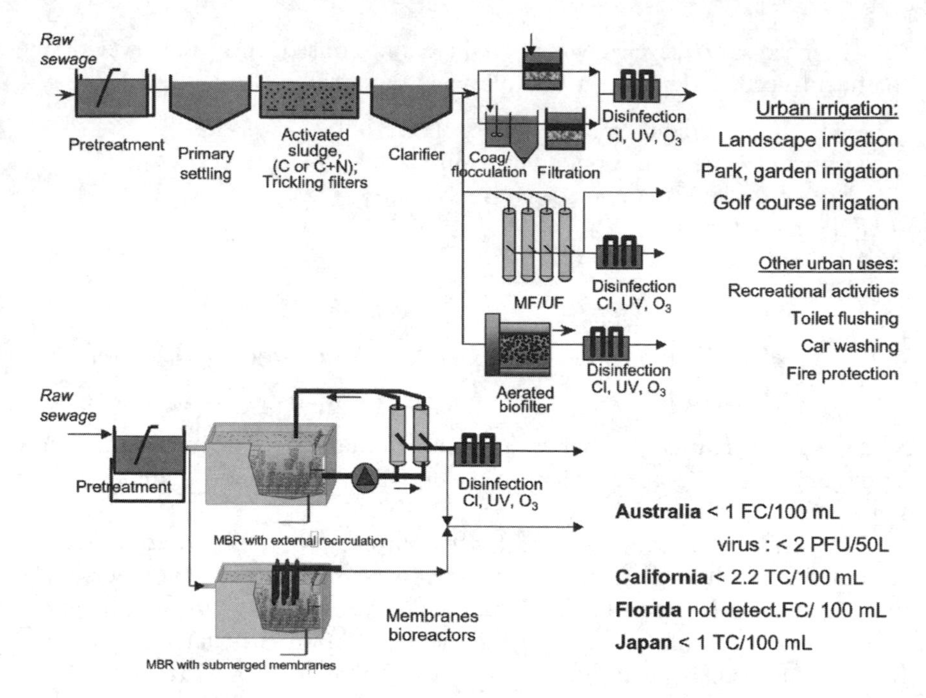

Figure 4.4 Typical treatment schemes for production of recycled water according to different requirements for landscape irrigation and other urban uses.

Another common treatment technology implemented in developing countries is stabilization ponds, which are low in cost, easy to operate, and generally have low maintenance requirements. When properly designed and operated, this treatment can meet the WHO[1] water quality requirements for both restricted and unrestricted irrigation (see also Chapter 7, §7.5). In the case of existing wastewater treatment plants, which, as a rule, include secondary treatment, additional polishing steps should be designed for disinfection with or without pretreatment (e.g., sand or multimedia filtration).

Compared to agricultural irrigation, the most appropriate treatment schemes for urban reuse include tertiary treatment and disinfection processes, which, in comparison to pond systems, have smaller footprints and greater efficiency and reliability of operation, such as aerated biofilters, membrane bioreactors, UV disinfection, and ozonation (Figure 4.4). Chapter 7 (§7.6) discusses the main principals of operation and design parameters of these treatment processes.

As a rule, the quality of recycled water used to irrigate recreational areas with free access to the public is subject to stringent quality requirements. Several hundred golf courses are irrigated with recycled water in the world, e.g., in France,[7] in South Africa,[8] in Spain,[9] and in the United States.[10] For such purposes, recycled water is tertiary treated in the municipal wastewater treatment or recycling plants or in satellite treatment units located at the golf course. Additional health-protection measures are often required (depending on treatment provided), such as night irrigation, use of microsprinklers, green belts around the golf course, and signage indicating the use of nonpotable recycled water (see also Chapter 6).

Water reuse for golf courses, greenbelts, and hotel gardens is often reported to be one of the most competitive water reuse applications, requiring optimization of both investment and operational costs. Advantages include reduced infrastructure needs if the golf courses are located in suburban areas near existing municipal wastewater treatment plants as well as a willingness to pay relatively high charges for recycled water (compared to agricultural irrigation) that would allow recovery of at least operation and maintenance costs.

4.2.2 Main Disinfection Processes Used in Water Reuse Systems

In terms of public health protection, the disinfection process has crucial importance for health safety. The most common processes used for disinfection of recycled water prior to reuse for irrigation, as shown in Figure 4.3 and Figure 4.4, are chlorination, UV irradiation, and maturation ponds (see also Chapter 7, §7.5 and §7.6). Table 4.1 summarizes the main design parameters, advantages and drawbacks of these technologies.

Polishing pond treatment (maturation ponds) is the simplest and least costly technology for the recycling of medium-quality effluent in small communities ($\leq 3000\,m^3/d$). This process produces no harmful by-products and provides storage capacity to accommodate variations in water demand. The major disadvantages of polishing pond treatment are low flexibility of the process, high evaporation losses, sludge removal, and the inability to remove or inactivate all pathogens if proper detention times are not employed. Over the last decade, an increased number of studies conducted in different countries have shown that stabilization pond systems in series can produce effluent with microbiological water quality suitable for unrestricted irrigation (WHO guidelines: < 1000 FC/100 mL and < 1 helm egg/L) for hydraulic residence times varying in the range of 20–90 days according the climate conditions and optimal lagoon depth of 1.0–1.5 m. For disinfection purposes, two types of maturation ponds can be used, aerated or nonaerated lagoons, as a function of land availability. Under optimal operating conditions, the disinfection efficiency is 3–5 log removal of the fecal coliforms, with reported maximum values of up to 5–6 log removal. In situations where requirements address only nematode removal (e.g., restricted irrigation according the WHO guidelines[1]), the retention time may vary from 10 to 25 days depending on climate conditions.

Chlorination is the most commonly used process for wastewater disinfection.[12] Chlorine has proven to be a convenient and efficient disinfecting reagent, but it also generates disinfection by-products that may be harmful to human health and the environment. Given the high chlorine doses that may be required for wastewater disinfection, these reactions are likely to occur to a broad extent in recycled effluents. There are also safety concerns associated

Table 4.1 Disinfection Processes Used for Health Protection in Water Reuse Schemes for Irrigation: Design Parameters, Advantages, and Drawbacks

Process	Design parameters	Advantages	Drawbacks
Polishing (maturation) ponds	High influence of climate Depth: 1 to 1.5 m Hydraulic residence time: 10–90 d, commonly between 15 and 30 d Surface area: 2.5 m^2/inhabitant Recommended for small and medium plants (< 50,000 inhabitants)	Provides additional storage capacity Easy to operate Robust to meet WHO guidelines Efficient for removal of bacteria and helminths Cost-efficient for small units	High evaporation losses Sludge accumulation with potential disposal problems requiring infrequent but intensive work Effluent quality is a function of hydraulic loads, climate, and season Can cause nuisance (odor, insects) Requires large footprints No disinfectant residual Additional contamination possible by presence of birds, etc.
Chlorination	Chlorine dose: 5–20 mg/L Hydraulic residence time: 20–90 min CT: 100– 450 mg.min/L Plug-flow contactors Recommended for all plant sizes	Presence of a disinfectant residual Easy to operate and control Little equipment needed Efficient on nitrified effluents	Formation of potentially harmful by-products Weak efficiency in case of incompletely nitrified effluent Heavy safety regulations in the case of chlorine gas Efficiency depends on water quality
UV irradiation	UV dose: 25–140 mJ/cm^2 Low- or medium–pressure UV systems	Low footprint No by-products High efficiency for inactivation of viruses, bacteria, and cysts (*Cryptosporidium*) Easy to operate Cost-efficient (similar to chlorination)	Dose difficult to measure Lamp aging and fouling difficult to assess No disinfectant residual Potential regrowth of bacteria in case of low applied dose Efficiency depends on water quality (removal of suspended solids required for complete disinfection)
Ozonation	Transfered ozone dose: 5–20 mg/L Contact time: 5–10 min Completely mixed contactors	Decreases color and odor of effluents High efficiency for inactivation of viruses, bacteria, and protozoa Dose easy to monitor Cost-efficient for large plants	Too expensive on a small scale (but very efficient for aquaculture)

Source: Adapted from Refs. 7, 11–13.

with transporting and storing chlorine gas. These safety and health concerns, along with cost and other factors, should be carefully evaluated prior to selection of a disinfection process. Even when other disinfectants are used, chlorine is often used as a finishing treatment to maintain chlorine residuals in distribution networks to minimize microbial regrowth (in-line injection and at a small dose) and odors, which is especially important in reuse systems for unrestricted urban irrigation. In such cases, typical residual chlorine concentrations at the outlet of water recycling plants are on the order of 1–2 mg/L (see Chapter 3).

Organic compounds, industrial wastes, and ammonia concentration (in particular the presence of nitrite or lack of ammonia) can strongly affect chlorine demand. Improving mixing characteristics of chlorine contactors and process control strategy can enhance the effectiveness of chlorination. Typical chlorine doses for municipal wastewater disinfection are about 5–20 mg/L with 30–90 minutes of contact time and usually allow for compliance with regulatory limits for conventional bacterial indicators such as coliforms.[12] Higher doses or contact times are required for low-quality wastewater, such as primary or trickling filter effluents or to meet stringent regulations such as California's Water Recycling Criteria[14] (CT > 450 mg.min/L). Bacteria are usually well destroyed by chlorination at relatively low dosages of the disinfectant, although the presence of suspended solids may affect the process; however, very high free chlorine concentrations may be needed to inactivate cysts and some viruses of concern. Chlorination is ineffective in inactivating some parasites, such as *Cryptosporidium*.

UV disinfection appears as a cost-competitive technology for a broad range of plant capacities.[12] The disinfection efficiency of UV light has been proven in the laboratory and at many full-scale facilities, with no known production of harmful compounds. Thus, UV radiation is the preferred technology for many wastewater disinfection projects. Disinfection data based on various types of secondary and tertiary treated wastewater indicated that 30–45 mJ/cm^2 doses of UV radiation were sufficient to ensure a 3–5 log removal of total and fecal coliforms and fecal streptococci.[11,15] The required doses are higher (up to 140 mJ/cm^2) where higher levels of disinfection are mandated. The required UV doses for a given pathogen log removal are significantly influenced by wastewater quality and especially by particulate matter and transmittance. UV systems can be recommended as a competitive solution for disinfection of secondary and tertiary effluents.

In large water reuse projects, the option of ozone disinfection should also be considered.[16,17] Ozone is a powerful disinfectant that improves the visual aspect and the odor of the recycled effluent. This can be decisive in gaining public acceptance. Also, ozone is only slightly more expensive than UV at high production rates. In case-specific studies, the cost of ozone could be decreased down to or lower than the cost of UV by implementation of different complementary technical options such as recycling of oxygen gas, production of oxygen from air or from adsorption techniques, etc.

A number of studies have demonstrated the feasibility of ozone to treat all effluents to moderate standards (like WHO recommendations[1]) for unrestricted irrigation or bathing zone protection.[11,15] Transferred ozone doses of around 15 mg/L would be required for poor-quality wastewater such as primary or secondary effluents produced by high-load activated sludge treatment. For a good-quality secondary effluent obtained after extended aeration, the ozone dose needed to meet the WHO recommendations may be as low as 3–5 mg/L. If stringent standards like the California Water Recycling Criteria[14] must be complied with, a tertiary filtration step is needed before ozonation to allow relatively low doses of ozone of about 8 mg/L to be effective. In the absence of a filtration step, very high doses of 40–50 mg/L of ozone would be necessary to meet stringent regulations.

4.2.3 Requirements for Recycled Water Storage and Distribution

Several modifications of recycled water quality can occur during the storage and distribution of recycled water (bacterial regrowth, additional contamination, etc.). For this reason, complementary measures should be considered by water-recycling managers in order to guarantee public health protection, particularly for landscape irrigation and other urban uses, including:

- Installation of separate storage and distribution systems separated from other types of water, especially from potable water
- Use of color coding to distinguish potable and nonpotable pipes and appurtenances
- Cross-connection and back-flow prevention devices and control measures
- Periodic use of tracer dyes to assure that there are no cross connections between recycled and potable supply lines
- Installation and appropriate maintenance of storage systems

Recycled water storage systems can be classified in two main groups: short-term and long-term (see Chapter 7, §7.7).

Short-term storage is typically required for landscape and agricultural irrigation. Detention time depends on peak irrigation demand and recycled water production with common values between 1 and 15 days. Both open and covered reservoirs can be used. The main advantages of covered reservoirs are that they reduce the potential for algal growth in the stored water and prevent external contamination from birds, animals, and surface runoff (Figure 4.5). As much as a 2–3 log increase in fecal coliform concentration has been reported in a number of open reservoirs and other water bodies, such as polishing lagoons, used for temporal storage of disinfected recycled water.

Long-term storage is applied mostly for seasonal storage of recycled water during the winter in order to satisfy the high irrigation demand during dry seasons. The most common forms of storage are open reservoirs or deep stabilization ponds. Long-term storage in aquifers is practiced in some locations. Surface spreading is the simplest, oldest, and most widely applied method of artificial recharge of aquifers. Recycled water percolates from

Figure 4.5 View of closed storage reservoirs in Santa Rosa, California, implemented for irrigation of vineyards.

spreading basins to an aquifer through a vadose zone. Direct groundwater recharge with injection of recycled water into aquifers requires a significantly higher water quality and, for this reason, is mostly applied for indirect potable reuse.

Significant improvement in water quality (physicochemical and biological), may take place during long-term storage (Chapter 7, §7.7). Coliform removal can reach 3–4 log units as orders of magnitude but depends greatly on hydraulic residence time and climate conditions.

4.2.4 Requirements for Reliability of Operation of Water Reuse Systems

The reliability of operation of water reuse systems is a great concern related to the health safety of irrigation with recycled water. Many water reuse regulations include specific requirements for water quality monitoring and enhancement of treatment reliability of water reuse schemes, especially for unrestricted irrigation (see Chapter 3).

The reliability of operation of reuse systems involves the capability to consistently deliver recycled water with the required quality. In other words, reliability represents the probability of adequate performance of a given system, measured by the percentage of time that the final product meets existing quality standards. Water reuse requires high operational reliability to maintain a high degree of health protection resulting from the use of recycled water.

The main engineering components of water reuse systems susceptible to failure are power supplies, mechanical equipment of treatment processes, distribution systems, etc. In the case of failure, backup systems are necessary to take over critical situations. During the last decade, a number of Failure Modes Effects and Criticality Analysis (FMECA) tools have been elaborated to identify the components of engineering systems most likely to cause failures. This approach is based on the evaluation of the major consequences of failures, including environmental pollution, exceedance of water quality limits, volume compliance, odors and flies, unsatisfactory sludge disposal, disinfection failures, health and safety, reputation and adverse public perception, financial impact (direct costs of repair, excess costs, loss of income), and contract failures. The resultant three-level scale (small, medium, large) relates to defined economic value. Failure modes are then prioritized as a function of their probability, severity, and detectability. Based on this prioritization and the associated technical-economic analysis of all potential failures of a given wastewater treatment plant, actions can be taken to prevent failure or to reduce the likelihood of failure. Finally, this analysis makes it possible to evaluate which investments would achieve a maximum decrease of risk at lower cost.

For water reuse, the most important and feasible safety features are:

- Alarm systems to provide warning of loss of power or failure of critical processes
- Implementation of emergency storage of inadequately treated effluent
- Automation
- Training of operators with establishment of operating manuals having emergency procedures and maintenance schedules

Redundancy should be provided in all water reuse systems for unrestricted urban irrigation to prevent any single component from becoming vital to a unit operation. This is particularly important if automatically actuated emergency storage or alternate disposal provisions are not provided for recycled water that does comply with requirements. The level of redundancy depends on whether the vital component is part of a critical or noncritical operational unit. Special attention should be given to power supply and control systems.

4.3 CONTROL OF RECYCLED WATER APPLICATION

The principal potential pathway of infection is from direct contact of humans with recycled water and crops or soil in the irrigated area. In this context, choosing the appropriate type and timing of irrigation can reduce health risks associated with pathogens.

Two main control measures for water application can be applied for health protection in the case of irrigation with recycled water:

1. Choice of appropriate irrigation method
2. Additional water application measures, such as irrigation timing and scheduling.

Table 4.2 Factors Affecting the Choice of Irrigation Method and Special Measures Required for Recycled Water Application

Irrigation method	Factors affecting choice	Special measures for irrigation with recycled water
Flood irrigation	Lowest cost Exact leveling not required Low water use efficiency Low level of health protection	Thorough protection of field workers, crop handlers, and consumers
Furrow irrigation	Low cost levelling may be needed Low water use efficiency Medium level of health protection	Protection of field workers, possibly of crop handlers and consumers
Sprinkler irrigation	Medium to high cost Medium water use efficiency Leveling not required Low level of health protection (aerosols)	Minimum distance 50–100 m from houses and roads Water quality restrictions Anaerobic wastes should not be used due to odor nuisance
Subsurface and drip irrigation	High cost High water use efficiency Higher yields Highest level of health protection	No protection measures required Water quality restrictions: filtration to prevent emitters from clogging

Source: Adapted from Ref. 1.

Table 4.2 summarizes the main factors affecting the choice of irrigation system in terms of health safety, costs, and other specific requirements. Surface irrigation involves less investment than other types of irrigation but exposes field workers to the greatest health risk.[1] If the effluent is not of the required water quality in terms of disinfection, sprinkler irrigation should not be used except for fodder crops, and border irrigation should not be used for vegetables, in particular those to be eaten uncooked. Subsurface and drip irrigation can provide the highest degree of health protection, as well as using water more efficiently and often producing higher crop yields. However, a high degree of reliable treatment is required to prevent clogging of emitters.

Risks associated with sprinkler irrigation are influenced by local weather conditions, e.g., they are higher in windy conditions. Irrigating during off-hours, e.g., night vs. day, helps to minimize potential human contact with recycled water. The minimal risk is attained when microirrigation devices are used, especially subsurface drippers.

For agricultural irrigation, microbial pathogens may present significant health risks to field workers, depending on the quality of the recycled water. The implementation of good cultivation practices is the only feasible means to ensure worker safety if low-quality water is used for irrigation. Table 4.3 lists

Table 4.3 Classification of Cultivation Practices as a Function of the Health Risk for Agricultural Workers

Low risk of infection	High risk of infection
Mechanized cultural practices	High dust areas
Mechanized harvesting practices	Hand cultivation
Crop dried prior to harvesting	Hand harvest of food crops
Long dry periods between irrigations	Moving sprinkler equipment
	Direct contact with irrigation water

Source: Adapted from Ref. 18.

some common situations associated with low and high health risks for farmers and field workers.

Other water application measures, such as ceasing irrigation for a certain period of time prior to harvesting (crop-drying), may help to reduce the risk of infection by allowing time for pathogens to die off. However, the use of this control measure will depend on the type of crops, the particular pathogens known to be present, and their survival times in the environment. For example, crop-drying does not result in lower, health risks associated with nematode eggs and cysts.

4.4 RESTRICTIONS ON CROPS AND PUBLIC ACCESS

In most cases, water reuse rules and regulations for irrigation specify the types of crop that can be irrigated with recycled water of a given quality, as well as public access restrictions. The microbiological quality of the water can directly affect the consumer because of the risk of infection. World Bank[19] defined three levels of risk in selecting a crop to be grown. They are presented in Table 4.4 in increasing order of public health risk.

Crop restriction is feasible and is particularly facilitated under the following conditions:

Existence of strong law enforcement
Public body control of water allocation
Irrigation project with strong central management
Adequate market demand for the crops irrigated with recycled water

Adopting crop restriction as a means of health protection in water reuse schemes requires a strong institutional framework and capacity to monitor and control compliance with regulations. Farmers need to be informed and assisted in developing adequate cropping practices.

The chemical quality of recycled wastewater needs also to be controlled, especially in relation to modifications that can be induced in soils and possible biomagnification related to the trophic chain (see also Chapter 5 and

Table 4.4 Levels of Risk Associated with Different Types of Crops Irrigated with Recycled Water

Lowest risk to consumer, but field worker protection still needed	Medium risk to consumer and handler	Highest risk to consumer, field worker, and handler
Agricultural irrigation		
Industrial crops not for human consumption (e.g., cotton, sisal) Crops normally processed by heat or drying before human consumption (grains, oilseeds, sugar beets) Vegetables and fruit grown exclusively for canning or other processing that effectively destroys pathogens Fodder crops and other animal feed crops that are sun-dried and harvested before consumption by animals	Pasture, green fodder crops Crops for human consumption that do not come into direct contact with wastewater, on condition that none must be picked off the ground and that sprinkler irrigation must not be used (tree crops, vineyards, etc.) Crops for human consumption normally eaten only after cooking (potatoes, eggplant, beets) Crops for human consumption, the peel of which is not eaten (melons, citrus fruits, bananas, nuts, groundnuts) Any crop not identified as high risk if sprinkler irrigation is used	Any crops eaten uncooked and grown in close contact with wastewater effluent (fresh vegetables such as lettuce or carrots, or spray-irrigated fruits) Spray irrigation regardless of type of crop within 100 m of residential areas or places of public access
Landscape irrigation		
Landscape irrigation in fenced areas without public access (nurseries, forests, green belts)	Golf courses with computer-guided irrigation	Golf courses with manual irrigation Landscape irrigation with public access (parks, lawns)

Source: Adapted from Ref. 19

Chapter 8). Salts and heavy metals are the main concern for soil-plant systems when reusing recycled water for irrigation.

The circumstances mentioned above lead to two types of restriction measures:

1. Access to the irrigated area, which can be applied either in the case of agricultural and landscape irrigation. The access restriction is related to contact of treated wastewater with risk groups, e.g., users, workers, and other persons likely to come in contact with recycled water, e.g., golfers.

2. Crop selection is a specific measure for agricultural practices. Crop selection is made according to water and soil properties, apart from the usual agronomic factors.

4.5 HUMAN EXPOSURE CONTROL

As mentioned previously, one of the main objectives when trying to reduce health hazards related to water reuse for irrigation is to prevent the population groups at risk from coming into direct or indirect contact with pathogens that may be present in the recycled water. Several groups of population may be considered to be at risk as a result of landscape and agricultural use of recycled water. The risk depends on several factors, including the microbiological quality of the recycled water used for irrigation. Examples of groups at risk, particularly if low-quality recycled water is used, are:

1. Agricultural workers and gardeners, as well as their families—high potential risk of disease through direct contact with recycled water, crops, and turf grass. Helminth infections present the highest risk in developing countries, mostly through contact with the skin, especially the feet if no appropriate shoes are used.
2. Crop handlers—the risk is high and similar to those for agricultural workers and gardeners. The principal infection pathways are direct contact with the skin of the hands or foot.
3. Consumers of crops, meat, and milk—relatively high risk in developing countries from bacteria and helminths infections, in particular from crops eaten uncooked.
4. People living near the areas irrigated with recycled water and those visiting the areas irrigated with recycled water (e.g., visitors of parks)—a potential risk is related to aerosols that may contain pathogens. Disinfection of recycled water greatly reduces the risk of disease transmission via aerosols, and there has been no epidemiological evidence to confirm the existence of significant risk from pathogens occurring in aerosols from sprinkler irrigation. Contact with turf, soil, or objects previously wet with recycled water may also present health risks if the recycled water is not adequately treated and disinfected.
5. Players in sport fields irrigated with recycled water (e.g., golfers)—the risks are similar to those of people visiting areas irrigated with recycled water.

Table 4.5[20] summarizes the theoretical risks of infection by various microorganisms for the main target groups in agricultural and landscape irrigation with recycled water compared to risks supported by epidemiological data. It is important to underline that using irrigation methods other than spraying can considerably reduce the risk of exposure to pathogens. In any case, education and information programs will be necessary.

Table 4.5 Theoretical and Actual Risks of Disease Caused by Different Types of Pathogens Associated with Agricultural Reuse of Recycled Water

	Risks from irrigation with recycled water		
Group at risk	**Crops for humans**	**Crops for animals**	**Nonconsumable crops**
Consumers of crops	Risk for virus, bacteria, protozoa and nematodes infections supported by epidemiological data	—	—
Consumers of animals	—	Potential risk for bacteria and cestodes infections (no epidemiological data)	—
Workers on sites irrigated with recycled water and crop handlers	No risk for virus infections, confirmed by epidemiological data; potential risk for protozoa and nematode infections (no epidemiological data); risk for bacterial infections, supported by epidemiological data		
People living near irrigated areas, park visitors, and golfers	No risk for virus and bacteria infections, confirmed by epidemiological data		

Source: Adapted from Ref. 20.

The main methods of exposure control for the risk groups during irrigation with recycled water are as follows:

1. Agricultural field workers and crop handlers:
   - Provision (and insistence on the wearing) of protective clothing
   - Maintenance of high levels of hygiene
   - Immunization against or chemotherapeutic control of selected infections (only if recycled water is not well disinfected)
2. Crops consumers:
   - Washing and cooking agricultural produce before consumption
   - High standards of food hygiene, which should be emphasized in the health education, associated with recycled water use schemes
3. Local residents:
   - Should be kept fully informed on the use of wastewater in agriculture so that they and their children can avoid these areas

Figure 4.6 Sign in a California park irrigated with recycled water.

- Sprinklers should not be used within 50–100 m of houses or roads, although there is no evidence to suggest that those living near wastewater-irrigated fields are at significant risk

4. Golfers and other athletes:
   - Should be notified that recycled water is used for irrigation

Some examples of measures for the protection of different risk groups are given in the most recent draft revision of the WHO guidelines.[21] For landscape irrigation, recycled water quality is often higher than the usual resources used, and thus risks for gardeners and visitors are reduced. People entering an area irrigated with recycled water should be informed by appropriate signs (Figure 4.6). Irrigation at night and special signs put on the irrigation systems help to avoid any contact with recycled water. Field workers and gardeners should wear protective clothes and avoid direct contact with recycled water.

Special care must always be taken in water reuse systems to ensure that workers and the public do not use recycled water for drinking or domestic purposes by accident or for lack of other alternative. All recycled water channels, pipes, and outlets should be clearly marked and preferably painted with a characteristic color (e.g., purple is used in California for all water reuse distribution networks). Wherever possible, outlet fittings should be designed and selected so as to prevent misuse.

REFERENCES

1. WHO, World Health Organization, *Health Guidelines for the Use of Wastewater in Agriculture and Aquaculture*, Report of a WHO Scientific Group, Technical Report Series 778, World Health Organization, Geneva, Switzerland, 1989.
2. Blumenthal, U.J., Strauss M., Mara, D.D., and Cairncross, S., Generalised model of the effect of different control measures in reducing health risks from water reuse, *Wat. Sci. Techn.*, 21, 567, 1989.
3. Lazarova, V., *Guidelines for Irrigation with Recycled Water*, Internal Report Suez-Environment-CIRSEE, France, 2003.
4. Lazarova, V., Wastewater reuse: technical challenges and role in enhancement of integrated water management. *E.I.N. Int.*, Dec. 99, 40–57, 1999.
5. Blumenthal, U., et al., Guidelines for the microbiological quality of treated wastewater used in agriculture: recommendations for revising WHO guidelines, *WHO Bull.*, 78, 9, 1104, 2000.
6. Jiménez, B., Chávez., A., Maya, C., and Jardines, L., Removal of microorganisms in different stages of wastewater treatment for México City, *Wat. Sci. Techn.*, 43, 10,155, 2001.
7. Lazarova, V. (ed.), Role of water reuse in enhancing integrated water resource management, Final Report of the EU project CatchWater, EU Commission, 2001.
8. Jagals, P. and Lues, J.F.R., The efficiency of a combined waste stabilisation pond/maturation pond system to sanitise wastewater intended for recreational re-use, *Wat. Sci. Techn.*, 33, 7, 117, 1996.
9. Mujeriego, R. and Sala, L., Golf course irrigation with reclaimed wastewater, *Wat. Sci. Techn.*, 24, 9, 161, 1991.
10. U.S. Golf Association, Wastewater Reuse for Golf Course Irrigation, Lewis Publishers, Boca Raton, FL, 1994.
11. Lazarova, V., et al., Advanced wastewater disinfection technologies: state of the art and perspective, *Wat. Sci. Techn.*, 40, 4/5, 203, 1999.
12. Lazarova, V., Wastewater disinfection: assessment of the available technologies for water reclamation, in: *Water Conservation*. Vol. 3. Water Management, Purification and Conservation in Arid Climates, ed. by M.F.A. Goosen and W.H. Shayya, Technomic Publishing Co. Inc., 2000, 171.
13. Brissaud, F., et al., Hydrodynamic behaviour and fecal coliform removal in a maturation pond, *Wat. Sci. Techn.*, 42, 10/11, 119, 2000.
14. State of California, Water recycling criteria, California Code of Regulations, Title 22, Division 4, Chapter 3. California Department of Health Services, Sacramento, CA, 2000.
15. Savoye, P., Janex, M.L., and Lazarova, V., Wastewater disinfection by low-pressure UV and ozone: a design approach based on water quality, *Wat. Sci. Tech.*, 43, 10, 163–171, 2001.
16. Lazarova, V., Technical and economic evaluation of UV and ozone for wastewater disinfection, in *Proc. IWSA World Water Congress*, Symp. IWA/AIDIS on Reuse of treated sewage effluents, September 18, Buenos Aires, 1999.
17. Xu, P., et al., Wastewater disinfection by ozone: main parameters for process design, *Wat. Res.*, 36, 4, 1043–1055, 2002.
18. Food and Agriculture Organization (FAO), Quality control of wastewater for irrigated crop production, Water Reports 10, by Westcot, D., Food and Agriculture Organization, Rome, 1997.

19. World Bank, Wastewater irrigation in developing countries: health effects and technical solutions. Technical Paper No. 51, Shuval H.I. et al., eds, Washington, DC, 1986.
20. World Bank, Sanitation and disease-health aspects of excreta and wastewater management, World Bank Studies in Water Supply and Sanitation 3, by Feachem, R. G., Bradley, D. J., Garelick, H. and Mara, D. D., published for the World Bank by John Wiley & Sons, Chichester, UK, 1983.
21. Carr, R., personal communication, 2002.

5

Code of Successful Agronomic Practices

Valentina Lazarova, Ioannis Papadopoulos, and Akiça Bahri

CONTENTS

5.1 Amount of water used for irrigation 104
5.2 General water quality guidelines for maximum crop production 105
5.3 Choice of management strategy of irrigation with recycled water 106
5.4 Selection of irrigation method 108
  5.4.1 Criteria for selection of an appropriate irrigation method 108
  5.4.2 Comparison of irrigation methods 113
  5.4.3 Final considerations for the choice of irrigation method 122
5.5 Crop selection and management 123
  5.5.1 Code of practices to overcome salinity hazards 123
  5.5.2 Code of practices to overcome boron, sodium and chloride toxicity 135
  5.5.3 Code of practices to overcome trace elements toxicity 137
5.6 Code of management practices of water application 142
  5.6.1 Leaching and drainage 142
  5.6.2 Using other water supplies 145
  5.6.3 Adjusting fertilizer applications 145
  5.6.4 Management of soil structure 146
  5.6.5 Management of clogging in sprinkler and drip irrigation systems 147
  5.6.6 Management of storage systems 148
References 149

1-56670-649-1/05/$0.00 + $1.50

Alternative water resources for irrigation include treated municipal wastewater, storm water runoff, and irrigation return flow. The principal concern with regard to irrigation using these waters is related to potential adverse effects on soil and crops as well as the management that may be necessary to control or compensate for water quality–related problems.

This section is specifically dedicated to irrigation with treated municipal wastewater and in particular the good agronomic practices enabling to improve public health safety and to reduce potential adverse impacts on crops, soil, and aquifers. In this respect, the term "recycled water" is used to indicate appropriately treated municipal effluents and should be differentiated from other types of recycled water such as drainage or storm water.

For irrigation with recycled water, parameters of agronomic significance, (See Chapter 2) are of concern depending on the characteristics of the systems where the plants grow. When the parameters vary outside a certain range determined by crop nature, soil, and agronomic practices, water may be unsuitable for agricultural use. Agricultural irrigation guidelines, which define this range of variation of irrigation water quality, have been developed to help farmers, operators, and decision makers. If recycled water quality does not match these guideline values, growers can either select more tolerant crops, manage soil characteristics, or change agronomic practices.

Several handbooks were developed concurrently with the main development of water reuse, but they were mainly based on the knowledge gained with traditional irrigation schemes[1,2] or prepared for specific local conditions and farmers.[3] Recently, more specialized handbooks have appeared, dealing mainly with the reuse of reclaimed water for specific purposes such as golf course irrigation.[4]

5.1 AMOUNT OF WATER USED FOR IRRIGATION

In irrigation, especially with recycled water, it is important to apply an appropriate, well-controlled quantity of water sufficient to meet the crop requirements and to prevent accumulation of salts in the soil. The use of insufficient water leads to decrease in crop production. However, excessive flooding can be more harmful, as it saturates the soil for a long time, inhibiting aeration, leaching nutrients, inducing salinization, and polluting underground water. Poorly managed irrigation has detrimental effects on land productivity, and the cost of land rehabilitation may be prohibitive.

Irrigation requirements depend on the crop, the period of plant growth, and climatic conditions (mainly precipitation and evapotranspiration). In all cases, the water needed for normal plant growth is equal to evapotranspiration (more than 99% of the water absorbed by plants is lost by transpiration and evaporation from the plant surface). Additional amount is required for leaching practices. Water application efficiency must be also considered.

Crop evapotranspiration is mainly determined by climatic factors and hence can be estimated with accuracy using meteorological data. Detailed

Table 5.1 Water Requirement of Selected Crops

Type of crop	Amount of irrigation water (mm/growing period)	Sensitivity to water supply
Alfalfa	800–1600	Low to medium-high
Banana	1200–2200	High
Bean	300–500	Medium-high
Cabbage	380–500	Medium-low
Citrus	900–1200	Low to medium-high
Cotton	700–1300	Medium-low
Groundnut	500–800	Low
Maize	500–800	High
Potato	500–700	Medium-high
Rice	350–700	High
Sunflower	800–1200	Low
Sorghum	450–650	Medium-low
Wheat	450–650	Medium-high

Source: Adapted from Ref. 3.

recommendations for the calculation of crop water use are provided by FAO.[5] A computer program, CROPWAT, can be downloaded from the FAO website to determine the water requirements of various crops from climatic data from almost all continents. Table 5.1 illustrates the water requirements of some selected crops.[3,4] Unlike agricultural crops or turfgrass, landscape ornamentals are composed of many species (groundcover, shrubs, trees, etc.) with different water requirements. In addition, the density of landscape planting and microclimate highly influence evapotranspiration and, consequently, water needs.

Water requirements should be calculated to adjust irrigation rate periodically, at least monthly, to ensure that the correct amount of recycled water is applied at the right time to meet crop requirements, taking into account climate variations. It is recommended to apply recycled water only under dry weather conditions, with regular inspections to avoid ponding of recycled or runoff water.

5.2 GENERAL WATER QUALITY GUIDELINES FOR MAXIMUM CROP PRODUCTION

The suitability of recycled water for irrigation is evaluated according to relevant criteria indicating the potential of creating soil conditions hazardous to crop growth and subsequently to animals and humans consuming these crops. These criteria are:

1. Soil permeability and tilth (coarse, cloddy, and compacted soil aggregates), evaluated by electrical conductivity and sodium adsorption ratio. The harmful effects result from excessive exchangeable sodium, high soil pH, and low electrical conductivity.

2. Recycled water and soil salinity, evaluated by the total dissolved solids content, chlorides, and electrical conductivity. Salinity affects crop transpiration and growth (fewer and smaller leaves).
3. Recycled water and soil toxicity, evaluated by the concentration of toxic ions. Plant growth, evidenced by leaf burn and defoliation, is affected mostly by boron, sodium, and chloride.
4. Nutritional imbalance in recycled water.

Traditionally, irrigation water is grouped into various classes in order to indicate the potential advantages as well as problems associated with its use and to achieve optimum crop production. Nevertheless, water quality classifications are only indicative guidelines, and their application must be adjusted to conditions prevailing in the field. The conditions of water use in irrigation are very complex and difficult to predict. Recently, for example, water quality criteria on salinity have been revised, allowing the use of salt affected water for irrigation. Consequently, the well-known classification of FAO[7] of irrigation water according to electrical conductivity and SAR is no longer considered as relevant.

In summary, even with the guidance on global water quality criteria, the suitability of water for irrigation will greatly depend on the climatic conditions, physical and chemical properties of the soil, the salt tolerance of the crop grown, and the management practices.

5.3 CHOICE OF MANAGEMENT STRATEGY OF IRRIGATION WITH RECYCLED WATER

Success in using treated wastewater for crop production will largely depend on adopting appropriate strategies aimed to optimize crop yield and quality, maintain soil productivity, and safeguard the environment. Several management alternatives are available, and their combination will offer an optimum solution for a given set of conditions. The choice of best management strategies for irrigation with recycled water becomes more limited with increasing salinity, sodicity, or toxic element concentration. For example, under such conditions, leaching and drainage must be increased.

The user should have information as to effluent supply and quality (Table 5.2) to ensure the formulation and adoption of an appropriate on-farm management strategy. The required information includes the quantity of available water, means and timing of its supply, as well as the main physical-chemical and microbiological water characteristics.

Basically, an on-farm strategy for using recycled water consists of a combination of the following measures:

1. Selection of the appropriate irrigation method
2. Proper selection of crops with adequate salt and specific ion tolerance
3. Adoption of appropriate management practices
   - Proper soil amendments and soil management
   - Leaching and sufficient drainage to dispose of excess water and salts

Table 5.2 Information on Recycled Water Supply and Quality Required for the Definition of Appropriate Management Strategy

Information on recycled water	Decisions on irrigation management
Effluent supply	
Total amount of effluent that would be made available during the crop-growing season	Total area that could be irrigated
Effluent available throughout the year	Storage facility during non–crop-growing period either at the farm or near wastewater treatment plant
Rate of delivery of effluent either as m^3/d or L/s	Area that could be irrigated at any given time, layout of fields and facilities, and system of irrigation
Type of delivery: continuous or intermittent, or on demand	Layout of fields and facilities, irrigation system, and irrigation scheduling
Mode of supply: supply at farm gate or effluent available in a storage reservoir to be pumped by the farmer	Need to install pumps and pipes to transport effluent and irrigation system
Effluent quality	
Total salt concentration and/or electrical conductivity of the effluent	Selection of crops, irrigation method, leaching, and other management practices
Concentrations of cations, such as Ca^{2+}, Mg^{2+}, and Na^+	Assess sodium hazard and undertake appropriate measures
Concentration of toxic ions, such as boron and Cl^-	Assess toxicities likely to be caused by these elements and take appropriate measures
Concentration of trace elements, particularly those suspected of being phytotoxic	Assess trace toxicities and take appropriate measures
Concentration of nutrients, particularly nitrate nitrogen	Adjust fertilizer levels, avoid over fertilization, and select crop
Level of suspended solids	Select appropriate irrigation system and measures to prevent clogging problems
Levels of intestinal nematodes and fecal coliforms	Select appropriate crops and irrigation system

Source: Adapted from Ref. 6.

Adequate timing for both irrigation and leaching
Proper use of fertilizers

Special attention should be paid to the selection of irrigation practices when recycled water is characterized by high salinity and sodicity. The most important recommendations are as follows:[1–3]

Verify that soil permeability and drainage are adequate
Determine initial salinity and sodicity of the soil and reclaim if necessary
Determine the chemical composition of irrigation water to assess potential soil and crops hazards
Leach to prevent salt accumulation
Fertilize and control weeds and insects, because healthy plants withstand salinity better

Furthermore, in arid and semi-arid areas where there are limited water supplies for irrigation, one option to overcome water scarcity problem could be alternate use of recycled water and the conventional sources of water by:

Blending conventional water with treated effluent
Using the two sources in rotation

In a more global view, it is also recommended to consider the policies undertaken by the agricultural authorities, such as, the European Union common policy on agriculture.

5.4 SELECTION OF IRRIGATION METHOD

The technical advances in irrigation materials and methods from the end of the twentieth century are becoming part of usual agricultural practice. Selection of irrigation method should be done taking into account these developments, which include computer models on water use optimization, developed for golf course irrigation. In addition to advanced irrigation technologies, low-cost and simple-to-use methods have been developed and implemented in emerging countries.

5.4.1 Criteria for Selection of an Appropriate Irrigation Method

The common irrigation methods could be classified in 3 main groups, as shown in Figure 5.1, depending on the location of water application on the soil surface, which can be on all, part of, or under the surface. Further distinction between irrigation methods comes from the type of water, which can be applied with low or high velocity or as spraying water.

The most common and widely used classification of irrigation methods is based on the practical experience, taking into account mostly the application of water (Table 5.3):

Surface irrigation, where water is applied directly on the soil surface by gravity
Overhead or sprinkler irrigation, where water is distributed over the soil surface under high pressure as small drops similar to rainfall
Localized or microirrigation, where small quantities of water is applied near the roots of crops as drops, tiny streams or mini-spray including a number of methods such as bubbler, trickle, micro-spray or subsurface drip irrigation.

Under normal conditions, the type of irrigation method selected will depend on water supply conditions, climate, soil, crops to be grown, cost of irrigation method, available irrigation material, and the ability of the farmer to manage the system (Table 5.3). However, when using recycled water as the source of irrigation, other factors such as contamination of plants and harvested product, farm workers or the environment, as well as salinity and

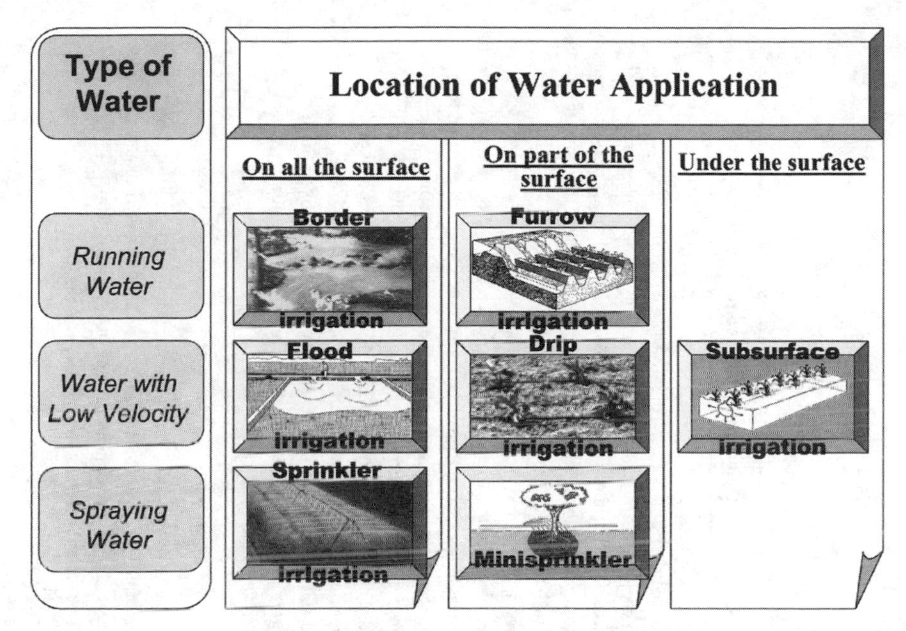

Figure 5.1 Classification of irrigation methods.

toxicity hazards need to be considered. There is considerable scope for reducing the undesirable effects of wastewater use in irrigation through the selection of appropriate irrigation methods.

The choice of irrigation method using recycled water is governed by the following technical factors:

Choice of crops
Wetting of foliage, fruits, and aerial parts
Distribution of water, salts, and contaminants in the soil
Ability to maintain high soil water potential
Efficiency of water application
Complexity of irrigation equipment
Potential of farm workers to suffer health problems derived from exposure to water components
Potential to contaminate the environment
Capital (installation) and operation costs, including energy requirements, labor availability, and maintenance costs

Table 5.4 presents an analysis of these factors in relation to four widely practiced irrigation methods: border, furrow, sprinkler, and drip irrigation. Common irrigation methods could be classified in three main groups, as shown in Figure 5.1, depending on the location of water application on the soil surface.

Water application efficiency becomes especially important in areas with water scarcity and high evaporation rates. In these conditions, mini-sprinklers and drip irrigation ensure the most effective water use (Figure 5.2). Moreover,

Table 5.3 Basic Features of Selected Irrigation Systems

Irrigation method	Topography	Crops	Comments and recommendations
Surface (flood) irrigation			
Widely spaced borders	Land slopes capable of being graded to less than 1% slope and preferably 0.2%	Alfalfa and other deep-rooted close-growing crops and orchards	The most desirable surface method for irrigating close-growing crops where topographical conditions are favorable. Even grade in the direction of irrigation is required on flat land and is desirable but not essential on slopes of more than 0.5%. Not suitable for sandy soils. Can be useful for salinity control purposes. *Water application efficiency 45–60%.*
Graded contour furrows	Variable land slopes of 2–25%, but preferably less	Row crops and fruit	Especially adapted to row crops on steep land, though hazardous due to possible erosion from heavy rainfall. Unsuitable for rodent-infested fields or soils that crack excessively. Actual grade in the direction of irrigation 0.5–1.5%. Ensure better health protection than border irrigation and can be useful for salinity control. *Water application efficiency 50–65%.*
Rectangular checks (levees)	Land slopes capable of being graded so single or multiple tree basins will be leveled within 6 cm	Orchards	Especially adapted to soils that have either a relatively high or low water intake rate. May require considerable grading. *Water application efficiency 30–60%.*
Subirrigation	Smooth-flat	Shallow-rooted (potatoes or grass) or hydrophilic (sugar cane, date palm) crops	Requires a water table, very permeable subsoil conditions, and precise leveling. Very few areas adapted to this method. For use of recycled water should be replaced by subsurface irrigation. *Water application efficiency 50–70%.*

Sprinkler irrigation			
Center pivot Lateral move Hand-move Solid set, etc.	Undulating 1–>35% slope	All crops, well suited for turf grass	Allows uniform, efficient and frequent application, as well as addition of chemicals and fertilizers. Good for rough or very sandy lands in areas of high production and good markets. Good method where power costs are low. May be the only practical method in areas of steep or rough topography. Good for high rainfall areas where only a small supplementary water supply is needed. High degree of automation. Capital costs typically 50 to 100% higher than surface irrigation, as well as operation and maintenance (O&M) costs. Affected by wind. High evaporation losses. *Water application efficiency 60–85%.*
Localized or microirrigation			
Mini-sprinklers	Wide range of terrain conditions	All crops	Point source application by a small spray on the soil surface, usually without overlapping. High efficiency and possibility of addition of chemicals. Especially suited for irrigation with recycled water enabling high-frequency and low-volume irrigation with easy scale up for small units. Capital costs typically over 100% higher than sprinkler irrigation, as well as O&M costs (+50%). *Water application efficiency 70–90%.*
Drip/trickle	Any topographic condition suitable for row crop farming	Row crops, fruits or vineyards	Perforated pipe on the soil surface drips water at base of individual vegetable plants or around fruit trees. High efficiency and possibility of addition of chemicals. Has been successfully used with saline irrigation water. Capital costs similar to mini-sprinkler irrigation and typically over 300–400% higher than surface irrigation. *Water application efficiency 70–95%.*
Subsurface	Field slope is limiting, in particular undulating terrain	Fruit trees, perennial row crops	Application of water below the soil surface through emitters or porous tube with discharge rates similar to drip irrigation. This system provides the highest health protection when using recycled water for irrigation. The main constraints are the high capital costs and needs of good design, operation and maintenance. *Water application efficiency 70–98%.*

Source: Adapted from Refs. 3, 6.

Table 5.4 Evaluation of Common Irrigation Methods in Relation to Use of Recycled Water

Parameter of evaluation	Furrow irrigation	Border irrigation	Sprinkler irrigation	Drip irrigation
Foliar wetting and consequent leaf damage resulting in poor yield	No foliar injury as crop is planted on ridge[a]	Some bottom leaves may be affected, but damage is not so serious as to reduce yield[a]	*Severe leaf damage can occur, resulting in significant yield loss*[b]	No foliar injury under this method of irrigation[a]
Salt accumulation in the root zone with repeated applications	*Salts tend to accumulate in ridge, which could harm crop*[b]	Salts move vertically downwards and are not likely to accumulate in the root zone[a]	Salt movement is downwards and root zone is not likely to accumulate salts[a]	*Salt movement is radial along the direction of water movement; a salt wedge is formed between drip points*[c]
Ability to maintain high soil water potential	*Plants may be subject to water stress between irrigations*[c]	*Plants may be subject to water stress between irrigations*[c]	*Not possible to maintain high soil water potential throughout growing season*[b]	Possible to maintain high soil water potential throughout the growing season and minimize the effect of salinity[a]
Suitability to handle brackish wastewater without significant yield loss	*Fair to medium; with good management and drainage, acceptable yields are possible*[c]	*Fair to medium; good irrigation and drainage practices can produce acceptable levels of yield*[c]	*Poor to fair; most crops suffer from leaf damage and yield is low*[b]	Excellent to good; almost all crops can be grown with very little reduction in yield[a]
Cost of equipments, operation and maintenance	Low costs; land preparation of furrows, 30–450 m long, 20–30 cm deep	Low costs; land preparation of field areas with downslope 0.1–0.4%	Very high cost of equipment, significant O & M costs, and need for periodic maintenance	High cost of equipment, moderate O&M costs, and need for maintenance

[a]Irrigation method is recommended.
[b]Irrigation system is not advisable.
[c]Irrigation method may cause problems.

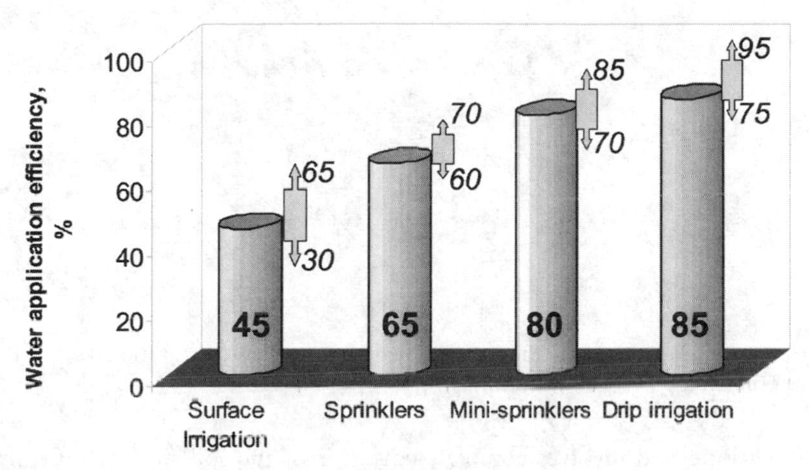

Figure 5.2 Theoretical water application efficiency of common irrigation systems: mean value and range of variation.

these systems provide the best health protection by lowering the probability of direct contact with recycled water.

From the health aspect point of view, the following points should be considered during the selection of the most appropriate irrigation method:

1. All irrigation methods are appropriate to be used for recycled water, which is in compliance with water reuse guidelines or regulations for unrestricted irrigation, provided that agronomic criteria are also met.
2. A number of existing regulations require higher water quality for sprinkler irrigation because of the possible disease transmission with aerosols.
3. Sprinkler irrigation with recycled water that does not meet the health criteria is possible in conditions of implementation of specific management practices such as crop selection (industrial crops and fodder), irrigation scheduling (irrigation during night), and other restrictions (no irrigation during windy conditions).

In addition to the technical and health considerations, socioeconomic conditions should also be taken into account.[8] In this context, an irrigation system considered most appropriate in one country or region may not be so in another.

5.4.2 Comparison of Irrigation Methods

5.4.2.1 Surface Irrigation Methods

Surface or flood irrigation methods account for about 95% of the world's irrigated area because of the low cost and simplicity of use and implementation.[3] In this case, irrigation water is simply flowing by gravity across the irrigated

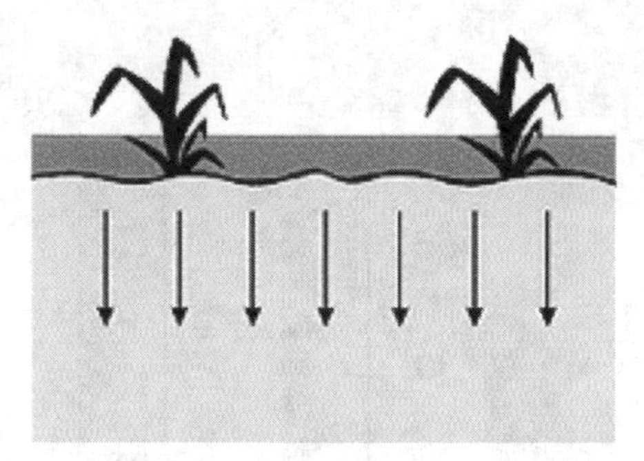

Figure 5.3 Flood irrigation: the entire root zone is wetted to saturation. (From Ref. 8, with permission.)

area. This method involves complete coverage of the soil surface with treated effluent (Figure 5.3) and is normally not an efficient method of irrigation.

Flood irrigation is not suitable on steep slopes, on soils with high hydraulic conductivity (sandy soils with too rapid infiltration), for shallow-rooted crop species, and at sites where it is critical to minimize deep drainage. As surface irrigation systems normally result in the discharge of a portion of the irrigation water from the site, some methods of tailwater return or pump-back may be required in areas where discharge is not permitted. To avoid surface ponding of stagnant effluent, land leveling should be carried out carefully and appropriate land gradients should be provided.

The efficiency of surface irrigation methods in general (borders, furrows) is not greatly affected by water quality, although the health risk inherent to these systems is most certainly of concern. Some problems might arise if the effluent contains large quantities of suspended solids, which settle out and restrict flow in transporting channels, gates, pipes, and appurtenances. The possibility of biofilm growth on the transport pipeline systems could be a concern.

Border Irrigation

In border irrigation systems, the irrigated land is divided into wide long bays bordered by earth mounds, also called strips or borders. Recycled water flows into each bay through a pipe or gate valve from the distribution channel. As a rule, this method is used on land of flat topography.

This system can lead to contamination of vegetable crops growing near the ground and root crops and will expose farm workers to the effluent more than any other method. Thus, from both health and water conservation points of view, border irrigation with recycled water is not very satisfactory, although it can occasionally be useful for salinity-control purposes.

Furrow Irrigation

Furrow irrigation does not wet the entire soil surface (Figure 5.4), because the irrigation water is distributed through shallow, narrow, gently sloping furrows

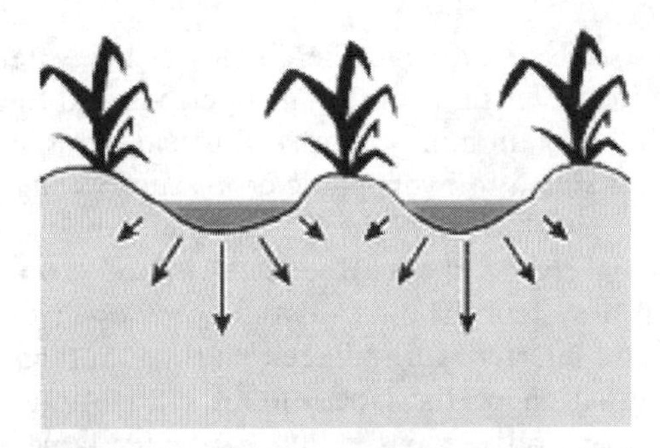

Figure 5.4 Furrow irrigation: the entire root zone is wetted to near-saturation if furrows are closely spaced. (From Ref. 8, with permission.)

Figure 5.5 Furrow irrigation of crops in Cyprus

(narrow ditches dug between the rows of crops). Typically, water is supplied to the furrows from gated pipes or siphons (Figure 5.5).

This method can reduce crop contamination, since plants are grown on ridges. However, complete health protection cannot be guaranteed. Contamination of farm workers is potentially medium to high, depending on automation. If the effluent is transported through pipes and delivered into individual furrows by means of gated pipes, risk to irrigation workers is minimal.

Subirrigation

This method, which has a limited application, consists in supplying water to the root zone of crops by artificially regulating the groundwater table. Open ditches are usually dug to a depth below the water table, and the level of the water is controlled by check dams or gates. In this manner, the ditches can serve either to drain excess water and thereby lower the water table during wet periods or to raise the water table during dry periods and thereby wet the root zone from below. Subirrigation may be used for field crops and pasture, as well as orchards. It is best suited to hydrophilic crops such as sugar cane and dates.

This system is applicable only where the water table is naturally high, as it frequently is along river valleys. Reduced risk of crop contamination is expected with such a system. The disadvantage of open ditches is that they interrupt the field and interfere with tillage, planting, and harvesting as well as take a significant fraction of the land out of cultivation. An alternative is to place porous or perforated pipes below the water table, with controllable outlets (subsurface irrigation).

5.4.2.2 Sprinkler Irrigation Methods

Sprinkler or overhead irrigation methods create artificial rainfall, and, thus, they are generally more efficient in terms of water use since greater uniformity of application can be achieved (Figure 5.6). Sprinklers are mounted on riser pipes and scattered throughout the irrigation area by blocks with overlap of 25–50% to achieve uniform wetting. The spraying is accomplished by using several rotating sprinkler heads or spray nozzles or a single gun-type sprinkler. In general, sprinkler systems are the most expensive and are less suitable for plantations harvested in short rotation, because of their vulnerability to damage by heavy machinery. Table 5.5 summarizes the most important advantages and disadvantages of sprinkler systems.[2,9] Sprinkler irrigation is especially suitable for continuous low vegetation (such as ground cover turf grass, decorative estates, some fodder crops), plants with shallow root systems, as well as any vegetation that benefits from overhead water and high humidity.

It is important to stress that sprinkler irrigation should not be used at sites where spray drift is undesirable for health or environmental reasons. These overhead irrigation methods may contaminate ground crops, fruit trees, farm workers, and other people not related to the facility. In addition, pathogens contained in aerosols may be transported downwind and create a health risk to nearby residents. Therefore, buffer zones or devices (tree barriers)

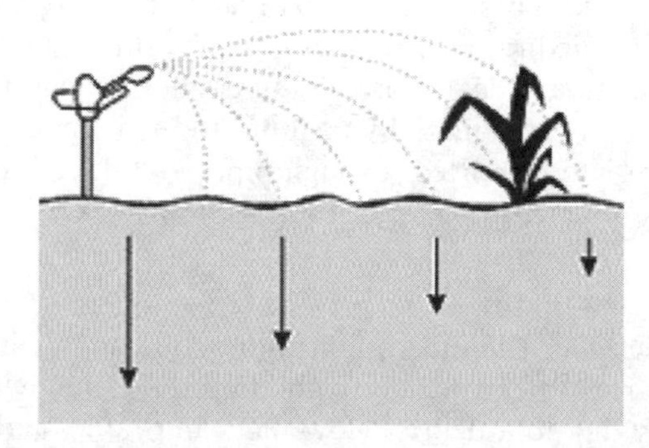

Figure 5.6 The wetting pattern of sprinkler irrigation. (From Ref. 8, with permission.)

Table 5.5 Advantages and Disadvantages of Sprinkler Irrigation Systems Relative to Surface Irrigation

Advantages	Disadvantages
Suitable to use on porous and variable soils	Initial cost can be high
Suitable to use on shallow soil profiles	Energy costs higher than for surface systems
Suitable to use on rolling terrain	Higher humidity levels can increase disease potential for some crops
Suitable to use on easily eroded soils	Sprinkler application of highly saline water can cause leaf burn
Suitable to use with small flows	Water droplets can cause blossom damage to fruit crops or reduce quality of some fruit and vegetable crops
Suitable to use where high water tables exist	Portable or moving systems can get stuck in some clay soils
Can be used for light, frequent applications	Higher levels of preapplication treatment generally required for sprinkler systems than for surface systems to prevent operating problems (clogging)
Control and measurement of applied water is easier	Distribution subject to wind distortion
Tailwater control and reapplication is minimized	Wind drift of sprays increases potential for public exposure to wastewater
	Vulnerability to damage by logging machinery during harvesting

Source: Adapted from Refs. 2, 9.

should be established around irrigated areas. In some systems (i.e., center pivot), the sprinkler nozzles may be dropped closer to the ground to reduce aerosol drift and thus minimize the buffer requirements. In all cases, wind must be considered as a limiting factor.

Labor requirements are usually low, with the exception of maintenance. As a rule, sprinklers and plastic risers need to be replaced about every 10 years. Generally, mechanized or automated systems have relatively high capital costs and low labor costs compared with manually moved sprinkler systems. Rough land leveling is necessary for sprinkler systems to prevent excessive head losses and achieve uniformity of wetting.

Sprinkler systems are more affected by water quality than surface irrigation systems, primarily as a result of the clogging of orifices in sprinkler heads, potential leaf burn, and phytotoxicity, in particular when water is saline and contains excessive toxic elements. Inadequate treatment, especially filtration, leads to clogging of sprinklers. The micro-spray systems are the most vulnerable to clogging. Sediment accumulation in pipes, valves, and distribution systems should also be taken into account. Secondary wastewater treatment has generally been found to produce an effluent suitable for distribution through sprinklers. Nevertheless, further precautionary measures are often adopted, such as additional treatment with granular filters or microstrainers and use of nozzle orifice diameters not less than 5 mm.

Homogeneous water distribution can be also disrupted by weeds, in particular with low-pressure systems. Recommended measures to avoid such constraints are periodic weed control by herbicides or by mounting the sprinklers on taller risers.

5.4.2.3 Micro-sprayer Irrigation

Micro-spray irrigation is a part of localized or microirrigation systems. Micro-sprayers, also called mini-sprinklers or spitters, ensure the greatest uniformity in effluent distribution and are similar in principle to drip systems, where water is applied only to a part of the ground surface (Figure 5.7). These systems eject fine water jets from a series of nozzles, mounted on moving spindles or fixed sprinkler arms (Figure 5.8). Compared to drip emitters, micro-sprayers can water a larger area of several square meters. In addition, clogging problems are reduced thanks to the larger nozzle orifices. The discharge rate is also greater, ranging from 20 to 300 L/h.

The pressure requirement is lower compared to conventional sprinklers, more than 2 atm, which still requires pumping or reservoir elevation of over 20 m. Another disadvantage relative to drip irrigation is the evaporation loss and wetting of foliage that can be damaged by brackish water.

It is important to stress that these systems, which are commonly used for trees, are especially suited for irrigation with recycled water because of their benefits, similar to drip irrigation, enabling high-frequency and low-volume irrigation with easy scale-up for small irrigation units.

5.4.2.4 Drip Irrigation

Drip irrigation is another microirrigation system, which often is considered as synonymous of trickle irrigation. The principle is the same, but the discharge rates of trickle systems could be higher than the upper limit of drip emitters,

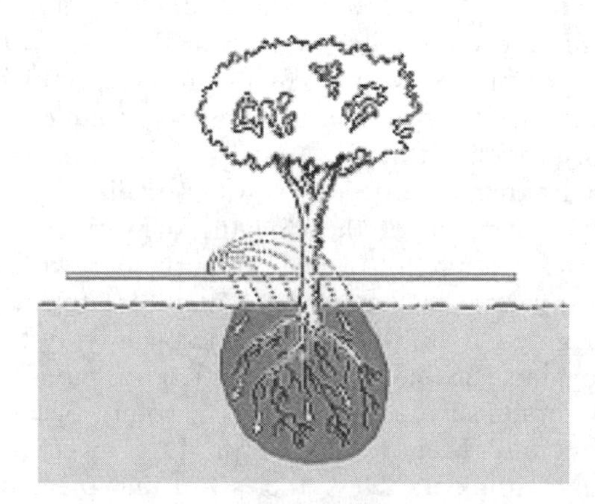

Figure 5.7 The wetting pattern of micro-spray irrigation. (From Ref. 8, with permission.)

Figure 5.8 Micro-spray irrigation of crops.

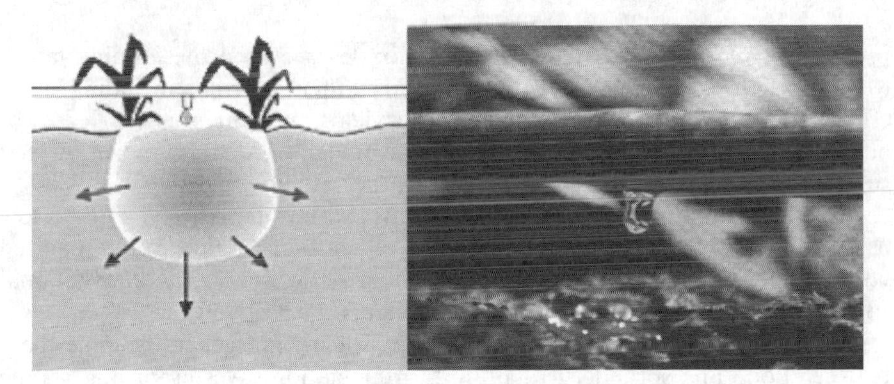

Figure 5.9 The wetting pattern of drip irrigation. (From Ref. 8, with permission.)

which is 12 L/h. In drip irrigation methods, water is applied through a network of small emitters placed on the ground surface (Figure 5.9 and Figure 5.10). At regular intervals of the distribution network, near the plants or trees, a hole is made in the tube and equipped with an emitter to supply water to the plants slowly, drop by drop. The saturated zone is usually less than 50% and depends on the density of the wetting points, the rate of application, and the soil properties. As a rule, drip emitters are designed to supply controlled water rate of 1–10 L/h. The operating pressure is in the range of 0.5–2.5 atm, preferably 1 atm at the dripper, which is dissipated to atmospheric pressure by the head loss of the emitters.

The most important advantage is that these systems can be used with all types of soils and topography without special land preparation, provided that

Figure 5.10 View of drip irrigation of crops in Cyprus.

pressure regulators or regulated dripper nozzles are used in slopping lands. Moreover, these systems provide numerous other benefits, including more efficient use of recycled water, higher crop yields, and the greatest degree of health protection for farm workers and consumers, particularly when the soil surface is covered with plastic sheeting or other mulch.

Drip irrigation is particularly suited to sites with water scarcity where a low irrigation rate is desirable. To achieve a given rate of irrigation, the required volume for drip systems is about 4-, 10- and 50-fold lower than conventional sprinkler, furrow, and border irrigation systems, respectively.

The capital and operation costs of drip systems fall generally somewhere between flood and sprinkler irrigation. Drip irrigation saves labor but has the highest installation cost. The successful operation of these systems requires good supervision and maintenance: regular inspection and cleaning of emitters, repair of leaks, flushing of the network (at least twice a year) with necessary acidification (to prevent precipitation of calcium carbonate), and monthly adjustment of the irrigation schedule for permanent plantings, according to seasonal weather and watering requirements.

One of the major constraints of drip irrigation is blockage of the emitters because of the very small orifices (0.1–2 mm). Thus, recycled water should be free of suspended solids. Some dissolved compounds such as Ca^{2+} and Fe^{2+} can precipitate.

From the point of view of water quality, drip irrigation systems have the advantage to allow using recycled water with high salinity or high BOD, because the entire root zone is not saturated, thus providing good soil aeration.

Drip irrigation is most suitable for row crops (vegetables, soft fruit) and tree and vine crops where one or more emitters can be provided for each plant. Generally only high-value crops are considered because of the high capital costs of installing a drip system. Nevertheless, several experiments and real plot application to extensive crops (wheat and alfalfa) have shown that it is possible to consider drip irrigation for these crops as well. It is important to stress that relocation of subsurface systems can be prohibitively expensive.

For irrigation of urban parks and other green areas, some dripper systems are buried underground to reduce the risk of human contact with recycled water. The capital and operating costs of such systems are, however, higher, and a very good filtration of effluent is required.

Bubbler irrigation, a technique developed for localized irrigation of tree crops, avoids the need for small emitter orifices, but requires careful setting for its successful application. In this case, water is allowed to "bubble out," even under a very limited pressure, from open thin-walled vertical tubes ($d = 1$–3 cm) connected to the buried lateral irrigation tubes ($d \leq 10$ cm). The main advantages of bubbler irrigation are its low cost, ease of installation and operation, as well as lower vulnerability to water quality.

When compared with other systems, the main advantages of drip irrigation are as follows:

Increased crop growth and yield achieved by optimizing the water, nutrients, and air regimes in the root zone

High irrigation efficiency: no canopy interception, wind drift, or conveyance losses and minimal drainage losses

Minimal contact between farm workers and effluent

Low energy requirements: the trickle system requires a water pressure of only 100–150 kPa (1–1.5 bar)

Low labor requirements: the trickle system can be easily automated and allows combined irrigation and fertilization

Accurate control of leaching and drainage (with adequate design, maintenance, and scheduling of irrigation)

Easier removal and reinstallation during harvesting periods compared to sprinklers

In summary, drip irrigation is best suited to areas where water is scarce, land is steeply sloping or undulating, water and/or labor are expensive, recycled water has high salinity, or the production of high-value crops requires frequent water applications.

5.4.2.5 Subsurface Irrigation

In subsurface irrigation, water is applied directly to the root zone via perforated or porous diffusers, placed 10–50 cm below the soil surface (Figure 5.11). The spreading pattern depends on the soil properties as well as on the distance between adjacent emitters and their discharge rates.

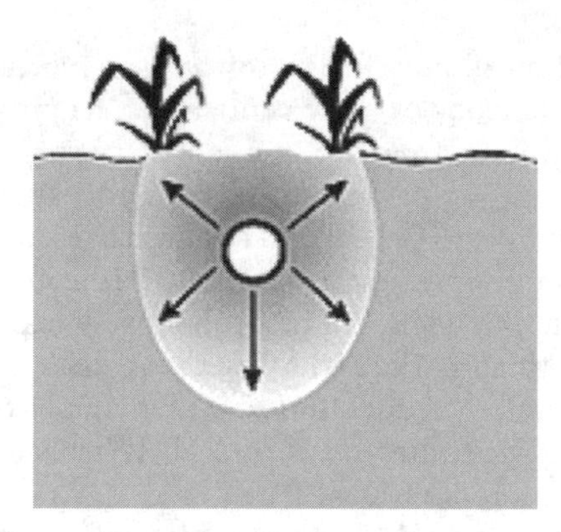

Figure 5.11 The wetting pattern of subsurface irrigation. (From Ref. 8, with permission.)

The main advantage of these systems is the health safety due to the absence of direct contact with recycled water.

A potential problem here is that the narrow orifices of the emitters may become clogged by roots, particles, algae, or precipitating salts. Such clogging is difficult to detect as readily as when the tubes are placed over the surface in above-ground drip irrigation. Injecting an acidic or herbicidal solution into the tubes may help to clear some types of clogging, though the problem may recur periodically. Special drippers with a specific chemical are available that do not allow roots to enter the dripper.

In underground drip irrigation, the delivery of water via the feeder tubes can be constant or intermittent. For uniformity of application, there should be some means of pressure control. If the lines are long or the land is sloping, there can be considerable differences in the hydraulic pressure and therefore in delivery rate, unless pressure-compensated emitters are used. However, such emitters are expensive and usually of higher variation in flow and as a consequence of lower uniformity in water application.

Existing experience (Israel, California) has shown that this method is feasible for fruit trees and other perennial row crops. The major constraint for their implementation is the high capital and operation costs and the high frequency of renewal.

5.4.3 Final Considerations for the Choice of Irrigation Method

The most important factor in decision making as to irrigation method appears to be financial, i.e., the irrigation system cost. Nevertheless, the health risks associated with the different methods, as well as water savings, should be also taken into account.

Ideally, recycled water should be applied closely to the root zone using micro-sprayers or drip emitters. Among the numerous benefits of these systems is the ability to apply high-frequency irrigation better adjusted to the crop irrigation requirements. In this case, storage capacity and soil capability to retain moisture are no longer decisive. The major constraint is the need for continuous operation, because any short-term interruption can quickly result in plantation damage.

Furrow irrigation is suitable while leaching demand is high. To avoid human contact and allow pressurizing, recycled water should be conveyed in closed conduits and distributed via resistant plastic tubes.

It is important to emphasize that the irrigation method is one of several possible health control measures, along with crop selection, wastewater treatment, and human exposure control. Each measure interacts with the others, and, thus, any decision as to irrigation system selection has an influence on wastewater treatment requirements, human exposure control, and crop selection (e g., row crops are dictated by trickle irrigation). At the same time, the feasibility of the irrigation technique depends on crop selection. The choice of irrigation system might be limited if wastewater treatment has been already implemented without any possibility for plant upgrading and water quality improvement.

5.5 CROP SELECTION AND MANAGEMENT

The selection of the crops and plant species to be irrigated with recycled water must be done after a complete evaluation of the recycled water quality. This evaluation should be done not only with average values of salinity and other physical-chemical parameters, but also taking into account the typical range of variation of each compound of interest. If there exist peaks in any of these parameters, it is necessary to know their usual duration. This information is necessary to select plant species, that can tolerate temporarily extreme conditions without suffering major damage.

Good management practices have a crucial role in the preservation of soil properties and crop productivity. It was reported[10] that correct management of irrigation with brackish water for 26 years did not cause detrimental changes to soil salinity; new equilibrium of soil chemical parameters took place.

5.5.1 Code of Practices to Overcome Salinity Hazards

The most important factor for crop selection is the salinity of the irrigation water. In general, TDS of >4000 mg/L or conductivity of >6 dS/m represent a significant quality problem for irrigation. In some cases, high chloride concentrations are the controlling parameter affecting reuse potential for irrigation. Good salinity management practices can allow irrigation with total dissolved solids up to 7000 mg/L. In some cases, application of saline water for irrigation of agricultural crops (conductivity range 2–7 dS/m) leads to improvement of fruit quality due to higher sugar content.[11]

The rate of salt accumulation in the soil depends upon the quantity of salt applied with the irrigation water and will increase as water is removed from the soil by evaporation and transpiration. The adverse effects of salinity are usually associated with an increase in soil salinity and the osmotic pressure in the soil solutions, and thereby with adverse effects on both crop and soil. Plants will use more and more energy to extract water from a saline soil solution and therefore put less into their growth. It is, therefore, important to prevent harmful concentrations of salts in the root zone of irrigated crops or at least to maintain a portion of the root below salinity levels that a given crop can tolerate.

The main management practices for the safe use of saline recycled water for irrigation are as follows:

1. Source control
2. Selection of crops or crop varieties that will produce satisfactory yields under the existing or predicted conditions of salinity or sodicity
3. Special planting procedures that minimize or compensate for salt accumulation in the vicinity of the seeds
4. Irrigation to maintain a relatively high level of soil moisture and to achieve periodic soil leaching
5. Land preparation to increase the uniformity of water distribution and infiltration, leaching, drainage, and removal of salinity
6. Special treatments such as tillage and addition of chemical amendments, organic matter, and growing green manure crops to maintain soil permeability and tilth

The crop grown, quality of irrigation water, rainfall pattern, climate, and soil properties determine to a large degree the kind and extent of management practices needed.

5.5.1.1 Source Control

Source control is the first and the most affordable measure for water reuse managers to address salinity issues. It is important to protect urban wastewater for beneficial reuse by treating or diverting poorer quality industrial and commercial brine waste streams in separate sewers or ocean outfalls (brine-disposal measures). It is important to underline that, as a rule, no limits on TDS concentrations exist for discharge of industrial wastewater in urban sewers.

Other measures for source control include the rehabilitation and repair of leaky sewers infiltrated by brackish water. Some restrictions can be made on the use of certain types of residential softeners and other products for domestic use that are major sources of salts in wastewater. Such salinity control measures offer significant economic and public benefits not only for crop production and increased life of plumbing and irrigation systems, but also for all urban and industrial users.

5.5.1.2 Crop Selection

Not all plants respond to salinity in a similar manner; some crops can produce acceptable yields at much higher soil salinity than others. This is because some crops are better able to make the needed osmotic adjustments, enabling them to extract more water from a saline soil. Plants capable of good growth in saline environments and of extracting salts from the natural environment should also be considered in these cases.

The ability of a crop to adjust to salinity is extremely useful. In areas where a build-up of soil salinity cannot be controlled at an acceptable concentration for the crop being grown, an alternative crop can be selected that is both more tolerant to the expected soil salinity and able to produce economic yields. The relative salt tolerance of most agricultural crops is known well enough to provide general salt-tolerance guidelines. For specific crops, local tests can also be carried out to study their salt tolerance. Crops can be divided into four groups depending on their salt tolerance at the root zone.[6,7]

Sensitive crops can be used when the water is suitable for irrigation of most crops on most soils with little likelihood that soil salinity will develop. Some leaching is required, but this occurs under normal irrigation practices, except in soils of extremely low permeability.

Moderately sensitive crops can be used if a moderate amount of leaching occurs. Plants with moderate salt tolerance can be grown in most cases without special practices for salinity control.

Moderately tolerant crops should be selected when the water cannot be used on soils with restricted drainage. Even with adequate drainage, special management for salinity control may be required.

Tolerant crops should be selected when the water is not suitable for irrigation under ordinary conditions, but may be used occasionally under very special circumstances. In this case, the soil must be permeable, drained adequately, and irrigation water applied in excess to provide considerable leaching.

Guidelines for the use of recycled water with varying salinity for irrigation of these four groups of crops are provided in Table 5.6. Table 5.7 summarizes the most relevant management practices for crop selection to avoid salinity hazards.

A list of crops classified according to their tolerance and sensitivity to salinity is presented in Table 5.8. Although this list is only a relative ranking, it provides a good comparison of the performance of different crops. For example, turf grass and a number of industrial crops are characterized by good salt tolerance. The most salt-sensitive crops are fruits and some vegetables.

The yield potential of some crops depending on water and soil salinity is summarized in Table 5.9. The salt tolerance data in this table are used for conventionally irrigated crops. However, drip irrigation can maintain a high level of humidity in the root zone and a salinity level almost similar to that of the irrigation water.

Salt tolerance also depends on the type, method, and frequency of irrigation. The prevalent salt tolerance data apply most directly to crops irrigated by surface methods and conventional irrigation management. Salt concentrations may differ severalfold within irrigated soil profiles, and they change constantly. At the same time, the plant is most responsive to salinity in the root zone, where water uptake occurs. For this reason, salt concentration

Table 5.6 Guidelines for the Classification and Use of Brackish Water for Irrigation

Crop class		Nonsaline	Slightly saline	Medium saline	Highly saline	Very highly saline
	TDS (mg/L)	<500	500–2000	2000–4000	4000–5000	>9000–30,000
	EC_W (dS/m)	<0.7	0.7–3.0	3.0–6.0	6.0–14.0	>14.0–42
I	**Sensitive crops**	No limitation on use No yield reduction	Slight to medium restriction on use Up to 50% yield reduction possible	For restricted use only More than 50% yield reduction possible	Not recommended for use with this class of crops	Not recommended for use
II	**Moderately sensitive crops**	No limitation on use No yield reduction	Slight restriction on use Up to 20% yield reduction possible	Medium restriction on use Up to 50% yield reduction	For restricted use only More than 50% yield reduction possible	Not recommended for use
III	**Moderately tolerant**	No limitation on use No yield reduction	No limitation on use No yield reduction	Slight to medium restriction on use Up to 20–40% yield reduction possible	Medium restriction on use 40–50% yield reduction possible	For restricted use only More than 50% yield reduction possible
IV	**Tolerant crops**	No limitation on use No yield reduction	No limitation on use No yield reduction	No serious limitation on use Practically no yield reduction	Slight to medium restriction on use 20–40% yield reduction possible	For restricted use only More than 50% yield reduction possible; suitable for halophytes

Source: Adapted from Refs. 12–16.

Table 5.7 Recommendations for Crop Selection to Overcome Salinity Hazards

Salinity EC_W (TDS)	Recommendations
< 0.7 dS/m (< 450 mgTDS/L)	Full yield potential should be achievable with nearly all crops.
0.7–3.0 dS/m (450–2000 mgTDS/L)	With good management, most fruits and vegetables can be produced. Full yield potential is still possible, but care must be taken to **achieve the required leaching fraction** in order to maintain soil salinity within the tolerance of the crops.
3.0 dS/m (< 2000 mgTDS/L)	The water might still be usable, but its use may **need to be restricted to more permeable soils** and **more salt-tolerant crops**, where high leaching fractions are more easily achieved. This is being practiced on a large scale in the Arabian Gulf states, where drip irrigation systems are widely used. 1. If crops are **salt-sensitive**, solutions are: ***Increase leaching*** to satisfy a leaching requirement greater than 0.25–0.30 (negative points: excessive amount of water is required). ***Select irrigation system*** with uniform application, high efficiency, and frequency of irrigation (drip irrigation and mini-sprinklers). ***Scheduling of irrigation***: more frequent irrigation with micro-irrigation systems enable to maintain lower levels of salinity in the plant vicinity. ***Drainage*** allows for the leaching of excess salts (in combination with irrigation scheduling). ***Soil conditioners*** are not recommended because the high price and low efficiency in certain periods and conditions. In such a case, consider **changing to a more tolerant crop** that will require less leaching to control salts within crop tolerance levels. 2. Selection of **salt tolerant crops** with the ability to absorb high amounts of salts

Source: Adapted from Refs. 3, 6.

should be measured in the root zone. The best performance of irrigation with highly saline water has been reported for drip irrigation.

Sprinkler-irrigated crops are potentially subject to additional damage caused by foliar salt uptake and desiccation (burn) from spray contact with the foliage. Susceptibility of plants to foliar salt injury depends on leaf characteristics affecting the rate of absorption and is not generally correlated with tolerance to soil salinity. The degree of spray injury varies with weather conditions, being especially high when the weather is hot and dry. Night sprinkling has been proved to be advantageous in a number of cases. The information concerning the susceptibility of crops to foliar injury from saline sprinkling water is limited and indicates the most sensitive crops to be fruits

Table 5.8 Classification of Crops as a Function of Salt Tolerance

Salt-tolerant crops	Moderately salt-tolerant crops		Moderately salt-sensitive crops		Salt-sensitive crops	
Fiber, seed, and sugar crops						
Barley, *Hordeum vulgare*	Cowpea, *Vigna unguiculata*	Soybean, *Glycine max*	Broadbean, *Vicia faba*	Groundnut/peanut, *Arachis hypogaea*	Guayule, *Parthenium argentatum*	Sesame, *Sesamum indicum*
Cotton, *Gossypium hirsutum*	Oats, *Avena sativa*	Triticale, *X Triticosecale*	Castorbean, *Ricinus communis*	Rice, paddy, *Oryza sativa*		
Jojoba, *Simmondsia chinensis*	Rye, *Secale cereale*	Wheat, *Triticum aestivum*	Maize, *Zea mays*	Sugarcane, *Saccarum officinarum*		
Sugarbeet, *Beta vulgaris*	Safflower, *Carthamus tinctorius*	Wheat, Durum, *Triticum turgidum*	Flax, *Linum usitatissimum*	Sunflower, *Helianthus annuus palustris*		
	Sorghum, *Sorghum bicolor*		Millet, foxtail, *Setaria italica*			
Vegetable Crops						
Asparagus, *Asparagus officinalis*	*Artichoke, Helianthus tuberosus*	Squash, zucchini, *Cucurbita pepo melopepo*	Broccoli, *Brassica oleracea botrytis*	Lettuce, *Latuca sativa*	Bean, *Phaseolus vulgaris*	Onion, *Allium cepa*
	Beet, red, *Beta vulgaris*		Brussel sprouts, *B. oleracea gemmifera*	Muskmelon, *Cucumis melo*	Carrot, *Daucus carota*	Parsnip, *Pastinaca sativa*
			Cabbage, *B. oleracea capitata*	Pepper, *Capsicum annum*	Okra, *Abel moschus esculentus*	Pea, *Pisum sativum*
			Cauliflower, *B. oleracea botrytis*	Potato, *Solanum tuberosum*		
			Celery, *Apium graveolens*	Pumpkin, *Cucurbita pepo pepo*		
			Corn, sweet, *Zea mays*	Radish, *Raphanus sativus*		
			Cucumber, *Cucumis sativus*	Spinach, *Spinacia oleracea*		
			Eggplant, *Solanum melongena esculentum*	Squash, scallop, *Cucurbita pepo melopepo*		
			Kale, *Brassica oleracea acephala*	Sweet potato, *Ipomoea batatas*		
			Kohlrabi, *B. oleracea gongylode*	Tomato, *Lycopersicon lycopersicum*		
				Turnip, *Brassica rapa*		
				Watermelon, *Citrullus lanatus*		

Grasses and Forage Crops

Alkali grass, *Puccinellia airoides*	Barley (forage), *Hordeum vulgare*	Ryegrass, perennial, *Lolium perenne*	Alfalfa, *Medicago sativa*	Cowpea (forage), *Vigna unguiculata*
Alkali sacaton, *Sporobolus airoides*	Brome, mountain, *Bromus marginatus*	Sudan grass, *Sorghum sudanense*	Bentgrass, *Agrostis stolonifera palustris*	Dallis grass, *Paspalum dilatatum*
Bermuda grass, *Cynodon dactylon*	Canary grass, reed, *Phalaris arundinacea*	Trefoil, narrowleaf birdsfoot, *Lotus corniculatus tenuifolium*	Bluestem, Angleton, *Dichanthium aristatum*	Foxtail, meadow, *Alopecurus pratensis*
Kallar grass, *Diplachne Fusca*	Clover, sweet, *Melilotus*	Trefoil, broadleaf, *Lotus corniculatus arvenis*	Brome, smooth, *Bromus inermis*	Grama, blue, *Bouteloua gracilis*
Saltgrass, desert, *Distichlis stricta*	Fescue, meadow, *Festuca pratensis*	Wheat (forage), *Triticum aestivum*	Buffelograss, *Cenchrus ciliaris*	Lovegrass, *Eragrostis sp.*
Wheatgrass, fairway crested, *Agropyron cristatum*	Fescue, tall, *Festuca elatior*	Wheatgrass, standard crested, *Agropyron sibiricum*	Burnet, *Poterium sanguisorba*	Milkvetch, Cicer, *Astragalus deer*
Wheatgrass, tall, *Agropyron elongatum*	Harding grass, *Phalaris tuberosa*	Wheatgrass, intermediate, *Agropyron intermedium*	Clover, alsike, *Trifolium hydridum*	Oatgrass, tall, *Arrhenatherum, Danthonia*
Wildrye, Altai, *Elymus angustus*	Panic grass, blue, *Panicum antidotale*	Wheatgrass, slender, *Agropyron trachycaulum*	Clover, Berseem, *Trifolium alexandrinum*	Oats (forage), *Avena saliva*
Wildrye, Russian, *Elymus junceus*	Rape, *Brassica napus*	Wheatgrass, western, *Agropyron smithii*	Clover, ladino, *Trifolium repens*	Orchard grass, *Dactylis glomerata*
	Rescue grass, *Bromus unioloides*	Wildrye, beardless, *Elymus triticoides*	Clover, red, *Trifolium pratense*	Rye (forage), *Secale cereale*
	Rhodes grass, *Chloris gayana*	Wildrye, Canadian, *Elymus canadensis*	Clover, strawberry, *Trifolium fragiferum*	Sesbania, *Sesbania exaltata*
	Ryegrass, Italia, *Lolium italicum multiflorum*		Clover, white Dutch, *Trifolium repens*	Siratro, *Macroptilium atropurpureum*
			Corn (forage) (maize), *Zea mays*	Sphaerophysa, *Spaerophysa salsula*
				Timothy, *Phleum pratense*
				Vetch, common, *Vicia angustifolia*

(*continued*)

Table 5.8 Continued

Salt-tolerant crops	Moderately salt-tolerant crops		Moderately salt-sensitive crops	Salt-sensitive crops	
Fruit and nut crops					
Date palm, *Phoenix* dactylifera	Fig, *Ficus carica*	Pineapple, *Ananas comosus*	Grape, *Vitis sp.*	Almond, *Prunus dulcis*	Orange, *Citrus sinensis*
	Jujube, *Ziziphys jujuba*	Pomegranate, *Punica granatum*		Apple, *Malus sylvestris*	Passion fruit, *Passiflora edulis*
	Olive, *Olea europaea*			Apricot, *Prunus armeniaca*	Peach, *Prunus persica*
	Papaya, *Carica papaya*			Avocado, *Persea americana*	Pear, *Pyrus communis*
				Blackberry, *Rubus sp.*	Persimmon, *Diospyros virginiana*
				Boysenberry, *Rubus ursinus*	Plum, Prune, *Prunus domestica*
				Cherimoya, *Annona cherimola*	Pummelo, *Citrus maxima*
				Cherry, sweet, *Prunus avium*	Raspberry, *Rubus idaeus*
				Cherry, sand, *Prunus besseyi*	Rose apple, *Syzygium jambos*
				Currant, *Ribes sp.*	Sapote, white, *Casimiroa edulis*
				Gooseberry, *Ribes sp.*	Strawberry, *Fragaria sp.*
				Grapefruit, *Citrus paradisi*	Tangerine, *Citrus reticulata*
				Lemon, *Citrus lemon*	
				Lime, *Citrus aurantifolia*	
				Loquat, *Eriobotrya japonica*	
				Mango, *Mangifera indica*	

Source: Adapted from Refs. 6, 7.

Table 5.9 Yield Potential and Salinity Tolerance of Selected Crops[a]

Crops	100%		90%		75%		50%		0% max[c]	
	EC_e[b]	EC_w	EC_e	EC_w	EC_e	EC_w	EC_e	EC_w	EC_e	EC_w
Field crops										
Barley (*Hordeum vulgare*)[d]	8.0	5.3	10	6.7	13	8.7	18	12	28	19
Cotton (*Gossypium hirsutum*)	7.7	5.1	9.6	6.4	13	8.4	17	12	27	18
Sugarbeet (*Beta vulgaris*)[e]	7.0	4.7	8.7	5.3	11	7.5	15	10	24	16
Sorghum (*Sorghum bicolor*)	5.8	4.5	7.4	5.0	8.4	5.6	9.9	6.7	13	8.7
Wheat (*Triticum aestivum*)[d,f]	5.0	4.0	7.4	4.9	9.5	6.3	13	8.7	20	13
Wheat, durum (*Triticum turgidum*)	5.7	3.8	7.6	5.0	10	6.9	15	10	24	16
Soybean (*Glycine max*)	5.0	3.3	5.5	3.7	6.3	4.2	7.5	5.0	10	6.7
Cowpea (*Vigna unguiculata*)	4.9	3.3	5.7	3.8	7	4.7	9.1	6.0	13	8.8
Groundnut (peanut) (*Arachis hypogaea*)	3.2	2.1	3.5	2.4	4.1	2.7	4.9	3.3	6.6	4.4
Rice (paddy) (*Oriza sativa*)	3.0	2.0	3.8	2.6	5.1	3.4	7.2	4.8	11	7.6
Sugarcane (*Saccharum officinarum*)	1.7	1.1	3.4	2.3	5.9	4.0	10	6.8	19	12
Corn (maize) (*Zea mays*)	1.7	1.1	2.5	1.7	3.8	2.5	5.9	3.9	10	6.7
Flax (*Linum usitatissimum*)	1.7	1.1	2.5	1.7	3.8	2.5	5.9	3.9	10	6.7
Broadbean (*Vicia faba*)	1.5	1.1	2.6	1.8	4.2	2.0	6.8	4.5	12	8.0
Bean (*Phaseolus vulgaris*)	1.0	0.7	1.5	1.0	2.3	1.5	3.6	2.4	6.3	4.2
Vegetable crops										
Squash, zucchini (courgette) (*Cucurbita pepo melopepo*)	4.7	3.1	5.8	3.8	7.4	4.9	10	6.7	15	10
Beet, red (*Beta vulgaris*)[f]	4.0	2.7	5.1	3.4	6.8	4.5	9.6	6.4	15	10
Squash, scallop (*Cucurbita pepo melopepo*)	3.2	2.1	3.8	2.6	4.8	3.2	6.3	4.2	9.4	6.3
Broccoli (*Brassica oleracea botrytis*)	2.8	1.9	3.9	2.6	5.5	3.7	8.2	5.5	14	9.1
Tomato (*Lycopersicon esculentum*)	2.5	1.7	3.5	2.3	5.0	3.4	7.6	5.0	13	8.4
Cucumber (*Cucumis sativus*)	2.5	1.7	3.3	2.2	4.4	2.9	6.3	4.2	10	6.8

(continued)

Table 5.9 Continued

Crops	100%		90%		75%		50%		0% max[c]	
	EC_e[b]	EC_w	EC_e	EC_w	EC_e	EC_w	EC_e	EC_w	EC_e	EC_w
Spinach (*Spinacia oleracea*)	2.0	1.3	3.3	2.2	5.3	3.5	8.6	5.7	15	10
Celery (*Apium graveolens*)	1.8	1.2	3.4	2.3	5.8	3.9	9.9	6.6	18	12
Cabbage (*Brassica oleracea capitata*)	1.8	1.2	2.8	1.9	4.4	2.9	7.0	4.6	12	8.1
Potato (*Solanum tuberosum*)	1.7	1.1	2.5	1.7	3.8	2.5	5.9	3.9	10	6.7
Corn, sweet (maize) (*Zea mays*)	1.7	1.1	2.5	1.7	3.8	2.5	5.9	3.9	10	6.7
Sweet potato (*Impomoea batatas*)	1.5	1.0	2.4	1.6	3.8	2.5	6.0	4.0	11	7.1
Pepper (*Capsicum annuum*)	1.5	1.0	2.2	1.5	3.3	2.2	5.1	3.4	8.6	5.8
Lettuce (*Lactuca sativa*)	1.3	0.9	2.1	1.4	3.2	2.1	5.1	3.4	9.0	6.0
Radish (*Raphanus sativus*)	1.2	0.8	2.0	1.3	3.1	2.1	5.0	3.4	8.9	5.9
Onion (*Allium cepa*)	1.2	0.8	1.8	1.2	2.8	1.8	4.3	2.9	7.4	5.0
Carrot (*Daucus carota*)	1.0	0.7	1.7	1.1	2.8	1.9	4.6	3.0	8.1	5.4
Bean (*Phaseolus vulgaris*)	1.0	0.7	1.5	1.0	2.3	1.5	3.6	2.4	6.3	4.2
Turnip (*Brassica rapa*)	0.9	0.6	2.0	1.3	3.7	2.5	6.5	4.3	12	8.0
Forage crops										
Wheatgrass, tall (*Agropyron elongatum*)	7.5	5.0	9.9	6.6	13	9.0	19	13	31	21
Wheatgrass, fairway crested (*Agropyron cristatum*)	7.5	5.0	8.5	5.6	11	7.2	15	9.8	22	15
Bermuda grass (*Cynodon dactylon*)[g]	6.9	4.6	8.5	5.6	11	7.2	15	9.8	23	15
Barley (forage) (*Hordeum vulgare*)[e]	6.0	4.0	7.4	4.9	9.5	6.4	13	8.7	20	13
Ryegrass, perennial (*Lolium perenne*)	5.6	3.7	6.9	4.6	8.9	5.9	12	8.1	19	13
Trefoil, narrowleaf birdsfoot[h] (*Lotus corniculatus tenuifolium*)	5.0	3.3	6.0	4.0	7.5	5.0	10	6.7	15	10
Harding grass (*Phalaris tuberosa*)	4.6	3.1	5.9	3.9	7.9	5.3	11	7.4	18	12
Fescue, tall (*Festuca elatior*)	3.9	2.6	5.5	3.6	7.8	5.2	12	7.8	20	13

Wheatgrass, standard crested (*Agropyron sibiricum*)	3.5	2.3	6.0	4.0	9.8	6.5	16	11	28	19
Vetch, common (*Vicia angustifolia*)	3.0	2.0	3.9	2.6	5.3	3.5	7.6	5.0	12	8.1
Sudan grass (*Sorghum sudanense*)	2.8	1.9	5.1	3.4	8.6	5.7	14	9.6	26	17
Wildrye, beardless (*Elymus triticoides*)	2.7	1.8	4.4	2.9	6.9	4.6	11	7.4	19	13
Cowpea (forage) (*Vigna unguiculata*)	2.5	1.7	3.4	2.3	4.8	3.2	7.1	4.8	12	7.8
Trefoil, big (*Lotus uliginosus*)	2.3	1.5	2.8	1.9	3.6	2.4	4.9	3.3	7.6	5.0
Sesbania (*Sesbania exaltata*)	2.3	1.5	3.7	2.5	5.9	3.9	9.4	6.3	17	11
Spaerophysa (*Sphaerophysa salsula*)	2.2	1.5	3.6	2.4	5.8	3.8	9.3	6.2	16	11
Alfalfa (*Medicago sativa*)	2.0	1.3	3.4	2.2	5.4	3.6	8.8	5.9	16	10
Lovegrass (*Eragrostis sp.*)[i]	2.0	1.3	3.2	2.1	5.0	3.3	8.0	5.3	14	9.3
Corn (forage) (maize) (*Zea mays*)	1.8	1.2	3.2	2.1	5.2	3.5	8.6	5.7	15	10
Clover, berseem (*Trifolium alexandrinum*)	1.5	1.0	3.2	2.2	5.9	3.9	10	6.8	19	13
Orchard grass (*Dactylis glomerata*)	1.5	1.0	3.1	2.1	5.5	3.7	9.6	6.4	18	12
Foxtail, meadow (*Alopecurus pratensis*)	1.5	1.0	2.5	1.7	4.1	2.7	6.7	4.5	12	7.9
Clover, red (*Trifolium pratense*)	1.5	1.0	2.3	1.6	3.6	2.4	5.7	3.8	9.8	6.6
Clover, alsike (*Trifolium hybridum*)	1.5	1.0	2.3	1.6	3.6	2.4	5.7	3.8	9.8	6.6
Clover, ladino (*Trifolium repens*)	1.5	1.0	2.3	1.6	3.6	2.4	5.7	3.8	9.8	6.6
Clover, strawberry (*Trifolium fragiferum*)	1.5	1.0	2.3	1.5	3.6	2.4	5.7	3.8	9.8	6.6
Fruit crops[j]										
Date palm (*Phoenix dactylifera*)	4.0	2.7	6.8	4.5	11	7.3	18	12	32	21
Grapefruit (*Citrus paradisi*)[k]	1.8	1.2	2.4	1.6	3.4	2.2	4.9	3.3	8.0	5.4
Orange (*Citrus sinensis*)	1.7	1.1	2.3	1.6	3.3	2.2	4.8	3.2	8.0	5.3
Peach (*Prunus persica*)	1.7	1.1	2.2	1.5	2.9	1.9	4.1	2.7	6.5	4.3
Apricot (*Prunus armeniaca*)[k]	1.6	1.1	2.0	1.3	2.6	1.8	3.7	2.5	5.8	3.8
Grape (*Vitus sp.*)[k]	1.5	1.0	2.5	1.7	4.1	2.7	6.7	4.5	12	7.9
Almond (*Prunus dulcis*)[k]	1.5	1.0	2.0	1.4	2.8	1.9	4.1	2.8	6.8	4.5

(*continued*)

Table 5.9 Continued

Crops	100%		90%		75%		50%		0% max[c]	
	EC_e[b]	EC_w	EC_e	EC_w	EC_e	EC_w	EC_e	EC_w	EC_e	EC_w
Plum, prune (*Prunus domestica*)[k]	1.5	1.0	2.1	1.4	2.9	1.9	4.3	2.9	7.1	4.7
Blackberry (*Rubus sp.*)	1.5	1.0	2.0	1.3	2.6	1.8	3.8	2.5	6.0	4.0
Boysenberry (*Rubus ursinus*)	1.5	1.0	2.0	1.3	2.6	1.8	3.8	2.5	6.0	4.0
Strawberry (*Fragaria sp.*)	1.0	0.7	1.3	0.9	1.8	1.2	2.5	1.7	4	2.7

EC_w, irrigation water salinity; EC_e, soil salinity.

[a]These data should only serve as a guide to relative tolerances among crops. Absolute tolerances vary depending upon climate, soil conditions, and cultural practices. In gypsiferous soils, plants will tolerate about 2 dS/m higher soil salinity (EC_e) than indicated, but the water salinity (EC_w) will remain the same.

[b]EC_e means average root zone salinity as measured by electrical conductivity of the saturation extract of the soil, reported in deciSiemens per metre (dS/m) at 25°C. The relationship between soil salinity and water salinity ($EC_e = 1.5\,EC_w$) assumes a 15–20% leaching fraction and a 40–30–20–10% water use pattern for the upper to lower quarters of the root zone.

[c]The zero yield potential or maximum EC_e indicates the theoretical soil salinity (EC_e) at which crop growth ceases.

[d]Barley and wheat are less tolerant during germination and seedling stage; EC_e should not exceed 4–5 dS/m in the upper soil during this period.

[e]Beets are more sensitive during germination; EC_e should not exceed 3 dS/m in the seeding area for garden beets and sugar beets.

[f]Semi-dwarf, short cultivars may be less tolerant.

[g]Tolerance given is an average of several varieties; Suwannee and Coastal Bermuda grass are about 20% more tolerant, while Common and Greenfield Bermuda grass are about 20% less tolerant.

[h]Broadleaf Birdsfoot Trefoil seems less tolerant than Narrowleaf Birdsfoot Trefoil.

[i]Tolerance given is average for Boer, Wilman, Sand, and Weeping Lovegrass; Lehman Lovegrass seems about 50% more tolerant.

[j]These data are applicable when rootstocks are used that do not accumulate Na^+ and Cl^- rapidly or when these ions do not predominate in the soil.

[k]Tolerance evaluation is based on tree growth and not on yield.

Source: Adapted from Refs. 12, 13, 17.

Table 5.10 Relative Tolerance[a] of Selected Crops to Foliar Injury from Saline Water Applied by Sprinklers

Na or Cl concentrations[b] (mEq/L) causing foliar injury[c]			
<5	5–10	10–20	>20
Almond	Grape	Alfalfa	Cauliflower
Apricot	Pepper	Barley	Cotton
Citrus	Potato	Corn	Sugarbeet
Plum	Tomato	Cucumber	Sunflower
		Safflower	
		Sesame	
		Sorghum	

[a]Susceptibility based on direct adsorption of salts through the leaves.
[b]The concentration of Na or Cl in mEq/L can be determined from mEq/L by dividing mg/L by the equivalent weight for Na (23) or Cl (35.5) (mEq/L = mg/L/equivalent weight).
[c]Foliar injury is influenced by cultural and environmental conditions. These data are presented only as general guidelines for daytime sprinkler irrigation.
Source: Adapted from Ref. 13.

and nuts, such as almond, apricot, citrus, and plum. Table 5.10 reports on the tolerance of some crops to foliar injury from saline water with different concentrations of sodium and chloride.

If the exact cropping patterns or rotations are not known for a newly used area, the leaching requirement must be based on the least tolerant crops adapted to the area. In those instances where soil salinity cannot be maintained within acceptable limits for the preferred sensitive crops, changing to more tolerant crops will raise the area's production potential. If there is any doubt about the effect of wastewater salinity on crop production, a pilot study should be undertaken to demonstrate the feasibility of irrigation and the outlook for economic success.

5.5.2 Code of Practices to Overcome Boron, Sodium and Chloride Toxicity

A toxicity problem is different from a salinity problem in that it occurs within the plant itself and is not caused by water shortage. Toxicity normally results when certain ions (boron, chloride, sodium, trace elements) are taken up by plants with the soil water and accumulate in the leaves or other parts of the plant during water transpiration to such an extent that the plant is damaged. A lack of equilibrium in the soil solution or solid components can cause problems of bioavailability or unbalanced absorption. The degree of damage depends upon time, relative concentration of toxic compounds, crop sensitivity, and crop water use. If the damage is severe enough, crop yield can be reduced.

The main toxic ions generally occurring in treated municipal effluents are the following (in order of toxic effect): boron (B) > sodium (Na) > chloride (Cl).

Table 5.11 Classification of Crops as a Function of Boron Tolerance

Very sensitive (< 0.5 mg/L)	**Moderately sensitive (1.0–2.0 mg/L)**
Lemon, *Citrus limon*	Pepper, red, *Capsicum annum*
Blackberry, *Rubus* spp.	Pea, *Pisum sativa*
Sensitive (0.5–0.75 mg/L)	Carrot, *Daucus carota*
Avocado, *Persea americana*	Radish, *Raphanus sativus*
Grapefruit, *Citrus X paradisi*	Potato, *Solanum tuberosum*
Orange, *Citrus sinensis*	Cucumber, *Cucumis sativus*
Apricot, *Prunus armeniaca*	**Moderately tolerant (2.0–4.0 mg/L)**
Peach, *Prunus persica*	Lettuce, *Lactuca sativa*
Cherry, *Prunus avium*	Cabbage, *Brassica oleracea capitata*
Plum, *Prunus domestica*	Celery, *Apium graveolens*
Persimmon, *Diospyros kaki*	Turnip, *Brassica rapa*
Fig, kadota, *Ficus carica*	Bluegrass, Kentucky, *Poa pratensis*
Grape, *Vitis vinifera*	Oats, *Avena sativa*
Walnut, *Juglans regia*	Maize, *Zea mays*
Pecan, *Carya illinoiensis*	Artichoke, *Cynara scolymus*
Cowpea, *Vigna unguiculata*	Tobacco, *Nicotiana tabacum*
Onion, *Allium cepa*	Mustard, *Brassica juncea*
Sensitive (0.75–1.0 mg/L)	Clover, sweet, *Melilotus indica*
Garlic, *Allium sativum*	Squash, *Cucurbita pepo*
Sweet potato, *Ipomoea batatas*	Muskmelon, *Cucumis melo*
Wheat, *Triticum eastivum*	**Tolerant (4.0–6.0 mg/L)**
Barley, Hordeum vulgare	Sorghum, *Sorghum bicolor*
Sunflower, *Helianthus annuus*	Tomato, *L. lycopersicum*
Bean, mung, *Vigna radiata*	Alfalfa, *Medicago sativa*
Sesame, *Sesamum indicum*	Vetch, purple, *Vicia benghalensis*
Lupine, *Lupinus hartwegii*	Parsley, *Petroselinum crispum*
Strawberry, *Fragaria* spp.	Beet, red, *Beta vulgaris*
Artichoke, Jerusalem, *Helianthus tuberosus*	Sugarbeet, *Beta vulgaris*
Bean, kidney, *Phaseolus vulgaris*	**Very tolerant (6.0–15.0 mg/L)**
Bean, lima, *Phaseolus lunatus*	Cotton, *Gossypium hirsutum*
Groundnut/Peanut, *Arachis hypogaea*	Asparagus, *Asparagus officinalis*

Source: Adapted from Refs. 1, 6, 13.

Boron is an essential element to plants, but in concentrations of > 0.5 mg/L could be toxic to some sensitive crops. Table 5.11 provides a list of some crops depending on the maximum boron concentrations tolerated in soil water without yield or vegetative growth reductions. For most ornamental crops, maximum quality of plants is obtained when irrigation water contains < 0.5 mg/L of boron. Less sensitive ornamentals exhibit increasing leaf scorch when boron content is > 1.0 mg/L.

The main aspects of boron toxicity that can be taken into account in water reuse projects are as follows:

Boron tolerance varies depending upon climate, soil conditions, and crop varieties.

Maximum concentrations in irrigation water are approximately equal to the reported values in wet soil or slightly less.
Fruit trees are more sensitive to boron than vegetables.
Generally, boron is tolerated by turf grasses better than other plants (up to 10 mg/L).
Boron toxicity is difficult to correct without changing the crop or water source.

The most important management practices recommended to overcome boron toxicity are as follows:

Leaching
Frequent irrigation
Appropriate choice of micro-irrigation systems

Chloride and sodium are less toxic than boron, and the implementation of proper irrigation management can help end-users to reduce significantly any potential toxic effect from municipal effluents. As for boron, not all crops are equally sensitive to these toxic ions.

Chloride concentrations in recycled water with adverse impacts on irrigation range from 100 mg/L (very sensitive plant species) to 900 mg/L (tolerant plant species). Some guidance as to the sensitivity of crops to sodium and chloride is provided in Table 5.12 and Table 5.13, respectively. Possibilities of synergistic circumstances should also be considered.

5.5.3 Code of Practices to Overcome Trace Elements Toxicity

As explained in Chapter 2, municipal wastewater may contain a number of toxic compounds, including trace organics and inorganics. Some of these may be removed during the treatment process, but others will persist and could present phytotoxic problems or accumulate in the crop with potential adverse health effects. Thus, municipal wastewater effluents should be checked for trace element toxicity hazards, particularly if contamination with industrial wastewater is suspected. With the high retention rate in the soil and the low use by plants, the maximum application rate should not exceed that which will allow normal crop growth and still not exceed any allowable concentration in the crop's production.

In general, the higher the exchange capacity of the soil, the more trace elements a soil can accept without potential hazards. A soil-plant system has a good self-potential to degrade certain pollutants, but attention should be paid to the accumulation of nonbiodegradable materials in soils treated with effluents. Trace elements accumulate in the upper part of the soil because of strong adsorption and precipitation phenomena. If a soil is highly permeable (sandy soil) or if the groundwater table is close to the surface, care must be taken to prevent groundwater pollution. Plant trace element uptake is generally small compared to soil build-up. For this reason the soil should be protected from trace element build-up.

Table 5.12 Classification of Crops as a Function of Sodium Tolerance

Sensitive	Semi-tolerant	Tolerant
Avocado, *Persea americana*	Carrot, *Daucus carota*,	Alfalfa, *Medicago sativa*
Deciduous fruits	*Clover, Ladino, Trifolium repens*	Barley, *Hordeum vulgare*
Nuts	Dallisgrass, *Paspalum dilatatu*	Beet, garden, *Beta vulgaris*
Bean, green, *Phaseolus vulgaris*	Fescue, tall, *Festuca arundinacea*	Beet, sugar, *Beta vulgaris*
Cotton (at germination), *Gossypium hirsutum*	Lettuce, *Lactuca sativa*	Bermuda grass, *Cynodon dactylon*
Maize, *Zea mays*	Bajara, *Pennisetum typhoides*	Cotton, *Gossypium hirsutum*
Peas, *Pisum sativum*	Sugarcane, *Saccharum officinarum*	Paragrass, *Brachiaria mutica*
Grapefruit, *Citrus paradisi*	Berseem, *Trifolium alexandrinum*	Rhodes grass, *Chloris gayana*
Orange, *Citrus sinensis*	Benjuí, *Mililotus parviflora*	Wheatgrass, crested, *Agropyron cristatum*
Peach, *Prunus persica*	Raya, *Brassica juncea*	Wheatgrass, fairway, *Agropyron cristatum*
Tangerine, *Citrus reticulata*	Oat, *Avena sativa*	Wheatgrass, tall, *Agropyron elongatum*
Mung, *Phaseolus aurus*	Onion, *Allium cepa*	Karnal grass, *Diplachna fusca*
Mash, *Phaseolus mungo*	Radish, *Raphanus sativus*	
Lentil, *Lens culinaris*	Rice, *Oryza sativus*	
Groundnut (peanut), *Arachis hypogaea*	Rye, *Secale cereale*	
Gram, *Cicer arietinum*	Ryegrass, Italian, *Lolium multiflorum*	
Cowpeas, *Vigna sinensis*	Sorghum, *Sorghum vulgare*	
	Spinach, *Spinacia oleracea*	
	Tomato, *Lycopersicon esculentum*	
	Vetch, *Vicia sativa*	
	Wheat, *Triticum vulgare*	

Table 5.13 Chloride Tolerance of Fruit Crop Cultivars and Rootstocks

Crop	Rootstock or cultivar	Maximum permissible Cl^- without leaf injury[a]	
		Root zone (Cl_e) (mEq/L)	Irrigation water $(Cl_w)^{b,c}$ (mEq/L)
Rootstocks			
Avocado, *Persea Americana*	West Indian	7.5	5.0
	Guatemalan	6.0	4.0
	Mexican	5.0	3.3
Citrus, *Citrus* spp.	Sunki Mandarin	25.0	16.6
	Grapefruit		
	Cleopatra mandarin		
	Rangpur lime		
	Sampson tangelo	15.0	10.0
	Rough lemon		
	Sour orange		
	Ponkan mandarin		
	Citrumelo 4475	10.0	6.7
	Trifoliate orange		
	Cuban shaddock		
	Calamondin		
	Sweet orange		
	Savage citrange		
	Rusk citrange		
	Troyer citrange		
Grape, *Vitis* spp.	Salt Creek, 1613-3	40.0	27.0
	Dog Ridge	30.0	20.0
Stone fruits, *Prunus* spp.	Marianna	25.0	17.0
	Lovell, Shalil	10.0	6.7
	Yunnan	7.5	5.0
Cultivars			
Berries, *Rubus* spp.	Boysenberry	10.0	6.7
	Olallie blackberry	10.0	6.7
	Indian Summer	5.0	3.3
	Raspberry		
Grape, *Vitis* spp.	Thompson seedless	20.0	13.3
	Perlette	20.0	13.3
	Cardinal	10.0	6.7
	Black Rose	10.0	6.7
Strawberry, *Fragaria* spp.	Lassen	7.5	5.0
	Shasta	5.0	3.3

[a]For some crops the concentration given may exceed the overall salinity tolerance of that crop and cause some reduction in yield in addition to that caused by chloride ion toxicities.

[b]Values given are for the maximum concentration in the irrigation water. The values were derived from saturation extract data (EC_e) assuming a 15–20% leaching fraction and $EC_d = 1.5\ EC_w$.

[c]The maximum permissible values apply only to surface-irrigated crops. Sprinkler irrigation may cause excessive leaf burn at values far below these.

Source: Adapted from Refs. 6, 13.

Table 5.14 Soil Boundary Concentrations Depending on Soil pH (mg/kg dry solids)

Element	EC Directive (1986) pH 6–7	USA (1993/1995)	Germany (1992) pH 5–6 clay < 5%	Germany (1992) pH > 6 clay>5%	Portugal (1996) pH ≤ 5.5	Portugal (1996) pH > 7	Spain pH < 7	Spain pH > 7	Sweden (1995)
As	—	20.5	—		—		—		—
Cd	1–3	19.5	1	1.5	1	4	1	3	0.4
Co	—	—	—		—		—		30
Cr	100–150[a]	1500	100		50	300	100	150	30
Cu	50–140	750	60		50	200	50	210	40
Hg	1–1.5	8.5	1		1	2	1		0.3
Mo	—	9	—		—		—		—
Ni	30–75	210	50		30	110	30	112	210
Pb	50–300	150	100		50	450	50	300	150
Se	—	100	—		—		—		100
Zn	150–300	1400	150	200	150	450	150	450	1400

[a]Planned.

Table 5.15 Annual Pollutant Loading Rates (kg/ha year)

	EC[a] Directive (1986)	Germany (1992)	Portugal (1996)	Sweden[b] (1995)	USA[c] (1993/1995)
As	—	—	—	—	2
Cd	0.15	0.016	0.15	0.00175	1.9
Cr	4[d]	1.5	4.5	0.1	150
Cu	12	1.3	12	0.6	75
Hg	0.1	0.013	0.1	0.0025	0.85
Mo	—	—	—	—	0.9
Ni	3	0.3	3	0.05	21
Pb	15	1.5	15	0.1	15
Se	—	—	—	—	5
Zn	30	2.5	30	0.8	140

[a]Mean value over a 10-year period.
[b]Maximum average annual application rates over a 7-year period.
[c]Mean value over a 20-year period.
[d]Planned.

Table 5.14 to 5.17 provide guideline information on soil boundary concentrations of some inorganic trace elements, annual and cumulative loading rates, and estimated irrigation duration to reach heavy metal loading limits.

Table 5.16 Cumulative Loading Limit for Land Application (kg/ha)

	EC Directive	France	Germany	Sweden[a]	USA[b]
As	—	—	—	—	41
Cd	1.25–6.25	5.4	8.4	0.012	39
Cr	—	360	210	0.7	3000
Cu	75–300	210	210	2.5	1500
Hg	2.25–3.5	2.7	5.7	0.0175	17
Mo	—	—	—	—	18
Ni	12.5–125	60	60	0.35	420
Pb	0–625	210	210	0.7	300
Se	—	—	—	—	100
Zn	175–550	750	750	5.6	2800

[a]Maximum loading over a 7-year period.
[b]With an application quantity of 10 t dry matter/ha year.

Table 5.17 Calculated Length of Time for Wastewater-irrigated Agricultural Soils to Reach Heavy-metal Loading Limits

Element	Typical concentration (mg/L)	Annual input (1.2 m/yr water depth)	Suggested loading (kg/ha) at soil CEC[a]			Time (in yrs) to reach soil loading limit at CEC[a]		
			<5	5–15	>15	<5	5–15	>15
Cd	0.005	0.06	5	10	20	82	167	333
Cu	0.10	1.2	125	250	500	104	208	416
Ni	0.02	0.24	125	250	500	521	1042	2083
Zn	0.15	1.8	250	500	1000	139	278	556
Pb	0.05	0.60	500	1000	2000	833	1667	3333

[a]Cation exchange capacity expressed in units of mEq/100 g soil.

According to FAO,[3] the major considerations for the management of trace element toxicity can be summarized as follows:

The concentrations of heavy metals and trace elements in municipal effluents are low, as a rule, and do not represent a major problem in water reuse projects for irrigation.

For moderate contamination with trace elements of recycled water, no particular management of irrigation is required for calcareous soils.

Under acid conditions in soils, heavy metals can be a problem and specific management measures should be implemented such as liming (addition of $CaCO_3$), limitation of the use of acid fertilizers, and selection of tolerant crops and plants that do not accumulate the heavy metals of concern.

5.6 CODE OF MANAGEMENT PRACTICES OF WATER APPLICATION

In addition to crop and irrigation method selection, water and soil management play an important role in the successful use of recycled water for irrigation.

5.6.1 Leaching and Drainage

Appropriate water-management practices must be followed to prevent salinization. If accumulated salts are not flushed out of the root zone by leaching and removed from the soil by effective drainage, salinity problems can build up rapidly. Consequently, leaching and drainage are two important water-management practices to avoid salinization of soils.

5.6.1.1 Leaching

Under irrigated agriculture, a certain amount of excess irrigation water is required to percolate through the root zone to remove the salts that have accumulated as a result of irrigation and evapotranspiration. This process of displacing the salts from the root zone is called leaching, and that portion of the irrigation water that mobilizes the excess of salts is called the leaching fraction. Salinity control by effective leaching of the root zone becomes more important as irrigation water becomes more saline.[6]

The leaching fraction (LF) is equal to the depth of water that passes down below the root zone divided by the depth of water applied at the surface of the soil. To estimate the leaching requirement (LR), both the irrigation water salinity (EC_W) and the crop tolerance to soil salinity (EC_e) must be known and used in the following equation:

$$LR = \frac{EC_W}{5(EC_e) - EC_W} \tag{5.1}$$

where:

LR = the minimum leaching requirement needed to control salts within the tolerance EC_e of the crop with ordinary surface methods of irrigation

EC_W = the salinity of the applied irrigation water, dS/m

EC_e = the average soil salinity tolerated by the crop as measured in a soil saturation extract (it is recommended to use in this calculation an EC_e value that can be expected to result in at least 90% or greater crop yield).

The total annual depth of water that needs to be applied to meet both the crop demand and leaching requirement can be estimated from Equation (5.2):

$$AW = \frac{ET}{1 - LR} \tag{5.2}$$

where:

AW = the depth of applied water, mm/year
ET = the total annual crop water demand, mm/year
LR = the leaching requirement expressed as a fraction (LF mentioned above)

Depending on the salinity status, leaching can be carried out at each irrigation, every other irrigation, or less frequently. With good-quality recycled water the irrigation application level will almost always apply sufficient extra water to accomplish leaching. With high-salinity irrigation water, meeting the leaching requirement is difficult and requires large amounts of water.

Soil leaching is needed for almost all crops when electrical conductivities of saturation extracts exceed 10 dS/m and for moderately tolerant crops for values over 3 dS/m. The salinity of the upper 0.6 m of soil is of greatest concern. In general, application of 10–20 cm of water before planting coupled with a similar irrigation immediately following planting is often sufficient. Leaching by such preirrigation can be achieved by flood, sprinkler, or trickle irrigation. Salinity level higher than 10 dS/m may require more leaching.

Rainfall must be considered in estimating the leaching requirement and in choosing the leaching method. In years with average or high rainfall, sufficient leaching may take place as a result of rainfall events. However, this is unlikely to happen in years with less than average rainfall. Rainfall seasonal distribution is also to be considered.

Furthermore, in dry years, the effluent may become more saline as a result of changed water use habits by contributors to the waste stream, greater concentration in storage ponds due to less dilution by rainfall, and greater water loss by evaporation. In order to ensure sufficient leaching, the soil must be sufficiently permeable. This is an essential selection criterion for a successful irrigation with most types of recycled water.

If irrigation with recycled water is managed in such a way that salt does not accumulate in the root zone, then a question need to be asked: What is the ultimate fate of the added salt? This will depend largely on the underlying stratigraphy and groundwater conditions. If the underlying material is sufficiently porous, more salt could be stored between the root zone and the water table. Nevertheless, this storage directly below the irrigation site is limited and not a long-term solution. Once it is filled, salt will reach the water table, lateral movement is probable, and the salt will move off-site.

5.6.1.2 Drainage

Drainage is defined as the removal of excess water from the soil surface (surface drainage) and below the surface (subsurface drainage) so as to permit optimum growth of plants. The importance of drainage for successful irrigated agriculture has been well demonstrated. It is particularly important in semi-arid and arid areas to prevent secondary salinization. In these areas, the water table will rise with irrigation when the natural internal drainage of the soil is not adequate. When the water table is within a few meters of the soil surface, capillary rise of saline groundwater will transport salts to the soil surface. At the surface, water evaporates, leaving the salts behind. If this process is not under control, salt accumulation will continue, resulting in salinization of the soil. In such cases, subsurface drainage can control the rise of the water table and, hence, prevent salinization.[6]

Salinity problems in many irrigation projects in arid and semi-arid areas are associated with the presence of a shallow water table. In this context, the role of drainage is to lower the water table to a desirable level, at which it does not contribute to the transport of salts to the root zone and the soil surface by capillarity phenomena.

The important elements of the total drainage system are as follows:

Ability to maintain a downward movement of water and salt through soils.
Capacity to transport the desired amount of drained water out of the irrigation scheme.
Facility to dispose of drained water safely.

Such disposal can pose a serious problem, particularly when the source of irrigation water is treated wastewater, depending on the composition of the drainage effluent.

5.6.1.3 Impact of Irrigation on Groundwater

Certainly, leaching is essential to prevent salinization of the root zone, but because there can be excess nitrate in this zone, leaching will result in the movement of nitrates and salts to the groundwater. Its impact will depend on a range of factors, listed below:

The depth of the water table will affect the time delay before the effects of effluent irrigation are seen at the water table. This may also provide the opportunity, in the case of nitrates, for further denitrification to occur if the soil is wet enough and there is a suitable carbon source.
The quality of groundwater prior to irrigation will determine whether or not the salt and/or nitrate reaching the groundwater will have a detrimental impact.
The extent of dilution of salt and nitrate when reaching the groundwater will depend on the rate of recharge and the rate of flow of groundwater beneath the irrigation site, which in turn depends on aquifer permeability and hydraulic gradient. The amount of dilution will also depend on the size of the irrigated area.
The proximity of the effluent irrigation site to discharge zones and to water supply wells determines the likelihood of contaminated water finding its way into rivers and drinking water supplies.

For these reasons, monitoring of groundwater beneath effluent irrigation is an essential indicator of environmental performance.

These circumstances should be considered in a different way if irrigation is localized near the coastline or over brackish water aquifers.

5.6.2 Using Other Water Supplies

5.6.2.1 Blending of Recycled Water with Other Water Supplies

One of the options available to farmers is the blending of treated effluent with conventional sources of water, canal water, or groundwater, if multiple sources are available. It is possible that a farmer may have saline groundwater and, if he has non–saline-treated wastewater, could blend the two sources to obtain a blended water of acceptable salinity level. The microbial quality of the resulting mixture after blending could be superior to that of the unblended recycled water.

5.6.2.2 Alternating Recycled Water with Other Water Sources

Another strategy is to use the treated wastewater alternately with canal water or groundwater, instead of blending. From the point of view of salinity control, alternate applications of the two sources is superior to blending. However, an alternating application strategy will require dual conveyance systems and availability of the effluent dictated by the alternate schedule of application.

5.6.3 Adjusting Fertilizer Applications

The fertilizer value of recycled water is of great importance. The typical concentrations of nutrients in treated wastewater effluent from conventional sewage treatment processes are nitrogen 50 (20–85) mgN/L, phosphorus 10 (4–15) mgP/L, and potassium 30 (10–35) mg/L.

In general, irrigation with recycled water (treated urban wastewater) at an application rate of 100 mm/ha would provide the following quantity of fertilizing elements:

Total nitrogen, N: 16–62 kg (in arid and semi-arid regions up to 90–300 kg)
Phosphorus, P: 4–24 kg
Potassium, K: 2–69 kg
Calcium, Ca: 18–208 kg
Magnesium, Mg: 9–110 kg
Sodium, Na: 27–182 kg (in arid and semi-arid regions up to 200–600 kg)

Assuming an application rate of 5000 m^3/ha year, the fertilizer contribution of the effluent would be:

N: 250 kg/ha year
P: 50 kg/ha year
K: 150 kg/ha year

Thus, all of the nitrogen and much of the phosphorus and potassium normally required for agricultural crop production would be supplied by the effluent. In addition, other valuable micronutrients and the organic matter contained in the effluent will provide complementary fertilizing benefits.

The fertilizing value of wastewater cannot be assessed using only the results of chemical analysis; the analysis should also consider the modifications occurring in the soil among the organic and mineral compounds. These modifications can change the nutrients taken up by the plants and depend on soil characteristics, climate, and type of crops. Additional fertilizers may be supplied during specific crop growth stages.

If the recycled water has high nutrient concentrations, it is desirable to choose plant species with a high demand for these elements to ensure that most of them will be assimilated by the plants. Good assimilation of nutrients by the plants will reduce the possibility of a deep nitrogen percolation and groundwater pollution.

As a rule, the phosphorus content in recycled water is too low during the early growth period to affect crop yield. Soil phosphorus gradually builds up with time, reducing the need for supplemental phosphorus fertilizers in future years. Excess phosphorus has not been a problem to date in reuse schemes, and no guideline value is given for phosphorus content in wastewater. However, it is recommended to check the phosphorus content in recycled water in conjunction with soil testing for fertilization planning. The use of recycled water for irrigation can only partially meet crop needs.

Farmers should take into account the fertilizer value of recycled water and save money by reducing consumption of fertilizers. They should also keep in mind that an excess of nitrogen can reduce the quality of the crops and that the nitrogen content of recycled water varies throughout the year, as do plant nitrogen requirements, which vary with development stage. For this reason, nitrogen content of water and soil should be carefully controlled, and measures such as denitrification or crop rotation should be considered when necessary.

Algae growth in storage systems induced by excess nitrogen can be minimized by screen filters or chemical control (e.g., copper sulfate).

5.6.4 Management of Soil Structure

Soil characteristics should be also studied and taken into account. The main soil parameters are as follows:

Soil profile: Depth of the soil will help to determine potential plant growth and water needs.

Soil texture: Relative proportions of sand, silt, and clay largely determine the soil physical properties (e.g., ability to hold nutrients and water, movement of air and water, and workability).

Soil structure: The way in which soil particles are grouped together affects moisture relationships, aeration, heat transfer, and mechanical impedance of root development.

Slopes: Containment of runoff and erosion are potential problems.

Soil structure in case of sodicity problems can be regenerated physically by specific cultivation practices: deep tillage, working organic residues back into the soil, or chemically by incorporating calcium either directly (gypsum), or indirectly (sulfuric acid) into the soil and then leaching. The addition of sulfuric acid has similar effects as gypsum, since it reacts with soil lime and release calcium. This is, however, a high-risk practice and should be performed under strict technical supervision.

Routine light applications of gypsum may be advantageous to avoid sodium problems. This technique is mainly used for golf courses or parks. The amount of gypsum or other amendments added to the soil can be estimated from the amount of exchangeable sodium to be replaced by calcium. It takes 0.4 ton/ha of gypsum to replace 1 mEq/100 g of exchangeable sodium to a depth of 0.2 m.[1,7] The amount of water required to dissolve 1 ton of gypsum ranges from 300 to 1200 m^3. Soil rehabilitation with gypsum may require annual application for several years until the soil is reclaimed to a depth of 0.6–0.9 m.

5.6.5 Management of Clogging in Sprinkler and Drip Irrigation Systems

Clogging of sprinkler and drip irrigation system is likely to occur. Various chemical, physicochemical and biochemical phenomena (Table 5.18) contribute to clogging, including biofouling (bacterial or algae growth), deposits (suspended solids), and scaling (chemical precipitation), which may be located in the sprinkler head, emitter orifice, or supply lines, causing plugging and decrease of irrigation efficiency.[18]

The most frequent clogging problems occur with drip irrigation systems. Such systems are often considered ideal, as they are totally closed systems and avoid the problems of worker safety and drift control. However, drip irrigation requires very good operation and maintenance.

Table 5.18 Physical, Chemical, and Biological Contributors to Clogging of Localized Irrigation Systems as Related to Irrigation Water Quality

Physical (suspended solids)	Chemical (precipitation)	Biological (bacteria and algae)
Sand	Calcium or magnesium carbonate	Filaments
Silt	Calcium sulfate	Slimes
Clay	Heavy metals hydroxides, oxides, carbonates, silicates, and sulfides	Microbial depositions:
Organic matter	Fertilizers;	Iron
	Phosphate	Sulfur
	Aqueous ammonia	Manganese
	Fe, Zn, Cu, Mn	Bacteria
		Small aquatic organisms:
		Snail eggs
		Larva

Source: Adapted from Ref. 18.

Table 5.19 Water Quality and Clogging Potential in Drip Irrigation Systems

		Degree of restriction on use		
Potential problem	**Units**	**None**	**Slight to moderate**	**Severe**
Physical				
Suspended solids	mg/L	< 50	50–100	100
Chemical				
pH		< 7.0	7.0–8.0	8.0
Dissolved solids	mg/L	< 500	500–2000	2000
Manganese	mg/L	< 0.1	0.1–1.5	1.5
Iron	mg/L	< 0.1	0.1–1.5	1.5
Hydrogen sulfide	mg/L	< 0.5	0.5–2.0	2.0
Biological	Maximum			
Bacterial populations	number/mL	< 10,000	10,000–50,000	50,000

Source: Adapted from Refs. 6, 19.

The guidelines presented in Table 5.19 could be used to evaluate the suitability of a given water for use in drip irrigation systems. These guidelines should be applied only for a preliminary and more general evaluation of water suitability for drip irrigation. Other factors, such as temperature, sunlight, emitter types, and flow rates, must also be taken into account because they greatly alter the degree of the clogging potential. It is important to stress that combinations of two or more factors affect irrigation efficiency more severely than a single factor acting alone but are more difficult to estimate and resolve. It is likely that a drip irrigation system compatible with wastewater use will become available with the development of new emitter designs.

For higher concentrations of suspended solids and nutrients, sand or gravel filtration of secondary treated effluent and/or regular flushing of lines have been recommended as an efficient measure.[20] Combined with flushing, chlorination effectively prevents biofouling.

5.6.6 Management of Storage Systems

A storage facility is, in most cases, a critical link between the wastewater treatment plant and the irrigation system. Storage facilities (both operational and seasonal) are needed for the following reasons:

- To equalize daily variations in flow from the treatment plant and store excess when average recycled water flow exceeds irrigation demands, including winter storage.
- To meet peak irrigation demands in excess of the average recycled water flow.
- To minimize disruptions in the operation of the treatment plant and irrigation system. Storage is used to provide insurance against the possibility of unsuitable recycled water entering the irrigation system

and to provide additional time to solve temporary water-quality problems.

To provide additional treatment. Oxygen demands, suspended solids, nitrogen, and microorganisms are reduced during storage. Managers of irrigation systems must be aware of the scope and nature of these changes in order to modify the agricultural procedures in an appropriate manner. The longer the recycled water is stored, the lower the nutrient concentrations in the water released from the irrigation system to the plants and the higher the proportion of phosphorus to nitrogen. Whereas in the case of nutrient-rich effluents it may be desirable to have an important reduction in nutrient concentrations in order to avoid or reduce overfertilization, it may simply be a waste of money if the recycled water has low nutrient concentration, because all the nutrients lost during the storage will have to be added afterwards to maintain good growth conditions.

REFERENCES

1. Pettygrove, G.S., and Asano, T., *Irrigation with Reclaimed Municipal Wastewater—A Guidance Manual*, Lewis Publishers Inc., Chelsea, 1985.
2. FAO, Food and Agriculture Organization of the United Nations, *Quality Control of Wastewater for Irrigated Crop Production*, Water reports 10, by Westcot, D., Rome, 1997.
3. FAO, Food and Agriculture Organization of the United Nations, *Users Manual for Irrigation with Treated Wastewater*, FAO Regional Office for the Near East, Cairo, Egypt, 2000.
4. USGA, *Wastewater Reuse for Golf Course Irrigation*, Lewis Publishers, Boca Raton, FL, 1994.
5. FAO, Food and Agriculture Organization of the United Nations, *Guidelines for Prediction of Crop Water Requirements*, Irrigation and Drainage Paper 24, by Doorenbos, J. and Pruitt, W.O., Rome, 1994.
6. FAO, Food and Agriculture Organization of the United Nations, Irrigation and Drainage Paper 47, *Wastewater Treatment and Use in Agriculture*, by Pescod, M., Food and Agriculture Organization, Rome, 1992.
7. FAO, Food and Agriculture Organization of the United Nations, *Water Quality for Irrigation, Irrigation and Drainage Paper 29*, by Ayers, R. S. and Wescot, D. W., Rome, 1985.
8. FAO, Food and Agriculture Organization of the United Nations, *Small-scale Irrigation for Arid Zones: Principles and Options*, W3094/E, Rome, 1997.
9. USEPA, *Guidelines for Water Reuse*, Manual, EPA/625/R-92/004 (United States Environmental Protection Agency) & USAID (United States Agency for International Development), Cincinnati, OH, 1992.
10. Bahri, A., Salinity evolution in an irrigated area in the Lower Medjerda Valley in Tunisia, *Science du Sol*, 31, 3, 125, 1993.
11. Oron, G., Agriculture, water and environment: future challenges, *Wat. Supply*, 3, 4, 51, 2003.
12. Maas, E.V. and Hoffman, G.J., Crop salt tolerance—current assessment, *J. Irrig. Drainage Div.*, ASCE, 103, IR2, 115, 1977.

13. Maas, E.V., Salt tolerance of plants, in *The Handbook of Plant Science in Agriculture*, ed. Christie, B.R., CRC Press, Boca Raton, FL, 1984.
14. Van Genuchten, M.T. and Hoffman, G.J., Analysis of crop salt tolerance, in *Soil salinity under irrigation: Process and management*, I. Shainberg and J. Shalhevet (eds.), Ecological Studies 51. Springer-Verlag, New York, 1984, pp. 225–271.
15. Meiri, A., Plant response to salinity: experimental methodology and application to field, in *Soil Salinity under Irrigation*, ed. Shainberd, I. and Shalveret, Springer-Verlag, New York, 284, 1984.
16. Abrol, I.P., Yadav, J.S.P., and Massoud, F.I., Salt-affected soils and their management, *FAO Soils Bulletin*, 39, 1988.
17. Kandiah, A., Environmental impacts of irrigation development with special reference to saline water use, in *Water, Soil and Crop Management Relating to the Use of Saline Water*, AGL/MISC/16, FAO, Rome, 152, 1990.
18. Bucks, D.A. and Nakayama, F.S., Injection of fertilizers and other chemicals for drip irrigation, in *Tech. Conf. Proc. Irrig. Assoc.*, 1980.
19. Nakayama, F.S., Water analysis and treatment to control emitter plugging, in *Proc. Irrig. Assoc. Conf.*, 21–24 February, Portland, Oregon, 1982.
20. Papadopoulos, I. and Stylianou, Y., Ongoing research on the use of treated sewage effluent for irrigating industrial and fodder crops, in *Treatment and Use of Sewage Effluent for Irrigation*, Pescod, M.B. and Arar, A., eds., Butterworths, Sevenoaks, Kent, 1988.

6

Codes of Practices for Landscape and Golf Course Irrigation

Bahman Sheikh

CONTENTS

6.1 Benefits of and constraints on the use of recycled water for landscape irrigation 152
6.2 Effects of recycled water on turfgrass 153
6.3 Best practices for golf course irrigation 153
6.4 Prevention of adverse effects of recycled water on turfgrass 156
6.5 Management of adverse effect of water reuse on soils 157
6.6 Recommendations to avoid adverse effects of water reuse on groundwater 157
6.7 Economic and financial aspects of landscape irrigation 158
6.8 Customer acceptance of recycled water for irrigation of landscaping and golf courses 160
References 161

In general, use of recycled water for irrigation on landscapes and golf courses and its impact on soils and plants follow the same general agronomic principles as those applicable to irrigation of agricultural crops discussed in Chapter 5. Therefore, in the following sections those principles are not repeated. Instead, guidelines, codes of good practice, and precautions specific to landscapes and golf courses are presented.

1-56670-649-1/05/$0.00 + $1.50

6.1 BENEFITS OF AND CONSTRAINTS ON THE USE OF RECYCLED WATER FOR LANDSCAPE IRRIGATION

The most important benefit of the use of recycled water for landscape irrigation is that it frees an equivalent volume of potable water for basic human needs. That water, already present in the urban setting, would be readily allocated to higher human needs assuming the necessary institutional arrangements can be made. In addition, the presence of macro- and micro-nutrients in recycled water reduces the need for fertilization to a significant degree. Reliability of supply of recycled water—in times of drought, when potable water may be withheld from use for landscape irrigation—becomes a particularly important benefit to those for whom maintenance of an attractive landscape at all times is critical. Some landscaping features represent large investments, which can be destroyed by drought if an alternate source of water is not available. Thus, another great advantage of recycled water for landscape use is that it is drought-proof, since wastewater flow is largely unaffected by drought restrictions.

Constraints on the use of recycled water depend on the mineral quality of the effluent and the level of treatment before use. Table 6.1 summarizes the most important negative impacts of some constituents in recycled water on plants, soils, and groundwater. Recycled water with a salinity level below 1.5 dS/m would not have an appreciable effect on the quality of most landscape materials. Higher levels of salinity may require more intensive management efforts and possibly a limited plant palette to avoid the most salt-sensitive plant materials. Sodium is a particular problem in turf irrigation, especially on golf greens where the grass is cut very low to the ground and foot traffic tends to be heavy. The typically higher levels of sodium in recycled water can cause a rise in the SAR index, which would indicate increasing levels of permeability reduction and water flow restriction within the soil profile. Also, if the recycled water is treated only to a secondary or lower quality level,

Table 6.1 Impact of High Constituent Levels in Irrigation Water on Plants, Soils, and Groundwater

Constituent causing impact	Impact on plants	Impact on soils	Impact on groundwater
High salinity	Lower yield	Salinization	Degradation of water quality
High sodium (SAR)	Toxicity, leaf burn	Lower permeability	—
High chloride	Toxicity	—	—
High boron	Toxicity	—	—
High levels of microorganisms	Farm worker and consumer health	—	Public health
Nitrogen	Better yield	—	Pollution

Source: Adapted from Ref. 2.

irrigation would be restricted to nighttime, buffer zones would be necessary, and public access to the site would have to be restricted during the irrigation period.

6.2 EFFECTS OF RECYCLED WATER ON TURFGRASS

Turfgrass irrigated with recycled water (if containing unusually high levels of sodium, hence SAR), as indicated in Table 6.2, may have patches of yellow, brown, or dead grass. On golf courses where the greens are not designed with proper drainage, such problems would appear more quickly and persist in spite of management practices to counter them. It should be noted that golf courses irrigated with secondary effluent for over 20 years in Tunisia continue to produce high-quality turfgrass with no indication of any adverse effects.[1] Figure 6.1 shows the El Kantaoui Golf Course, on the west coast of Tunisia, which has been continuously irrigated with secondary treated recycled water for the past 20 years.

6.3 BEST PRACTICES FOR GOLF COURSE IRRIGATION

Best practices for golf course irrigation with recycled water are generally similar to those recommended for agricultural irrigation (see also Chapter 5). However, most golf courses are intensively maintained with a view to maximum aesthetic value.

The high level of care and expertise practiced on most golf courses provides for opportunities to integrate the special needs of using recycled water successfully, such as:

Good drainage, especially on the greens and other intensely played surfaces.
Provision of an adequate leaching fraction.
Balanced fertilization, accounting for the fertilizer value present in reclaimed water.

Table 6.2 Typical Landscape Plant Species and Their Sensitivity/Tolerance to Salt

Sensitive (EC_w 1–2)	Moderately sensitive	Moderately tolerant	Tolerant ($EC_w > 10$)
Star jasmine	Yellow sage	Weeping bottlebrush	Brush cherry
Pyrenees	Orchid tree	Oleander	Evergreen pear
Cotoneaster	Southern Magnolia	European fan palm	Bougainvillea
Oregon grape	Japanese boxwood	Blue decaena	White/Purple iceplant
Photinia	Xylosma	Rosemary	Croceum iceplant
Tulip tree,	Cherry plum	Aleppo pine	
Crape myrtle		Sweet gum	

Source: Adapted from Ref. 2.

Figure 6.1 View of the El Kantaoui Golf Course on the west coast of Tunisia, which has been irrigated with secondary treated recycled water for the past 20 years.

Close attention to variations in daily evapotranspiration (ET_o) and soil water depletion, preferably using on-site instrumentation.

Use of amendments (gypsum, lime, sulfur, etc.), as necessary, to adjust the SAR of soil moisture, countering potentially higher irrigation water SAR.

Use of buffer zones where lower-quality (secondary and lower) reclaimed water is used.

Warning signs to alert golfers not to drink water from irrigation sources without arousing unnecessary fear of recycled water.

Blending with lower-SAR or lower-TDS water sources, if necessary and feasible, through seasonal use of alternate sources of water or real-time blending with such waters.

Dual plumbing to irrigate greens with potable water in cases of excessively high SAR or high salinity, as a last resort.

Selection of drought-resistant, salt-tolerant grass species for use on fairways and particularly on the greens and tees.

Adaptation of irrigation schedule and depth of irrigation to the quality of irrigation water from a recycled water source.

The most serious potentially adverse impacts of use of recycled water are listed in Table 6.1.

The yield reduction impact of high salinity of irrigation water on landscape plants is illustrated in Figure 6.2.[2] Sensitivity of the landscape palette (plant species from which landscape designers select) to salt levels in the irrigation water varies considerably, as illustrated on Table 6.2.

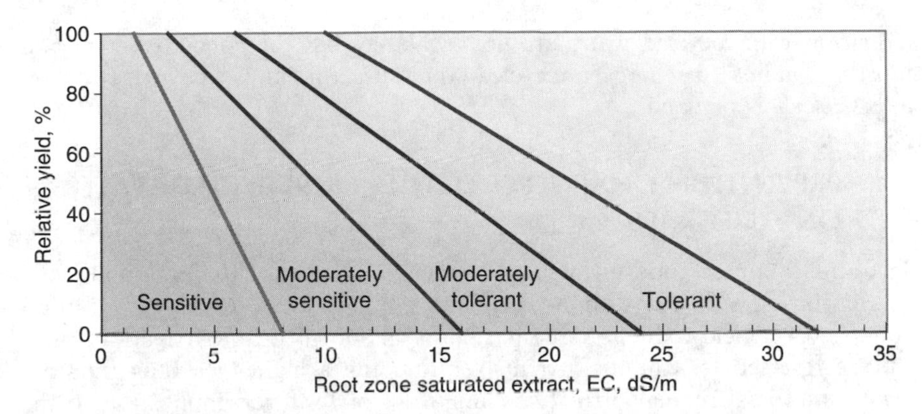

Figure 6.2 Sensitivity and yield reduction of plant materials as a function of salinity in the root zone.

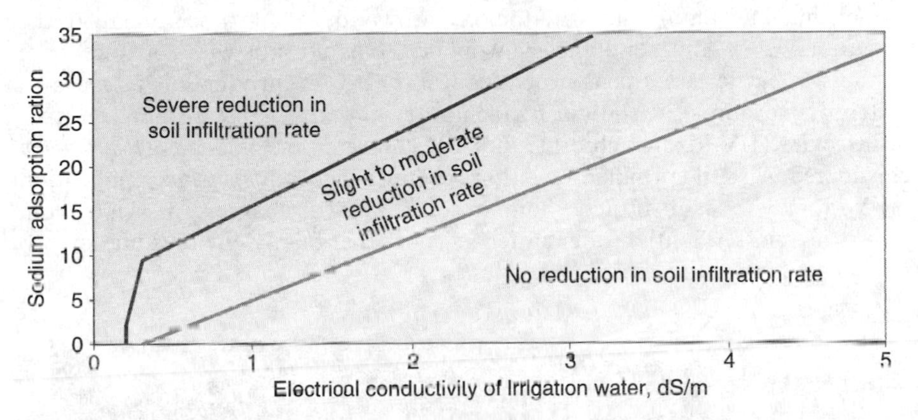

Figure 6.3 Combined effect of irrigation water SAR and EC on soil infiltration rate.

The soil permeability impact of high-SAR irrigation water is dependent also on the salinity, as shown in Figure 6.3.[3] Irrigation water with a relatively low salinity would more likely cause an SAR-induced infiltration rate reduction than one with higher salinity. Since most recycled water sources tend to have higher salinities than their original potable water counterparts, it would be a rare situation in which a soil permeability problem might be caused by irrigation with recycled water. Nonetheless, the problem is potentially serious enough to merit close attention and appropriate remedial measures, if necessary.

Groundwater pollution with nitrogen (especially in the nitrate form) is a serious threat due to the conservative nature of nitrates in the anoxic environment of the saturated aquifer. Examples abound in areas where long-term farming with excess nitrogen fertilization has occurred (such as in Orange County, California) and where highly concentrated livestock feedlots have been in operation over a long period of time. Examples directly related to irrigation with recycled water are not available, but the concern is real and must be addressed wherever recycled water use for landscape irrigation on a large scale is planned.

6.4 PREVENTION OF ADVERSE EFFECTS OF RECYCLED WATER ON TURFGRASS

The application of good management practices allows maintaining uniformly green playing surfaces with no patchiness, unevenness, or color variation (Figure 6.4). Yield reduction due to high levels of salt in the irrigation water can be reversed to a limited extent by providing a higher leaching fraction, deeper and less frequent irrigation, limitation of fertilizer application to the extent of the available fertilizer value in the reclaimed water, and below-ground application of irrigation water with drip irrigation.

The adverse impact of sodium can be countered by judicious applications of gypsum, lime, and/or sulfur to the soil or by adding such amendments to the irrigation water. Another approach would be to avoid salts with a sodium base in the wastewater treatment processes leading to the production of recycled water. For example, if sodium hypochlorite is used for disinfection, another disinfectant (UV, ozone, chlorine dioxide, chloramines, chlorine gas) may be considered. Also, if sprinkler irrigation is being practiced, a surface application method may be substituted. Source control of discharges of brine from industries and residential water softeners is another highly effective measure to reduce sodium at the source.

Figure 6.4 Golf course in Pebble Beach, California, irrigated since 1986 with recycled water.

Boron toxicity, especially in sensitive plants, where recycled water contains more than 0.5 mg/L of boron is preventable by blending with a water source having a lower boron concentration. In extreme cases where the boron concentration of recycled water is in excess of 3 mg/L, the only solution may be to avoid using the water from that source, especially in the germination and early growth stages of crops and ornamentals.

6.5 MANAGEMENT OF ADVERSE EFFECT OF WATER REUSE ON SOILS

The approaches listed above to mitigate adverse impacts on plants are equally effective in soils. In addition, alternate irrigation with water sources of lower salinity and lower sodium (hence SAR) would help counter such adverse impacts. This is particularly important in soils with higher levels of clay minerals, where reduced permeability can become a barrier to further efforts to remedy the situation. Sandy soils, on the other hand, offer greater flexibility and easier remediation.

6.6 RECOMMENDATIONS TO AVOID ADVERSE EFFECTS OF WATER REUSE ON GROUNDWATER

The most important potential adverse effect on groundwater, nitrogen build-up, can be avoided by balancing the fertilizer value of irrigation water in the overall fertilizer application program to avoid exceeding plant uptake capability. This problem is not unique to the use of recycled water. Farmers are often tempted to overapply fertilizers in order to increase crop yield with minimal increase in investment. Use of recycled water can exacerbate the problem if the fertilizer value inherent in recycled water is not accounted for (i.e., subtracted from planned chemical fertilizer applications).

Salinity impact on groundwater quality is a long-term issue, and the extent and immediacy of the problem can be calculated with simple mathematical modeling of the groundwater column beneath the irrigated perimeter. The larger the irrigation area, the greater the leaching fraction. The higher the irrigation water TDS and the purer the initial quality of the groundwater, the greater and the more immediate would be the impact. In case the impact is significantly adverse, it may be advisable to adopt measures to reduce its salinity. Salinity reduction measures include blending, partial desalinization, balanced irrigation with rainfall, and alternate irrigation with waters of lower salinity. If the natural recharge of the aquifer with high-quality rainwater (with very low salinity) is so much greater than the leached fraction of applied recycled water, the overall impact may be calculated to be negligible, even after several hundred years of irrigation. With sound knowledge of local groundwater and conservative assumptions about aquifer characteristics and local hydrology, it is possible to compute reliable estimates of the possible range of impacts under a variety of application scenarios.

Figure 6.5 Oak Hill cemetery in San Jose, California, irrigated with recycled water.

6.7 ECONOMIC AND FINANCIAL ASPECTS OF LANDSCAPE IRRIGATION

When the monetary value of all the benefits associated with a water reuse project are calculated and added up to a single dollar figure, traditional engineering cost-benefit analysis can serve to compare the project to its alternatives. When all the benefits are assigned to a single agency, such an analysis can also be used to determine the economic feasibility of the project. Unfortunately, neither of these situations normally prevails. In the first place, the benefits of water reuse include watershed protection, local economic development, improvement of public health, energy conservation, environmental protection, and other factors, which are not readily quantified by traditional cost-benefit techniques. Second, the benefits are usually fragmented among a number of agencies and the general public is not easily assigned. It is not always possible to allocate costs proportionately to the beneficiaries. As a result, water reuse projects are often undervalued when compared to other projects, and significant opportunities for beneficial reuse are lost.

The economic value of water recycling projects is routinely underestimated during the planning process due to the nonmonetary nature of their many benefits. The traditional engineering economic analysis favors strict formulas and regimented calculation to arrive at an objective estimation of a project's worth. Such an approach is embraced by public agencies, lending institutions, and regulatory bodies, which share a sincere desire for precision and which may be reluctant to make subjective judgments not required by policy. However, since certain benefits are not readily quantified—especially those with regional or global scope—this approach has the effect of setting the value of these "big picture" benefits equal to zero. To the resource economist using analytical techniques commonly employed by that discipline, these benefits may be more valuable than the more narrowly focused benefits accounted for in rigorous detail through traditional cost-benefit calculations. Nevertheless, from the standpoint of the project planners, they usually do not exist.

The effect is compounded by the fact that local agencies already tend to view their responsibilities as service providers narrowly. As the directors and managers of these agencies see it, their fiduciary responsibility to their constituents (ratepayers or stockholders) requires them to provide their core service at the lowest possible cost. Despite the gradual trend towards viewing themselves as "trustees of public resources," water supply agencies still by and large define their mission in terms of supplying water, and wastewater agencies have as their goal the treatment and discharge of wastewater. With respect to water reuse, neither entity takes into account more than the current cost of the project and the benefit of the water produced or the diversion of effluent accomplished. Such a narrow perspective may avoid the introduction of subjective scalars, but it provides decision makers with a very limited tool with which to evaluate alternatives.

This situation will only change as the decisions based on such simplistic cost-benefit analysis are challenged and more appropriate evaluations are offered in their place. This transition has already occurred in the arena of solid waste recycling. Three decades ago the value of recycling solid waste was also in dispute. Advocates were viewed as extremists, and their position was called impractical by opponents citing narrow, locally focused economic analyses. But with dwindling supplies of non-renewable resources and shrinking landfill space came the realization that solid waste recycling had benefits beyond short-term local economics, and efforts were made to identify and quantify those benefits. As a result, today the practice of solid waste recycling is widespread throughout the United States.

Even when all benefits are identified and evaluated, a water reuse project may not prove feasible as long as the benefits are distributed among entities that do not bear their cost. For example, a water reuse project may be sponsored by a wastewater agency to reduce the amount of effluent discharged into its receiving water ecosystem. Proposed as an effluent diversion alternative, the project will also provide a benefit as a new reliable local source of water. This benefit, however, will not accrue to the wastewater agency, but to the local water utility, which is responsible for meeting local water supply needs. As a result, the wastewater agency may not factor the water supply value of reuse into its planning process, even if the water agency is actively seeking new sources of water and is willing to serve recycled water to its customers. Or it may accept a value equal only to the current market price of water, without regard for the additional supply-related benefits of reliability or local control.

By the same token, if a water agency were to sponsor a water-recycling project without the collaboration of the local wastewater entity, any benefits accruing to the sewerage agency would likely be either ignored or undervalued.

Such disparity of interests is quite common in areas where the responsibility for water supply is distinctly separate from the authority to manage wastewater. An equally common result is that projects whose total value to the community exceed their cost are not implemented and opportunities for reuse are lost.

This division of interests is universal and stems from the traditional division of labor between those responsible for various aspects of water use, especially water supply and wastewater treatment. Even a centrally governed society like China maintains institutional barriers between water and wastewater agencies. As a result, these barriers have prevented implementation of water reuse projects. By contrast, in agencies where the entire water cycle is managed under the same administrative umbrella, water reuse gains ready acceptance because its diverse benefits are all appreciated. For example, the Irvine Ranch Water District, a California agency responsible for both water and wastewater utility management, has become a pioneer and leader in water reuse. However, when responsibilities are not consolidated, the challenge is to bring the stakeholders together to identify the benefits of water recycling, add them up, and assign their costs fairly to all beneficiaries.

6.8 CUSTOMER ACCEPTANCE OF RECYCLED WATER FOR IRRIGATION OF LANDSCAPING AND GOLF COURSES

New projects in areas without a prior history of use of recycled water can pose special challenges to the agency attempting to convert to using recycled water. These special challenges can come from members of the community who feel threatened by the change or consider the risk of using recycled water (no matter how low) to be unacceptable. Another source of opposition encountered is from the greenskeepers and superintendents of golf courses who may believe the quality of recycled water, particularly its salt content, inappropriate for golf course irrigation.

In the case of community members' opposition, the best approach is early and intensive involvement of the public in decision making in tandem with a comprehensive public information program. The public information program must define the problem being solved (usually lack of sufficient water) in clear and tangible terms. It must also explore alternative solutions (including those not involving use of recycled water), their advantages, costs, and disadvantages to the community.

In the case of golf courses with a management reluctant to use recycled water, the best approach is to start with a pilot project in which portions of the course are initially irrigated with recycled water and an opportunity is provided to observe the results and compare with areas irrigated with potable water. Field visits to other golf courses already using recycled water may also be useful, although there is often doubt expressed about the applicability of results from one place to another. Weaning existing golf courses from potable water after a switch to recycled water can continue to be a tough challenge until there is a change in administration and younger managers come up through the ranks to replace recalcitrant ones.

In all cases where there is a potential for a public acceptance challenge to the use of recycled water, the agency promoting recycled water should consult experts in the area of public relations. Together with such experts, the agency should develop positive branding strategies, emphasize the water quality improvement, environmental, and other values and benefits arising from the use of recycled water, and generally approach the public with complete openness and respect. Technical data and cost-benefit analyses alone are not sufficient to win over a suspicious public for whom the *status quo* represents a comfort zone not to be disturbed by the prospects of an unfamiliar source of water.

REFERENCES

1. Bahri, A., Basset, C., Oueslati, F., and Brissaud, F., Reuse of reclaimed water for golf course irrigation in Tunisia, in *Proc. IWA 1st World Water Congress*, Paris, July 3–7, 2000.
2. Maas, E.V., Crop salt tolerance, in *Agricultural Salinity Assessment and Management*, Tanji, K.K., ed., ASCE Manual and Reports on Engineering Practice, No. 71, ASCE, New York, 1990, Chapter 13.
3. Hanson, et al., *Agricultural Salinity and Drainage*, University of California, Davis, 1993.

7 Wastewater Treatment for Water Recycling

Valentina Lazarova

CONTENTS

7.1 Introduction 164
7.1.1 Choice of appropriate treatment 166
7.1.2 Main treatment processes used for wastewater treatment 166
7.1.3 Influence of sewer configuration on water quality 168
7.2 Physicochemical treatment of wastewater 168
7.2.1 Screening 168
7.2.2 Primary sedimentation 171
7.2.3 Coagulation/Flocculation 171
7.2.4 Flotation 172
7.3 Biological wastewater treatment processes 173
7.3.1 Activated sludge 173
7.3.2 Trickling filters 174
7.3.3 Rotating biological contactors 176
7.4 Advanced biofilm technologies 176
7.5 Nonconventional natural systems 178
7.5.1 Lagooning 178
7.5.2 Wetlands 183
7.5.3 Infiltration-percolation 184
7.5.4 Soil-aquifer treatment 186
7.6 Advanced tertiary treatment and disinfection 188
7.6.1 Tertiary filtration 189
7.6.2 Chlorination 194
7.6.3 Chlorine dioxide 198
7.6.4 UV disinfection 198

1-56670-649-1/05/$0.00 + $1.50

7.6.5 Ozonation 208
7.6.6 Membrane filtration 215
7.6.7 Membranes bioreactors 216

7.7 Storage and distribution of recycled water 219
7.7.1 Short-term storage 220
7.7.2 Long-term storage 220
7.7.3 Management of recycled water storage reservoirs 224
7.7.4 Control of water quality in distribution systems 225

7.8 Criteria for selection of appropriate polishing process before irrigation 226
7.8.1 Cost of additional treatment and reuse 226
7.8.2 Main criteria for selection of disinfection process 228

References 231

This chapter provides basic information on the design and operational parameters of wastewater treatment with specific attention to processes applied in water-recycling systems. Details provided should be used only as illustration and guidelines for the choice and preliminary design of water reuse treatment trains. It is important to stress that a number of parameters greatly influence wastewater treatment design and associated costs. For this reason, the proper design of water reuse treatment facilities requires professional judgment, operational experience, and in-depth knowledge of specific local conditions.

7.1 INTRODUCTION

As shown in Chapter 4, wastewater treatment is the most effective way to reduce the health risks associated with the use of recycled water for irrigation. The choice of treatment schemes depends on water quality requirements, type of irrigated crops, irrigation method, public access, and potential adverse impacts on soils and crops.

Figure 7.1 illustrates the water quality groups of concern for water reuse applications, i.e., irrigation, groundwater recharge, urban, recreational, industrial, and potable reuse.[1] In this figure, pretreated wastewater enters from the center. The first parameter of concern in primary wastewater treatment is suspended solids, which, if present in the effluent, can plug irrigation systems or soils and protect microorganisms, thus decreasing the disinfection efficiency of most treatments. The second important objective of wastewater treatment is the secondary treatment for carbon removal. Even if carbon removal is not strictly required for the reuse of effluents for irrigation, this process is of great importance for the reduction of the regrowth potential of residual microorganisms in distribution systems. The next parameter of interest

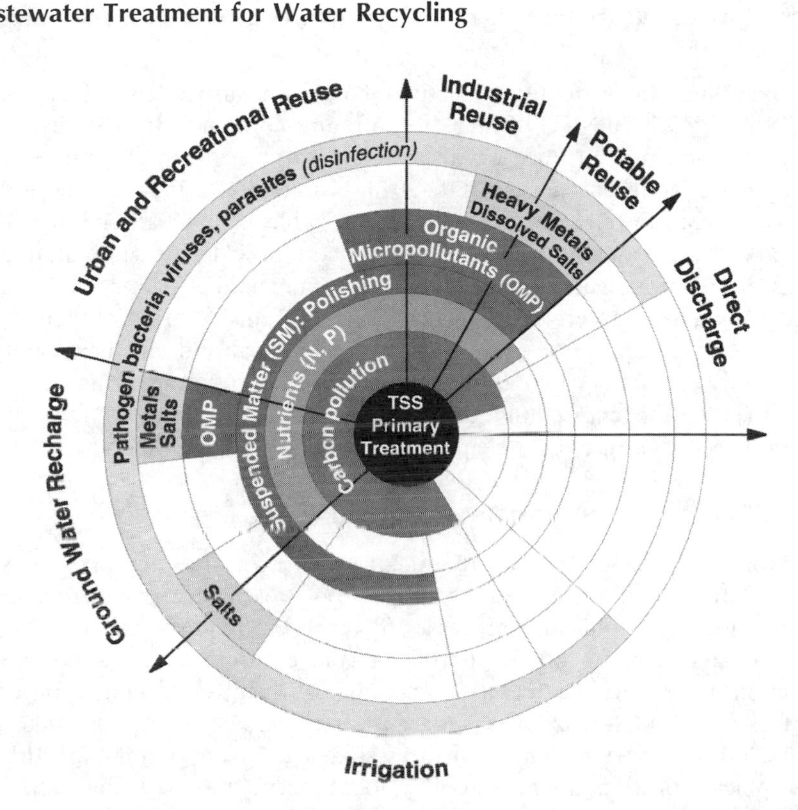

Figure 7.1 Treatment levels for target pollutants that need to be removed for different types of reuse purposes.

is nutrients. Generally recommended for agriculture, their presence in groundwater recharge effluents and most industrial applications is not desirable. Further treatment includes tertiary polishing, such as removal of residual suspended solids, some specific compounds like organic micropollutants, salts, and heavy metals, and, finally, disinfection.

Disinfection is the most important treatment step for almost all water reuse applications, and specifically for irrigation with recycled water, as shown in Chapter 4. The removal of organic micropollutants and, more specifically, synthetic organic compounds (SOCs) such as pesticides, PCBs, and PAHs, which are also known to be potentially harmful for human health, is generally required only for potable reuse applications. In some cases heavy metals and dissolved salts can also be of health and environmental concern. These last two groups of parameters are of particular interest only for potable reuse applications and industrial uses. In addition to microbiological pollution, salinity, sodium, and boron are some of the most important parameters for the use of recycled water for irrigation (see also Chapter 2).

The principal contaminants of urban wastewater are suspended solids with a typical concentration range of about 100–300 mg/L and organic matter expressed by a chemical oxygen demand (COD) of about 200–1000 mg/L. Accordingly, the treatment of municipal wastewater is typically designed to meet recycling water quality objectives based on suspended solids (<5–30 mgTSS/L or <0.1–10 NTU of turbidity), organic content (<10–45 mgBOD/L), biological indicators (total or fecal coliforms, *E. coli*, helminth eggs, enteroviruses), nutrient levels (<10–20 mgN/L and <0.1–2 mgP/L) and, in some cases, chlorine residual (>1–5 mg Cl/L). Table 3.1 in Chapter 3 provides a summary of water quality parameters of concern in water reuse regulations as well as their approximate range in secondary municipal effluents.[1,2]

As shown in Chapter 4, the starting point for consideration of water reuse for any specific application is ensuring the biological and chemical safety of using recycled water by applying appropriate treatment technologies. Consequently, the choice and design of treatment scheme to meet water quality objectives is a critical element of any water reuse system.

7.1.1 Choice of Appropriate Treatment

The choice of wastewater reuse treatment scheme includes at least four steps (Figure 7.2).[1] The first step is the evaluation of treatment performance of the available technical tools and processes to reach the required effluent quality. The next step consists of an analysis of the existing standards and other restrictions to identify the operation reliability of the given treatment processes and the need for storage. The choice of treatment scheme must take into account other important criteria such as size of the plant, climate conditions, geographical, social, political, and other local specificities. The final step is the technical-economic evaluation, including the analysis of the existing infrastructure (available land, labor requirements, electrical power, distance to public housing units, etc.), existing equipment (sewage system, wastewater treatment, need for plant extension or retrofitting, recycled effluent distribution and storage systems, need for satellite treatment, etc.) and the financial resources for the capital and operating costs.

7.1.2 Main Treatment Processes Used for Wastewater Treatment

Figure 7.3 summarizes the main treatment steps and individual processes used in wastewater treatment and reuse.[3] As a rule, a combination of physical,

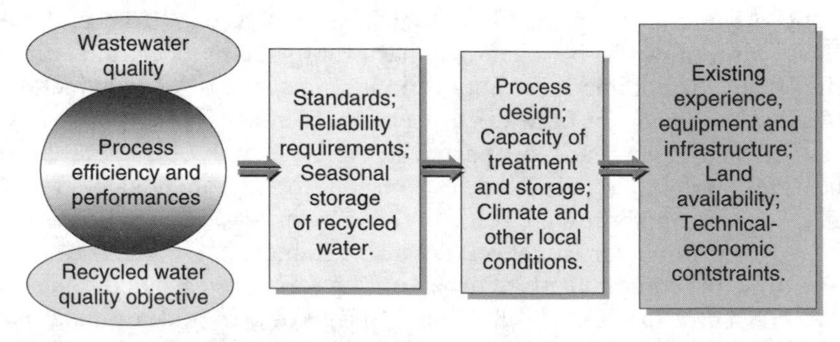

Figure 7.2 The main steps and criteria for selecting the most appropriate wastewater treatment for reuse purposes.

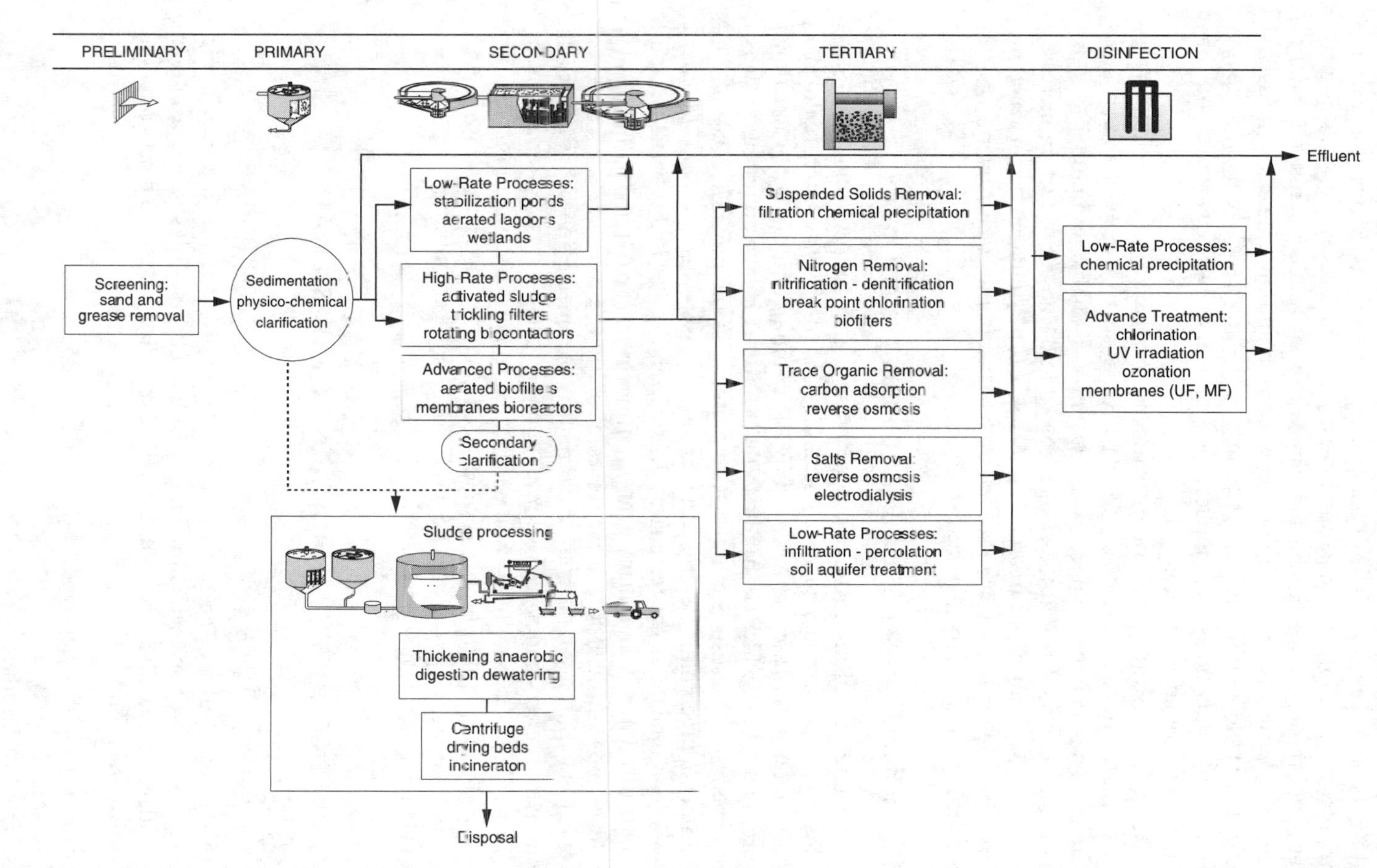

Figure 7.3 Typical combinations of wastewater treatment processes for water reuse and discharge in sensitive areas. (Adapted from Ref. 3.)

chemical, and biological technologies is used for the removal of the main pollutants of suspended solids, organic matter, pathogens, and in some cases nutrients, salt, or organic micropollutants. The main steps of wastewater treatment are primary treatment for removal of suspended solids, secondary biological treatment for the elimination of carbon and in some cases nitrogen, and tertiary treatment for effluent polishing or disinfection. The main removal mechanisms and fields of application for these wastewater treatment technologies are given in Table 7.1.[1]

Table 7.2 illustrates how some treatment processes can be combined to achieve water quality objectives prior to reuse.[1,2,4,5] Although almost any treatment level can be attained with currently available technologies, it is important to remember that the costs involved and sludge production increase almost exponentially as the treatment level rises. It is common for construction costs to double from primary to secondary treatment and to rise again by 50% from secondary to more advanced treatment.

7.1.3 Influence of Sewer Configuration on Water Quality

Sewer configuration and detention time have a great impact on wastewater quality and must be taken into account in the design of wastewater treatment. Combination sewers are characterized by the combination of wastewater flow and rainfall runoff, which contribute to the addition of some pollutants such as heavy metals and resuspension of settled material in the pipes. Because of the variability in precipitation events, drainage area, wastewater sources, and other factors, combined wastewater characteristics tend to be highly variable and difficult to predict. Consequently, stringent water reuse quality requirements would be more difficult to achieve for combination sewers.

Typical variations in the main water quality parameters in wastewater from combination sewers compared to conventional municipal wastewater are given in Table 7.3.[4] Lower BOD values and microbial contamination can be observed for high runoff flows. However, first runoff flush and low-rate runoff can be characterized by organic and microbial pollution as high as in conventional municipal wastewater.

7.2 PHYSICOCHEMICAL TREATMENT OF WASTEWATER

7.2.1 Screening

With the technical advances in treatment devices, the use of fine screens for grit removal as replacements for primary sedimentation tanks is increasing, in particular in some advanced water-recycling schemes. The three most common types of screens are:

Inclined self-cleaning type
Rotary drum type
Rotary disk screen

Table 7.1 Main Mechanisms and Applications of Treatment Processes Used for Wastewater Treatment and Reuse

Process	Description	Application	Comments
Conventional wastewater-treatment processes			
Sedimentation	Elimination of suspended solids, including particles, sand, and flocs	Pretreatment, primary and secondary settling	Colloidal solids should be transformed into suspended solids
Aerobic biological treatment	Transformation of organic matter to CO_2, H_2O, and biomass by micro-organisms in presence of oxygen	Wastewater containing significant quantities of organic matter	Several systems are available: activated sludge, oxidation ditches, biofilm reactors
Anaerobic biological treatment	Transformation of organic matters to CO_2, H_2O, and biomass by micro-organisms in absence of oxygen	Wastewater containing high quantities of organic matter or excess biomass from aerobic systems (sludge treatment)	Several systems are available: anaerobic lagoons, anaerobic activated sludge, USBG and biofilm reactors
Combination of aerobic/ anaerobic/ anoxic systems	Combination of different types of microorganisms to reduce the amount of N and P	Usually applied as advanced treatments, not for irrigation	
Advanced treatment processes used for additional treatment before reuse			
Coagulation-flocculation	Increase of solids size through addition of chemicals and particle aggregation	Usually before specific disinfection systems	Chemicals destabilize particles
Filtration	Removal of particle and colloidal solids by retention in granular media	Removal of particles above a certain level defined by media characteristics	Mainly sand and activated carbon used as filtration media
Disinfection processes	Removal or inactivation of pathogens using heat, lime and other chemicals, physical separation, UV light, etc.	Crucial process for health protection before reuse	Chlorination is most common method; UV disinfection is rapidly growing, as is ozonation
Membrane processes	Pressure-driven membrane processes based on size exclusion or molecular diffusion	Removal of impurities: bacteria, viruses, dissolved salts, colloids, etc. and production of high-quality recycled water	Main systems: microfiltration, ultrafiltration, nanofiltration, and reverse osmosis; important energy requirements

(*continued*)

Table 7.1 Continued

Process	Description	Application	Comments
Extensive and low-tech processes used in water-reuse schemes			
Extensive processes in liquid media	Biological degradation of organic matter in natural systems and disinfection by sunlight; macrophytes and/or algae could be a part of the system	Appropriate low-cost treatment, in particular for small treatment plants; high residence time and need for land	Main systems include stabilization and maturation ponds (lagooning), wetlands, and algae ponds; lagooning has the additional advantage of ensuring storage of recycled water
Extensive treatments using low-rate infiltration	Biological degradation of organic matter and retention of solids and pathogens in solid media	Low-cost treatment that can be used for small (IP) or large (SAT) units	Main systems include soil-aquifer treatment (SAT) and infiltration-percolation (IF)

Source: Adapted from Ref. 1.

Table 7.2 Treatment Levels Achievable with Some Typical Treatment Trains

Secondary treatment	Additional treatment	Suspended solids (mg/L)	Turbidity (NTU)	BOD (mg/L)	COD (mg/L)	N_{tot} (mg/L)	$P\text{-}PO_4$ (mg/L)
Activated sludge	None (secondary effluent)	10–30	5–15	15–25	40–90	10–50	6–15
	Granular media filtration	< 5–10	< 0.5–5	< 5–10	30–70	10–35	4–12
	Filtration + GAC	< 5	< 0.5–3	< 5	5–20	10–30	4–12
	Coagulation/flocculation	< 5–7	< 10	< 5–10	30–70	10–30	< 1–5
	Coagulation + filtration	< 1	< 0.5–2	< 5	20–40	< 5–25	< 1–2
	Aerated biofilters	2–10	0.5–5	< 5–15	20–50	10–30	4–12
	Maturation ponds	20–120	—	< 5–35	40–150	5–25	2–6
Trickling filters	None (secondary effluent)	20–40	5–15	15–35	40–100	15–60	6–15
	Granular media filtration	10–20	10	15–35	30–70	15–35	6–15
	Coagulation + filtration	< 5–10	0.5–5	< 5–10	30–60	10–30	4–12
MBR	None (secondary filtered effluent)	< 1	< 0.1–0.5	< 5	5–50	< 5–20	< 0.1[a]–10

[a]with chemical addition.
Source: Refs. 1, 2, 5.

Fine screens can achieve a significant degree of suspended solids removal from 20 to 35%.[5,6] Openings of fine screens commonly vary from 0.02 to 3 mm. Full-scale operation of the first two processes indicate that grit removal efficiency can reach up to 80–90% with, respectively, BOD removal of 15–25%

Table 7.3 Comparison of Water-quality Characteristics of Municipal Wastewater, Runoff, and Combined Wastewater Flow

Parameter	Rainfall	Stormwater runoff	Combined wastewater	Municipal wastewater
Suspended solids, mg/L	< 1	65–100	270–600	100–350
BOD_5, mg/L	1–15	8–10	60–220	110–400
COD, mg/L	9–16	40–73	260–460	250–1000
Total nitrogen, mgN/L	low	low	4–17	20–85
Total phosphorus, mgP/L	low	low	1.2–2.8	4–15
Fecal coliforms, cfu/100 mL		10^3–10^4	10^5–10^6	10^5–10^8
Metals, mg/L:				
Copper		0.02–0.09		0.007–0.06
Lead	0.03–0.07	0.03–0.35	0.14–0.6	0.003–0.45
Zinc		0.135–0.54		0.006–0.55

Source: Adapted from Ref. 4.

and solids removal of 15–30%. Rotary disks are characterized by the greatest efficiency and can achieve 40–50% removal of suspended solids.

Coarse screens consist of vertical or inclined bars spaced in equal intervals and are used primarily as a pretreatment to protect downstream treatment devices (pumps, aerators, biofilters, etc.).

7.2.2 Primary Sedimentation

This process is commonly used in wastewater treatment before biological treatment. The main goal is the removal of suspended solids through gravity settling of particles that are denser than water. Circular or rectangular sedimentation tanks with vertical or horizontal flow have been developed.

The basic design parameter of settling tanks is surface hydraulic loads, with average and peak values varying from 30 to 50 m^3/m^2 d and from 80 to 110 m^3/m^2 d, respectively.[2,4–6] The typical hydraulic residence time is 2 hours (1.5–2.5 h). Sludge storage and water depth (3–4 m) are also key design issues. In addition, inlet and outlet turbulence should be minimized.

Suspended solid removal efficiency of primary settling is about 50%. About 40% of the BOD and COD are also removed. Removal of some heavy metals can reach 35–50% (Cd, Cr, Pb, Zn) with a few exceptions, such as mercury (Hg, 11%).

7.2.3 Coagulation/Flocculation

Physicochemical treatment is an alternative approach to biological treatment of wastewater with high industrial effluent content or to achieve maximum fertilization capacity of recycled water. This technology has been implemented in numerous water-recycling schemes, both as primary and tertiary treatments.

The major advantage of this process is its ability to efficiently remove suspended solids (80–90%, including colloid particles) and carbon (50–80% of

BOD), its disinfection properties (80–90% of coliforms and helminth eggs), as well as phosphorus removal, with short hydraulic residence times. The main constraints for application in municipal wastewater treatment are the additional cost for chemicals, handling, and disposal of the great volumes of sludge. Lime, alum, ferric chloride, or ferric sulfate can be used as chemicals. Depending on the treatment objectives, the required chemical dosage and application rates are determined from bench- or pilot-scale tests.

Figure 7.4 illustrates the application of lime clarification for production of high-quality recycled water in the West Basin Water Recycling Plant in California.[7] Lime promotes coagulation by increasing wastewater pH and reacting with alkaline component and phosphorus, forming the precipitants of calcium carbonate, calcium hydroxyapatite, and magnesium hydroxide. After the first precipitation step and pH adjustment by recarbonation, wastewater is passed through a granular media filter to remove any residual flocs, after which reverse osmosis removes dissolved salts and organic compounds. Because of the high pH generated during lime clarification, high disinfection credit (4 log credit units for total coliforms removal, i.e., extended disinfection) has been given to this process by the California health authorities.

7.2.4 Flotation

In wastewater treatment, flotation is used principally to remove suspended matter and to concentrate biological sludge. This process could be used in water reuse schemes as a pretreatment before disinfection, when carbon and

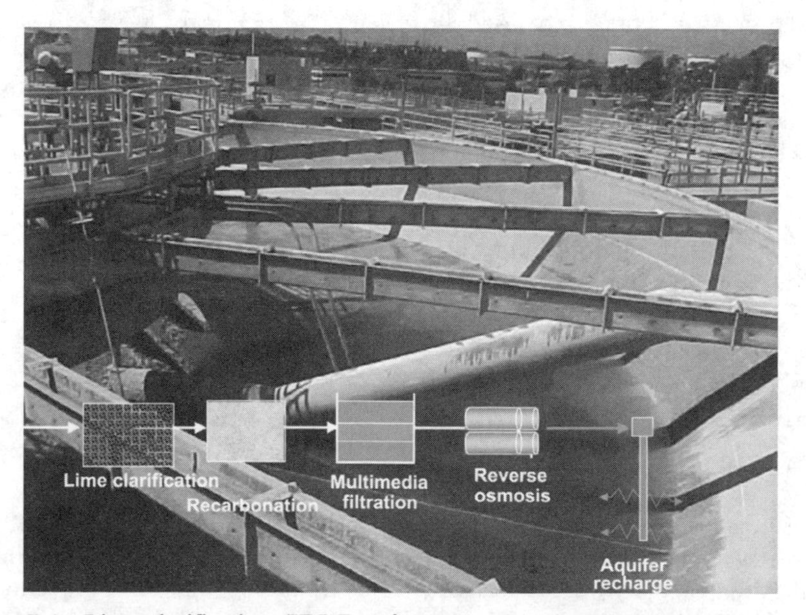

Figure 7.4 Lime clarification (IDI Degrémont) and its integration in the treatment scheme for aquifer recharge with recycled water in the West Coast Basin Barrier, California.

nitrogen removal is not required, e.g., for agricultural irrigation. This process has the advantage of high surface-loading rates and high removal of grease and floatable materials, which is beneficial for efficient removal of helminth eggs from recycled water. The main advantage compared to coagulation-flocculation is less sludge production. The main constraint is the high energy consumption.

Flotation is a separation process involving the introduction in the liquid phase of fine bubbles that attach to particles and solid phase removal resulting from the buoyant force of the combined particles–gas bubbles. Particles of higher and lower density than water can be removed.

Flotation processes can be classified according to the method of generation of air bubbles:

1. Dissolved air flotation: air is injected into the liquid, which is under pressure, followed by a release of pressure to generate fine air bubbles.
2. Air flotation: aeration is performed under atmospheric pressure.
3. Vacuum flotation: saturation with air occurs at atmospheric pressure, followed by application of a vacuum to the liquid.

In all these systems, removal efficiency can be enhanced by the use of various chemical additives (inorganic chemicals such as aluminum or ferric salts and activated silica, as well as organic polymers). From practical experience, it appears that the addition of 2–3% air yields satisfactory results.

7.3 BIOLOGICAL WASTEWATER TREATMENT PROCESSES

As a rule, secondary wastewater treatment is required for environmental protection and reduction of wastewater discharge. It is important to stress, however, that even if recycling water treatment is implemented after secondary treatment, the efficiency and reliability of the downstream processes highly influence recycling water quality and must be taken into account during water reuse project design. When municipal wastewater treatment facilities are initially designed with the objective of water reuse, the selection and design of biological treatment processes must consider some specific treatment requirements, in particular, improved reliability of operation and influence on disinfection efficiency.

The most commonly used biological processes are activated sludge and oxidation ditches. New treatment combinations and various types of innovative biological processes have been developed for carbon removal, nitrification, pre- and postdenitrification, and phosphorus removal. The major concern in water reuse systems is the removal of carbon pollution in order to reduce the regrowth potential of pathogens in distribution systems and, thus, guarantee efficient disinfection.

7.3.1 Activated Sludge

Activated sludge is a common method for providing secondary treatment of municipal wastewater. It can be used as a final treatment process in water reuse schemes that may be followed only by disinfection. For stringent water quality requirements, the activated sludge process is an intermediate step followed by processes that provide a higher level of treatment.

There are many variations on the basic conventional activated sludge process. Modifications are designed to optimize organic and nutrient removal on the basis of air or oxygen feed rate, hydraulic loading, flow pattern, and detention time. This process is very flexible and can be adapted to almost any type of biologically degradable wastewater. The most common variations of the activated sludge process are the conventional, tapered aeration, continuous-flow stirred tank, step aeration, contact stabilization, and extended aeration systems. The last process provides an effluent with very good quality that requires relatively low disinfection doses.

Typical average removal efficiency values of activated sludge processes are 45% for phosphorus, 70% for COD and $N\text{-}NH_4$, 80% for suspended solids and TOC, 85% for turbidity, and 90% for BOD.[4–6] High removal rates can be achieved for some heavy metals, e.g., 70–80% of chromium, copper, iron, lead, and silver, and about 50% of zinc and cadmium. Recent R&D studies indicated the high removal efficiency of biological treatment of some emerging pollutants, such as endocrine disruptors, pharmaceutical products, and organo-tin compounds.

Typical process loading for activated sludge processes ranges as follows:

High loading rate processes—organic loads (food-to-microorganism ratio) $C_m > 0.4\,kgBOD_5/kgSS$ d and solid detention time 3–5 days

Medium loading rate processes—organic loads $0.15 < C_m < 0.4\,kgBOD_5/kgSS$ d and solid detention time 5–15 days

Low loading rate processes (extended aeration)—organic loads $C_m < 0.15\,kgBOD_5/kgSS$ d and solid detention time 15–30 days

The oxidation ditch is essentially an extended aeration process. It consists of a ring-shaped channel about 0.9–1.5 m deep. An aeration rotor is placed across the ditch to provide aeration and circulation (0.3–0.6 m/s). Other configurations are the Carrousel process, with vertical aerators, and the Kraus process, with aeration of the supernatant from the sludge digesters.

In respect to treatment efficiency and reliability of activated sludge processes, the design of sedimentation units for suspended solids removal is critical to the overall system performance, especially for disinfection. A major portion of BOD in plant effluent can be attributed to the loss of suspended solids. For this reason, good separation of the solids in the sedimentation step is essential for producing high-quality recycling effluents.

7.3.2 Trickling Filters

Trickling filters with rock media (1–3 m) are the first biological systems applied to wastewater treatment since 1880s. Due to the relatively low reaction rates, low specific area for biofilm development, and clogging problems, plastic-medium trickling filters were developed and have been used for combined carbon removal and nitrification. Consequently, almost all trickling filters constructed since the 1980s have been equipped with plastic media.

The use of high-rate plastic media (Figure 7.5) allowed increasing filter depth to 4.3–6.4 m, improving process performance and reducing clogging problems. It is important to underline that different plastic media geometries affect the wastewater treatment efficiency.

Typical process loading rates for trickling filters range as follows [4–6]:

High loading rate: organic loads 0.7–0.8 $kgBOD_5/m^3\,d$ and hydraulic loads $> 0.7\,m^3/m^2\,h$

Low loading rate: organic loads 0.08–0.15 $kgBOD_5/m^3\,d$ and hydraulic loads $< 0.4\,m^3/m^2\,h$

In general, high trickling filter rates are used for secondary treatment, whereas intermediate- and low-rate processes are mostly implemented as tertiary treatments in combination with activated sludge or a high-rate trickling filter.

The main advantages of trickling filters are their simplicity and low operating costs (energy consumption and maintenance). This process is recommended for small and medium-size water reuse treatment schemes (< 20,000 p.e.) and for treatment of "light" raw wastewater with low organic pollution. The main disadvantage, however, is the limited treatment efficiency of trickling filters and their strong dependence on carbon and hydraulic loads. Several other disadvantages, such as high sensitivity to temperature, generation of odors, and mass transfer limitations, limit their application essentially to carbon removal for small treatment facilities.

To better regulate organic and hydraulic loading, it is recommended to implement a recirculation system. This mode of operation lets one maintain sufficient wetting rates for plastic media. Although trickling filters are fairly

Figure 7.5 Trickling filter distribution system and plastic media.

simple secondary treatment processes, influent to the filter must be monitored and controlled because changes in the raw wastewater characteristics can affect the growth and sloughing of biofilms. Adequate ventilation is also an important factor that can affect process performance. In hot and humid areas, forced ventilation may be required to provide adequate air circulation.

Effluent quality of trickling filters, as well as the removal of trace metals and inorganics, is usually not as high as that achieved by an activated sludge process. The average performances of rock filters are 60% for suspended solids and COD, 70% for BOD, less that 40–50% for copper, lead, and zinc, and no effect on cadmium and chromium.[4] Removal efficiency can approach 90% for carbon oxidation with plastic media operated at low hydraulic loading.

7.3.3 Rotating Biological Contactors

Rotating biological contactors (RBC) are another conventional attached-growth secondary treatment process that has been used for small to medium-sized facilities. The process consists of large-diameter plastic media mounted on a horizontal shaft and submerged at 40% in wastewater. These structures are ordinarily arranged in treatment stages with the possibility for nitrification in the latest stages. Shearing forces exerted on the biofilm (typical size 1–4 mm) by rotation cause excess biomass to be stripped into the mixed liquor. Secondary sedimentation normally follows.

This process requires relatively little O&M attention. Because of continuous growth of biofilm, sludge recycling is not required, simplifying the process operation. Typical removal rates are 90% for BOD and suspended solids, 70% for ammonia, and up to 70–90% for chromium, copper, iron, lead, mercury, and zinc.[4–6]

7.4 ADVANCED BIOFILM TECHNOLOGIES

The importance of biofilm reactors for municipal wastewater treatment and reuse has become widely recognized. The main factors promoting the development of new intensive biofilm technologies for wastewater treatment and reuse are the increasing volume of wastewater, limited space availability, and progressively tightening standards and water quality control.[8] Biofilm reactors are, in general, less complex to operate than activated sludge systems, eliminating sludge settling and recycling as well as the problems of sludge bulking and rising. Recently developed biofilm reactors are an attractive technological solution, offering numerous features and advantages.

During the last decade, granular fixed-bed biofilters have been widely implemented for the secondary and tertiary treatment of municipal wastewater. The advantage of this technology for water reuse schemes is its performance of both biological treatment and suspended solid removal simultaneously.[8] It is thought that granular biofilters can ensure volumetric nitrification rates up to 3.5 and 8 times higher than RBC and trickling filters (plastic media), respectively. For a given degree of treatment, granular biofilters require 3 times less aeration volume than activated sludge units and 20 times less than trickling filters. Specific energy consumption is similar to that of activated sludge tanks (approximately 1.0–1.4 kWh/kg COD removed), but the loading rates are much higher. These reactors are competitive especially when land is limited and/or tertiary treatment is required. The main constraint for operation is the frequent biofilter backwashing (one each 24 or 48 h) due to rapid clogging of the bed. However, it is important to stress that all backwashing procedures are fully automatic.

To avoid clogging problems and ensure better mass transfer and auto-control of biofilm thickness, several types of moving bed biofilm reactors have been developed in recent years,[8] including fluidized bed, moving bed, and air-lift systems.

Table 7.4 lists design loads reported for carbon and nitrogen removal in some commercial innovative biofilm reactors demonstrated on the basis of full-scale experience. It is worth stressing the significantly higher hydraulic and organic loads of these processes compared to conventional trickling filters and rotating biological contactors. High hydraulic loads up to 35 $m^3/m^2/h$ have been reported for the Biofor® process compared with the average design value of 0.4 $m^3/m^2/h$ for trickling filters. Higher volumetric loading and removal rates have also been reported in innovative fixed and moving bed reactors for treatment of municipal wastewater as well as for the aerobic and anaerobic treatment of industrial wastewater.

Table 7.4 Loading Criteria for Innovative Biofilm Reactors in Full-Scale Operations for Municipal Wastewater Treatment and Reuse

Type of biofilm reactor *(manufacturer)*	Type of treatment	Maximum installed capacity[a] (m^3/h [*p.e.*])[b]	Mass loading rates (kg/m^3 d)	Hydraulic loading rates (m^3/m^2 h)
Submerged biofilter	Carbon removal		18 (COD)	16
Biofor®	Nitrification	43,200	1.5 ($N\text{-}NH_4$)	20
(Degrémont)	Postdenitrification	[1,200,000]	6 ($N\text{-}NO_3$)	35
Submerged biofilter	Carbon removal			10
Biostyr®	Combined	43,200		
(OTV)	Nitrification/	[1,200,000]	1 ($N\text{-}NH_4$);	6
	Denitrification		5 (COD)	10
	Nitrification		1.5 ($N\text{-}NH_4$)	
Moving bed	Carbon removal		2.5–8 (BOD_7)	NA[c]
MBBR (Kaldness	Nitrification	3,585 [70,000]	0.15–0.35 ($N\text{-}NH_4$)	
Miljøtechnology)	Predenitrification		0.3 ($N\text{-}NO_3$)	
	Postdenitrification		0.7 ($N\text{-}NO_3$)	

[a]Peak wet weather flow.
[b]Population equivalents.
[c]NA, not applicable.
Source: From Ref. 8, with permission.

The relatively high capital and operation costs are the main constraints for the implementation of advanced biofilm reactors in water reuse schemes. For this reason, such processes are applied mostly for production of high-quality water for industrial purposes or indirect potable reuse.[9]

7.5 NONCONVENTIONAL NATURAL SYSTEMS

Most of the countries involved in wastewater reuse are located in the subtropical zone. These areas are generally characterized by alternating dry and wet seasons and average temperature values greater than 20°C, which enables them to use nonconventional natural systems, also called extensive or low-tech technologies. When sufficient land of suitable character is available, these technologies can often be the most cost-effective option for both construction and operation.

Extensive technologies are usually suitable for small communities and rural areas. The common element is the use of natural treatment mechanisms (vegetation, soil, microorganisms). In addition to having good treatment efficiency, such systems generally have low energy and maintenance requirements.

Natural systems range from small individual home units, with a capacity of about 1 m^3/d, to large plants up to 270,000 m^3/d, such as the soil-aquifer treatment in Dan Region, Israel (Table 7.5).[10]

7.5.1 Lagooning

Waste stabilization pond treatment, also called lagooning, is a low-tech system that has been practiced for more than 3000 years. It reproduces in a controlled environment the natural purification and disinfection processes found in lakes and streams. Different pond types have been developed to remove pathogens, organic matter, nutrients, colorants, and heavy metals:

Anaerobic ponds are heavily loaded systems used as a pretreatment step, with high depth and retention time.

Table 7.5 Main Categories and Characteristics of Natural Treatment Systems

	System characteristics	
Type of natural system	**Country (number of plants)**	**Maximum capacity**
Ponds or lagooning	USA (>7600); France (>200), Argentina, Morocco, Tunisia, South Africa, Jordan	240,000 m^3/d (Nairobi, Kenya)
Infiltration-percolation	France (>10), Spain (>10)	—
Soil-aquifer treatment	USA, Israel, France, Australia	270,000 m^3/d (Tel Aviv, Israel)
Wetlands	USA (>140), France, Spain	—

Source: Adapted from Refs. 1, 10.

Table 7.6 Design Parameters of Stabilization Ponds

Type of pond	Effective depth (m)	Detention time (d)	BOD loads[a] (kg/ha d)	Effluent suspended solids[b] (algal) (mg/L)	Comments
Anaerobic	2.5–5	20–50	220–560	80–160 (0–5)	Pretreatment
Facultative	1.5–2.5	25–180	50–200	40–60 (5–20)	C and N removal
Aerated	3–6	7–20		80–250	C and N removal, less space, less odor
Maturation	0.5–1.5	15–30	<17	10–30 (5–10)	Polishing disinfection step, storage function

[a] Temperature range 0–30°C; optimum temperature 20°C.
[b] Typical values; much higher values of algal growth can be observed in maturation ponds.
Source: Adapted from Refs. 1, 10, 11.

Facultative ponds, or oxidation ponds, are most commonly used for carbon removal. Aerobic conditions exist in the upper layers with anaerobic zones near the bottom.
Aerated ponds with mechanical floating aerators make possible increased loading rates commonly used in facultative ponds.
Maturation ponds are relatively shallow systems used as a polishing disinfection step.

Table 7.6 summarizes the main design parameters of pond systems.[1,10,11]

Maturation ponds (Figure 7.6) are commonly used in water reuse schemes in combination either with other ponds in series or with intensive biological treatment (activated sludge, trickling filters, etc.). This technology efficiently removes helminth eggs and ensures effluent disinfection by the direct action of UV light.

7.5.1.1 Main Characteristics of Maturation Ponds

Polishing pond treatment is the most simple and cheap technology for the recycling of secondary-quality effluent in small communities (≤3000 m^3/d). This treatment produces no harmful by-products and provides the storage capacity to accommodate variations in water demand. Table 7.7 illustrates the main advantages and disadvantages of maturation ponds, as well as the main design parameters.

Maturation ponds can be used as polishing, storage, and disinfection steps to achieve the maximum water quality requirements of WHO[12] (<1000FC/100 mL; <1 helm.egg/L) in small communities with a population of less than 20,000–50,000 p.e.[1] One of the main disadvantages of this technology is its restricted operation flexibility, especially with respect to flow and seasonal variations.[10] Another disadvantage of lagooning, in particular for dry and windy zones, is the high water loss due to evaporation.

Figure 7.6 Maturation ponds at the treatment plant Wadi Mousa near Petra, Jordan (agricultural irrigation).

Table 7.7 Main Characteristics of Maturation Ponds

Design parameters	Advantages	Drawbacks
Depth: 0.5–1.5 m Hydraulic residence time: 15–30 d Surface area: 2.5 m^2/inh Recommended for small and medium plants (<50,000 inhabitants) ⇒ *High influence of climate*	Provides additional storage capacity Easy to operate Robust to meet WHO guidelines Efficient for removal of bacteria and helminths Cost-efficient for small units	Sludge accumulation with potential disposal problems Sludge disposal requires infrequent but intensive work Effluent quality is a function of the hydraulic loads, climate, and season Can cause nuisance (odor, insects) Requires large footprints No disinfectant residual Additional contamination is possible by the presence of birds, etc.

Source: Adapted from Refs. 1, 10, 11.

As a rule, maturation ponds are constructed entirely of earth.[10] Traditionally, the basins are lined with compacted natural materials such as clay. With current groundwater concerns, synthetic barriers become the most common protection layer. Lagoons that are greater than 4 ha or are located in windy areas should be protected from wave erosion (0.3 m height minimum). The inlet structure for small lagoons is in the center, whereas large lagoons employ inlet diffusers with multiple outlet points to distribute water over a large area. Transfer and outlet structure should permit lowering of water level at a rate of 300 mm/week.

7.5.1.2 Main Operational Problems of Maturation Ponds

Poor effluent quality may be caused by overloading, short-circuiting, low temperature, toxic materials, blockage of light by plants, high turbidity or algal growth, loss of liquid volume by evaporation or leakage, etc. In addition, erosion of the dikes, the presence of animals, or the growth of rooted woody plants can destroy the structural stability of the lagoon walls and thereby create hazardous conditions.

Table 7.8 presents the most common problems and some possible solutions.[1,10,13]

7.5.1.3 Application of Lagooning in Water Reuse Systems

As mentioned in Chapter 4, during the last decade an increased number of studies conducted in different countries[1] have shown that stabilization pond systems in series can produce effluent with microbiological water quality suitable for unrestricted irrigation in compliance with the WHO[12] guidelines (3–5 log removal of fecal coliforms, hydraulic residence times from 20 to 90 days, and lagoon depth 1.2–1.5 m). For lower water quality requirements, e.g., restricted irrigation with only nematode parasite removal (WHO guidelines category B[12]), retention times are lower (10–15 days).

Stabilization ponds are used as common treatment systems on the France Atlantic and Mediterranean coast, in Argentina, Kenya, Tanzania, and have recently begun to be implemented in Morocco. This technology is recommended in Tunisia by ONAS for cities of between 2000 and 5000 inhabitants.

Nonaerated maturation ponds are used in a number of French plants for wastewater treatment and/or refining before reuse for irrigation. This technology, often designed for additional storage, is predominantly used in Atlantic coastal areas (Melle, Coullons, and the islands Noirmoutier, Ré, Mont-Saint-Michel and Porquerolles). On Noirmoutier Island, four maturation ponds with an overall volume of 0.193 Mm^3 have been installed for storage and disinfection before irrigation of high-quality potatoes. Another example is the tertiary lagooning in Clermont-Ferrand, used for polishing and disinfection of secondary effluent before irrigation of more than 750 ha of maize.

Other international examples of the integration of lagooning in water reuse systems include[14]:

In the city of Mindelo, Cape Verde, 2250 m^3/d, lagooning is applied for treatment and disinfection before agricultural reuse.

Maturation ponds are used for tertiary treatment of the urban sewage in Windhoek, Namibia.

Table 7.8 Some Common Operational Problems of Stabilization Ponds and Possible Solutions

Indicator	Probable cause	Solutions
High algal suspended solids in pond effluent	Weather or temperature that favors algae population	Draw off effluent from below the surface by use of a good baffling arrangement. Intermittent submerged rock filters or sand filter may be used. Provide shading, increase dissolved oxygen, add plankton or algae-eating fishes. Use multiple ponds in series. In some cases, alum dosages of < 20 mg/L have been used in final cells.
High weed growth	Poor maintenance	Periodic removing is the best method. Spray with approved weed-control chemicals. Lower water level to expose weeds, then burn. Increase water depth to above top of weeds. To control duckweed, use rakes or push a board with a boat, then physically remove duckweed from pond.
Scum formation	Pond bottom is turning over with sludge floating to the surface, poor circulation, wind	Use rakes, a portable pump to get a water jet, or motor boats to break up scum; broken scum usually sinks.
Odors	Overloading, poor circulation, weeds	Recirculate pond effluent to the pond influent to provide additional oxygen (1 : 6 ratio). Install supplementary aeration such as floating aerators. Apply chemicals such as sodium nitrate (1.3–4% per 1000 m^3) to introduce oxygen, repeat at a reduced rate on succeeding days.
Mosquitoes, midges	Poor circulation and maintenance	Keep pond clear from weeds and allow wave action on bank to prevent mosquitoes from hatching. Keep pond free of scum. Stock pond with *Gambusia* (mosquito fish). Spray with larvacide as a last resort; check with regulatory officials for approved chemicals.
Decreasing trend in pH	Overloading, long period of adverse weather, *Daphnia* feeding	pH should be on the alkaline side, preferably about 8–8.4. Apply recirculation. Check possible short-circuiting. Look of possible causes for algae die-off.

Source: Adapted from Refs. 10, 13.

One of the largest pond systems in Africa treats a dry weather flow of 80,000 m^3/d (peak capacity 240,000 m^3/d) from the city of Nairobi, Kenya. The treatment complex comprises eight parallel series of lagoons, including a primary facultative lagoon followed by a sequence of three maturation ponds. The effluent quality meets WHO guidelines for unrestricted irrigation ensuring >90% BOD_5 removal and >6 log reduction in fecal coliforms.

Another large system is operating in Mendoza, Argentina, where more than 140,000 m^3/d urban wastewater has been treated by lagoons with a total area of 290 ha before effluent reuse for irrigation of more than 2000 ha of forest, vineyards, fruits, and different crops.

The wastewater treatment plant of Al Samra, Jordan, provides recycled water for irrigation by treatment in series of stabilization ponds with daily flow rate of 120,000 m^3/d and total area of 181 ha.

Israel has extensive experience in the use of lagooning for agricultural reuse, both as seasonal long-term storage (Haifa) and wastewater treatment (Dan Region, Tel Aviv).

Some recent studies have indicated, however, that series of stabilization ponds were not completely efficient in removing pathogens from irrigation water, in particular when stringent water quality criteria are required for unrestricted irrigation.

7.5.2 Wetlands

Wetlands are another environmentally sound solution implemented in rural areas, sometimes in natural wetlands (marshes, bogs, swamps), and more frequently in constructed wetlands. The four general classes of constructed wetlands are[1,5]:

1. Surface flow marshes, very popular in the United States, with a mean surface of 2–5 ha and up to 400 ha
2. Vegetated subsurface flow beds, or reed beds, widely accepted throughout Europe, Australia, and South Africa
3. Submerged aquatic beds, which are less frequently used
4. Floating aquatics

The literature suggests that gravel-filled subsurface flow wetlands successfully remove protozoa and helminth parasites. For appropriate design and operation, the WHO[12] guidelines for unrestricted irrigation (category A) may be attained.[14] However, several field studies carried out in constructed wetlands for secondary treatment in Spain, as well as in Egypt, Uganda, and the United States, indicated that pathogen reduction (1–3 log reduction of fecal coliforms and coliphages) is not sufficient to satisfy water quality standards for either unrestricted and restricted irrigation.

It is important to stress that the spread of the use of constructed wetlands to developing countries has been depressingly slow, despite the favorable

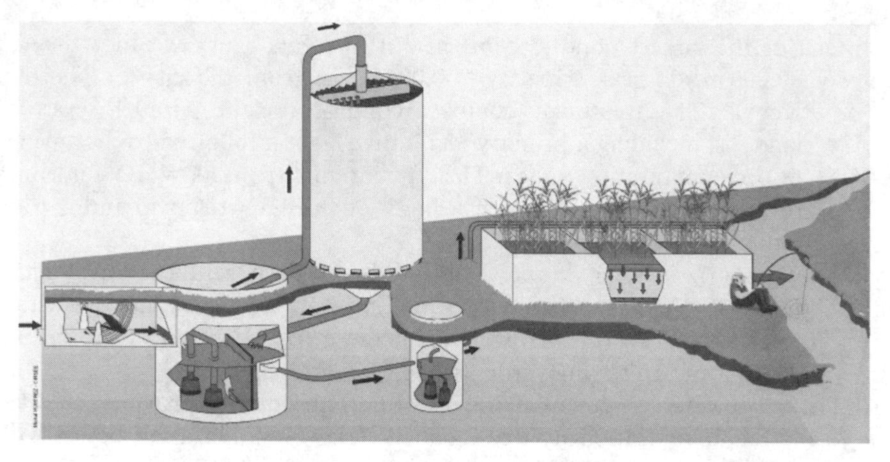

Figure 7.7 Integration of wetlands (Rhizopur system) as a polishing step in small-scale water reuse systems (Courtesy of Suez Environnement).

climate conditions. Designs appropriate to tropical and subtropical zones must be developed. A number of benefits can be achieved by implementing such technologies: more healthful environment, better water quality, sustainable cropping of plant biomass, waste recycling in agriculture, and irrigation. A good example is the use of constructed wetlands alone (rhizofiltration) or in combination with trickling filters for treatment of municipal wastewater in France at a small scale or rural areas. The optimized reed beds system Rhizopur® (Figure 7.7), developed by Suez Environment, can be used either as a main treatment or a polishing step, with surface areas of 2–2.5 m^2/inhabitant or 1–2 m^2/inhabitant, respectively.[15] A minimum of three reed beds is required for each stage of treatment. Plant density should be 4–5 plants/m^2. The recommended water feed frequency is maximum 4 floodings per day. The average capital costs reported are about 230 $/$m^2$, and operation costs vary from 15 to 25 $/inhabitant/year.

Wetlands need pretreatment, the operation of which is not easy to control. Odor and mosquito control is of primary concern. In areas where malaria occurs, wetland types with free water surface should not be implemented.

7.5.3 Infiltration-Percolation

Infiltration-percolation (Figure 7.8) is an efficient wastewater treatment process ensuring high treatment and disinfection efficiency with lower residence time—a few hours or days.[16] This treatment process acts as an aerobic biological filter and removes suspended solids, organic matter, nutrients, and microorganisms from the influent. It is based on flooding-drying cycles of deep sand infiltration basins. Disinfection efficiency is a function of the media, temperature, pH, and, most important, its saturation level and the hydraulic residence time.

Figure 7.8 An infiltration-percolation system in Pénestin, France.

Table 7.9 Main Characteristics of Infiltration-Percolation

Design parameters	Advantages	Drawbacks
Depth: 1–3.5 m	Easy to operate	Surface and media clogging
Surface area: 0.3–3 m²/inh	Robust to meet WHO guidelines	Effluent quality a function of the hydraulic loads, climate, and season
Minimum 2 units	Efficient for removal of helminths	Low tolerance of overloading (hydraulic and organic)
Intermittent water feed	Cost-efficient for small units	Filter performances highly influenced by sand quality (sand cleaning and strong selection required)
Recommended for small plants (<10,000 inhabitants)		Disinfection efficiency highly influenced by filter operation
Filter media: sand $d_{50} = 0.2$–0.7 mm with permeability 1.5–4.10^{-4} m/s		

Source: Adapted from Refs. 10, 11, 16.

Table 7.9 provides the main design parameters, as well as the advantages and disadvantages, of infiltration-percolation systems.[10,11,16] Enhanced disinfection is achieved if the filter bed remains unsaturated for a residence time of at least 30 hours (i.e., minimal time for a secondary effluent to run through a 4 m sand dune filtering bed at 0.5 m/d). Capital costs are typically in the range of 20–50 $/inhabitant, depending on plant capacity. The main advantage of infiltration-percolation is its simplicity of operation and the relatively low capital and operating costs. It requires less space than tertiary lagooning or soil-aquifer treatment. The most important disadvantage is the sensitivity of this process to variations in flow rates and water quality, especially suspended solid concentration.

In France, Morocco, and Spain, infiltration-percolation plants are small, with an average capacity of 600 m^3/d for secondary treatment plants and 3600 m^3/d for tertiary treatment plants. Larger plants exist in Israel and the United States.

7.5.4 Soil-Aquifer Treatment

Soil-aquifer treatment (SAT) provides wastewater purification during flow through unsaturated soils and the aquifer itself to recovery wells (Figure.7.9). In comparison with direct injection into the aquifer, this method ensures the treatment of lower-quality effluents without microbial contamination of groundwater. The removal efficiency for different pollutants (suspended solids, nutrients, pathogens, trace metals, and organics) varies widely with the type of soil, loading rate, and temperature. The hydraulic residence time varies from several months to one year.

As a rule, secondary to tertiary disinfected effluents are discharged to the unsatured vadose zone of permeable soil through recharge basins via intermittent wetting-drying operation mode (Figure 7.10). Complex treatment

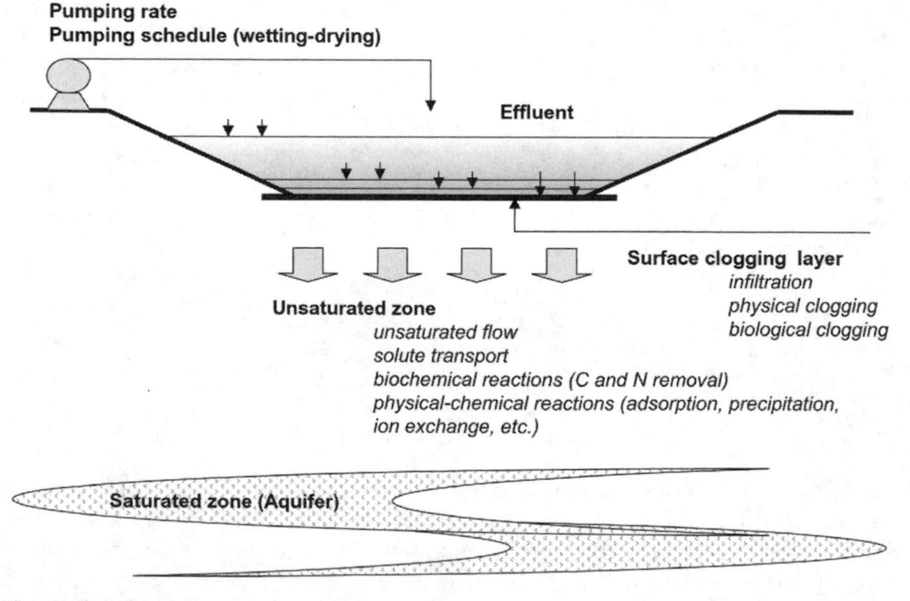

Figure 7.9 Soil-aquifer treatment schematic.

Figure 7.10 A recharge basin at the beginning of the filling cycle and a soil treatment machine: the Dan Region recycling plant, Israel (agricultural irrigation).

Table 7.10 Main Characteristics of Soil-Aquifer Treatment

Design parameters	Advantages	Drawbacks
Soil type: sand and sand-muddy soils	Provide large storage capacity	Surface and soil clogging
Infiltration rates: 0.15–0.3 m/d	Efficient for polishing and disinfection	Highly influenced by soil properties
Aquifer depth: >30 m	Easy to operate	Need for good knowledge of aquifer behavior
Infiltration basin depth: 0.2–3 m	Relatively low costs	
Infiltration basin bottom: sand		
Recharge cycles: 1–2 d recharge + 1–2 d drying to 2–3 weeks recharge + 2–4 weeks drying		

Source: Adapted from Refs. 1, 10.

mechanisms take place during the infiltration period, which include chemical precipitation, adsorption, ion exchange, biological degradation, nitrification, and denitrification. Water recovered after soil-aquifer treatment is of high quality and can be used for unrestricted agricultural irrigation.[17]

Table 7.10 summarizes the main characteristics of SAT: design parameters, advantages, and disadvantages. The most important operating parameter is the infiltration rate, which can be controlled by the duration of recharge cycles, periodic cleaning of the bottom layer of recharge basins, and periodic renewal of the upper soil layer. Low-duration cycles of operation (1–2 days for each step) limit system clogging and maintain aerobic conditions in the upper soil layers.

One of the largest water-reclamation plants with SAT treatment is implemented in the Dan Region, Israel,[18] serving a total population of about 1.3 million with an average wastewater flow of 270,000 m^3/d. Following activated sludge treatment, suspended solids are reduced by 98% to 7 mg/L, BOD is reduced by 99% to 6 mg/L, and ammonia is reduced by 78% to 8.2 mg/L. The polishing SAT step results in the total disinfection and drinking water quality of effluents. The main water quality characteristics of the SAT in the Dan Region are as follows[1,18]:

The concentrations of trace elements are below the maximum recommended limits for irrigation water used continuously on all soils, as well as below the maximum permissible limits for toxic substances in drinking water.

The bacteriological analysis of the water after SAT treatment shows no fecal or total coliforms and only 600 total bacteria per mL.

Two treatment lines have been constructed in the Dan Region: Soreq (1970, stabilization ponds in series and recharge, no longer in operation, US$170 million capital costs) and Yavne (1987, activated sludge and recharge, US$144 million capital costs). The operation of the recharge basins involves 1–2 days of wetting and 2–3 days of drying. The operation and maintenance costs for the actual treatment and reuse facility are 0.14 $/$m^3$ of treated effluents with an additional 0.233 $/$m^3$ amortization costs.

7.6 ADVANCED TERTIARY TREATMENT AND DISINFECTION

Advanced treatment processes (biological, physicochemical, membrane) are at the center of R&D programs for water reuse, their main objective being to ensure high and reliable water quality—chiefly removal of suspended particulate matter, viruses, and pathogens. Further improvement of water quality after conventional secondary treatment is achieved through tertiary treatments such as sand filtration, activated carbon filtration, coagulation-flocculation, low-pressure membrane filtration, electrodialysis, and reverse osmosis. In most cases, depending on the type of reuse application and the treatment previously applied, additional disinfection is necessary to satisfy wastewater reuse requirements. Bacteria, parasites, and viruses are among the key constituents that need to be removed or deactivated from the recycled water (see also Chapters 2–4).

Two types of process are commonly used for wastewater disinfection:

1. Physicochemical disinfection with processes such as chlorination, ozonation, and UV irradiation.

2. Membrane filtration processes, such as microfiltration, ultrafiltration, nanofiltration, reverse osmosis, and membrane bioreactors. These technologies guarantee high water quality and satisfy all current disinfection standards, including standards for virus removal (except for microfiltration), but require substantial investments.

The most common method of wastewater disinfection is chlorination.[19] However, the potential toxicity of chlorination by-products makes this process unattractive. UV irradiation has emerged in France and the United States as a viable alternative to chlorination, with a comparable and often more effective disinfection efficiency for control of viruses and bacteria. Ozone disinfection is becoming increasingly popular for wastewater treatment due to its high effectiveness in bacteria and virus inactivation.

Effluent toxicity, in particular disinfection by-products and other emerging micropollutants, is an issue of ever-increasing importance and could be included in future water reuse regulations (see Chapter 8). Traditional treatment methods have achieved limited success in the removal of toxic substances and emerging water quality parameters such as endocrine disruptors and pharmaceutical compounds. New advanced processes and new combinations of treatment technologies must be developed for complete removal of such trace compounds.

7.6.1 Tertiary Filtration

Filtration, a key step in producing high-quality recycled effluents, combines physical and chemical processes to remove solids from liquid phase. This technology has been used both as a final stage preceding disinfection before reuse or as one of a series of tertiary treatment processes. As a rule, filtration is used when the effluent maximum concentration limit is $\leq$10 mgSS/L. The principle of filtration consists of passing wastewater through a bed of granular media. For the removal of solids retained in the media and to avoid clogging, backwash flushes are used.

The direct application of water filtration technologies in water reuse systems has been unsuccessful because of the distinctive characteristics of wastewater solids. For this reason, specific filtration equipment has been developed for tertiary wastewater treatment and reuse using mono, dual, and multimedia filter beds. Some common combinations include anthracite and sand, activated carbon and sand, resin bed and sand, etc. The choice of filter vessel, either gravity or pressure, is generally determined by the role of filtration in the water reuse scheme in terms of interactions with other processes, as well as space availability and plant capacity.

A combination of filtration with biological oxidation in biofilm reactors has increasing application in water-reuse schemes. This chapter addresses only the application of filtration technology for the removal of suspended solids with or without the addition of chemical coagulation.

7.6.1.1 Treatment Efficiency of Tertiary Filtration

The performance of tertiary filtration is affected by many factors. For this reason, accurate design can only be achieved by means of pilot plant studies. It is important to stress that upstream secondary treatment greatly influences filtration efficiency.

Table 7.11 shows the treatment levels that can be reached by tertiary filtration in combination with the most common biological treatment processes.[4–6] The presence of algae impedes filtration of lagoon effluents. Chemical pretreatment is considered a good practice for such cases. In addition to these general parameters, tertiary filtration makes possible up to 50–90% removal of some heavy metals such as chromium, iron, manganese, and selenium. Preliminary coagulation may increase this removal rate by 20–38% for cadmium, copper, and lead.

With the increased regulatory requirements for unrestricted irrigation, filtration performances are generally evaluated using turbidity instead of

Table 7.11 Treatment Levels of Tertiary Filtration of Secondary Effluents

Treatment scheme	TSS (mg/L)	Turbidity (NTU)
Extended aeration + filtration	< 5	< 1–3
Extended aeration + coagulation + dual or multimedia filtration	< 2–5	< 0.5–2
Conventional activated sludge + filtration	5–10	< 1–5
Conventional activated sludge + coagulation + dual or multimedia filtration	< 5	< 0.5–5
Conventional activated sludge + filtration + activated carbon	< 2–5	< 0.5–3
High-rate trickling filter + filtration	10–20	—
High-rate trickling filter + coagulation + dual or multimedia filtration	< 5–10	< 2–5
Two-stage trickling filter + filtration	6–15	—
Two-stage trickling filter + coagulation + dual or multimedia filtration	< 5–7	< 1–5
Aerated/facultative lagoons + filtration	10–50	—
Aerated/facultative lagoons + coagulation + dual or multimedia filtration (low reliability possible because of algae presence)	< 5–30	—

Source: Adapted from Refs. 2, 5.

suspended solids (easy on-line control). Low values of turbidity (<2 NTU) can be achieved only when influent turbidity is low, 5–7 NTU, which corresponds to about 10–17 mg/L of total suspended solids (TSS). The following relationships[2] between turbidity and TSS can be used for settled secondary effluents:

$$\text{TSS (mg/L)} = 2.0 - 2.4 \times \text{turbidity (NTU)} \tag{7.1}$$

and for filtered effluents:

$$\text{TSS (mg/L)} = 1.3 - 1.5 \times \text{turbidity (NTU)} \tag{7.2}$$

7.6.1.2 Process Design

The design of tertiary filtration includes the following main steps[4,6]:

1. Pretreatment to enhance filterability: add inorganic or organic coagulants both upstream of the secondary clarifier and to the filter influent.

   Typical dosages of organic polyelectrolytes vary from 0.5 to 1.5 mg/L and 0.05 to 0.15 mg/L to settled or filtered influents, respectively. Use of proper coagulant dosage can be expected to provide effluent suspended solid concentrations of about 5 mg/L or less with proper media sizing and depth.
   In all cases, jar tests are recommended to determine optimum coagulant dosage. Overdosage may significantly impair operation. For daily influent concentrations of suspended solids of more than 30–50 mg/L, a preliminary treatment process of coagulation, flocculation, and sedimentation or flotation is recommended.

2. Choice of filter type and loading rates: in terms of driving forces, either gravity or pressure filters may be used. Multiple filters units are needed to allow continuous operation.

   Gravity filters are preferable for large plants with adequate capital resources. This process operates with lower filtration rates of 5–15 m^3/m^2 h and lower terminal head losses, typically 2.4–3 m.
   Pressure filters (Figure 7.11) can be operated with filtration rates of 25 m^3/m^2 h and terminal head losses up to 9 m without solid breakthrough. This technology is well appropriate for small plants and severe space limitations. This type of filter requires complex inspection, operation, and maintenance.

3. Media selection and characteristics: medium selection concerns media size, shape, composition, density, hardness, and size.

Figure 7.11 Pressurized tertiary sand filter (Courtesy of Degrémont).

The depth of solids penetration into a filter depends principally on the size of the filter medium. Poor filtrate results if the medium is too large, and surface clogging occurs for small media.
The effective size and required depth are interrelated. As a good practice, the minimum depth of the finest medium is at least 0.15 m, with typical design values of 0.5–0.9 m of sand and anthracite for conventional sand filters and 0.15–0.6 m layers of sand and anthracite for dual-media filters.
The minimum size d_{50} is 0.35 mm with typical design values of 0.4–0.8 and 0.8–2 mm for sand and anthracite, respectively. Tertiary filters rarely have a coarser medium than 2 mm.
Specific gravity of typical materials is 4.2 g/cm^3 for garnet sand, 2.6 g/cm^3 for silica sand, and 1.6 g/cm^3 for anthracite.

Slow sand filtration is no longer used for tertiary treatment. Rapid sand filters typically incorporate up to 0.6 m of sand supported by a gravel bed and are characterized by filtration rates in the range of 5–15 m^3/m^2 h and effluent suspended solids concentrations of 5–7 mg/L. Conventional rapid sand filtration is avoided in wastewater applications because of its susceptibility to rapid clogging. Deep-bed, coarse-medium filters typically involve single-medium bed depths of 1.2–1.8 m with typical sand diameter of 1.48 mm. A highly uniform sand (1.1 uniformity) is best for allowing full use of the bed. Flow rates up to $50\,m^3/m^2\,h$ have been reported for tertiary treatment, with nominal values of $15–20\,m^3/m^2\,h$. This technology is ideally suited for pressure applications, but a major disadvantage is that it requires deeper filter compartments, exceeding an overall depth of 3.6 m even in gravity applications.

Early problems with effective bed cleaning and economical volumes of backwash water seem to be solved by using an air backwash at $100–150\,m^3/m^2\,h$ with a water rinse at $15–20\,m^3/m^2\,h$. Total backwash water consumption is $6–8\,m^3/m^2$. Moving bed continuously operated filters (up or downflow moving beds or pulsed beds) have been developed specifically for wastewater treatment and reuse purposes (Figure 7.12).

7.6.1.3 Operational Problems of Tertiary Filtration

The most common operational problems of tertiary filtration and some solutions are given in Table 7.12[13]. Regular backwashing of filter media is the key to maintaining good filtration efficiency.

Figure 7.12 Tertiary moving bed sand filter (DynaSand) filter installed before ozonation of recycled water: demonstration plant for agricultural irrigation and aquifer recharge near Thessaloniki, Greece.

Table 7.12 Some Common Operational Problems of Tertiary Filtration and Possible Solutions

Indicator	Probable cause	Solutions
High effluent turbidity	High hydraulic and organic loading Improper upstream coagulation Algal growth	Backwash filter. Run jar tests and adjust coagulant dosage. High turbidity is not associated with a corresponding increase of head loss: chlorine addition is the solution in this specific case.
Short filter runs due to head loss	Surface clogging Inadequate filter cleaning	Increase surface wash cycle length. Reduce solids loads by improving upstream treatment. Replace sand media with dual media; bump filter with short backwash. Increase backwash duration and rate.
Mud ball formation on filter surface	Inadequate backwash and surface wash	Increase backwash duration and rate and length of surface wash cycle.
Loss of media during backwashing	Backwash rate too high Washwater troughs not level Surface wash cycle too long Uneven distribution of backwash water	Reduce backwash rate. Adjust washwater troughs Reduce length of surface wash so that it goes off at least 1 minute before backwash is complete Clean filter underdrains.
Excessive amount of backwash water (>5%)	Too high solids content. Surface wash system has failed. Inadequate surface wash. Excessive length of backwash	Improve upstream treatment (coagulation and settling). Repair surface wash system. Increase duration of surface wash. Reduce backwash cycle length.

Source: Adapted from Ref. 13.

7.6.2 Chlorination

Chlorination, the most universally practiced wastewater disinfection method since the late 1940s, plays a major role in preventing waterborne infectious diseases throughout the world. This disinfection process is also commonly used in water-recycling facilities (Figure 7.13). According to recent studies,[20] in 1990 78% of U.S. publicly owned treatment works were disinfecting treated

Figure 7.13 Plug-flow chlorine contact basins in Irvine Ranch, California (agricultural and landscape irrigation).

wastewater with chlorine or hypochlorite, and 28% of these practiced dechlorination. Use of dechlorination avoids by-product toxic effect but increases disinfection costs by up to 20–30%.

Several chlorine derivatives can be applied, such as gaseous chlorine, hypochlorite, or chloramine compounds. Numerous U.S. facilities have replaced gaseous chlorine with hypochlorite in order to improve operator safety and decrease operation and maintenance costs.

A number of theories have been proposed to explain the disinfection effect of chlorine and its related compounds, including oxidation of protoplasm, protein precipitation, modification of cell wall permeability, and hydrolysis. While all these effects may be operative, the predominant mechanism must depend on the type of organism, its physiological state, and specific conditions such as wastewater quality and environmental conditions (pH, T°C).

7.6.2.1 Design and Efficiency of Chlorination

The effectiveness of chlorine for disinfection depends on numerous variables, including temperature, pH, contact time, chlorine dose, turbidity, presence of interfering substances, and degree of mixing.

Facilities are normally designed to provide a minimum of 15–30 minutes of contact time at peak flow. For water reuse, the required contact time can be up to 90–120 minutes, as recommended, for example, by U.S. regulations.[21,22] Under these conditions, good inactivation of total and fecal coliforms occurs (4–6 log removal), as well as good virus removal.

Lower disinfection doses and contact times are characteristic of good-quality effluents, e.g., filtered activated sludge effluent. Organic and humic compounds or nitrites increase chlorine demand. Suspended solids also affect chlorine disinfection efficiency, because microorganisms are shielded by or embedded in particles. Typical chlorine doses for municipal wastewater disinfection are about 5–20 mg/L and 30–60 minutes of contact time[19] and usually allow for compliance with permits on conventional bacterial indicators (coliforms, *E. coli*). Higher doses are required for low-quality wastewater, such as primary or trickling filter effluents.

The required chlorine doses for disinfection of recycled water for irrigation purposes fall in the following ranges:

Primary effluents: 8–30 mg/L
Chemical precipitation: 2–6 mg/L
Trickling filter effluents: 5–15 mg/L
Activated sludge effluents: 2–9 mg/L
Nitrified effluents: 2–6 mg/L
Tertiary filtered effluents: 1–6 mg/L

As mentioned previously, wastewater chlorine requirements vary considerably depending on effluent quality. Organic compounds, industrial wastes, and ammonia concentration can strongly affect chlorine demand. Improving mixing characteristics of chlorine contactors and well-adapted process control strategy can enhance the effectiveness of chlorination.

Very high free chlorine concentrations may be needed to inactivate cysts and some viruses of concern. It has been observed that both ciliates and amoebae could withstand free chlorine residuals of 4 and 10 mg/L (pH 7.0, 25°C), respectively, and were still motile in chlorine solutions with lower residual concentrations after 30–60 minutes of exposure.[23] It was found that free chlorine produced no measurable inactivation of *Cryptosporidium parvum* oocysts by 4 or even by 24 hours.[24]

The main disadvantage of chlorine disinfection is the generation of toxic by-products (DBP). This phenomenon has been discussed since the 1970s, when naturally occurring organic materials in some water sources were found to react with chlorine and form carcinogenic trihalomethanes and other compounds (haloacetic acids and dissolved organic halogens). More recently, DBPs are becoming a great concern in enhancement of environmental and public health protection (see also Chapter 8). DBPs have been identified as potential human carcinogens and harmful for the environment at very low concentrations, as low as 0.1 mg/L.[25,26] It is important to stress that the presence of small concentrations of residual ammonia and low DOC (dissolved organic carbon) reduces by-product concentration.

Another important concern is the impact of chlorinated effluents on receiving water ecosystems. The major source of acute toxicity in chlorinated effluents is residual chlorine: chlorine residue at concentration levels as low as 0.002 mg/L is highly toxic for aquatic systems. For this reason, dechlorination is often requested for such effluents. Historically, most municipal wastewater treatment plants used sulfur dioxide gas to dechlorinate effluents. In addition to being a toxin, sulfur dioxide has been proven to be a carcinogen, and many plants are now using other sulfur-based reactants. Sodium bisulfite solutions are now commonly applied and have proved to provide fair dechlorination. However, after considerable research and full-scale demonstrations, some unsolved problems persist. Although most toxicological studies have shown that dechlorination reduces the toxicity of chlorinated effluents, some residual chlorine exceeding EPA regulatory limits may remain in dechlorinated effluents.[27] Moreover, dechlorination consumes dissolved oxygen, and reaeration becomes sometimes necessary. For this reason, some water reuse regulations include requirements for oxygen saturation (see Chapter 3, Table 3.7).

Disinfection efficiency and the possibility of maintaining residual chlorine are the main advantages of chlorine disinfection. Chlorination systems are reliable and flexible and the equipment is not complex.

The main disadvantages of chlorine disinfection can be summarized as follows:

- Production of toxic by-products
- Poor inactivation of spores, cysts, and some viruses at the low dosages used for coliform removal
- Stringent safety regulations leading to high investments for scrubbing systems and other safety equipment
- Need for dechlorination, increasing disinfection costs by 20–30%

7.6.2.2 Chlorine Residual

Depending on the type of reuse, a chlorine residual can be required. The two types of chlorine residuals that can be monitored are free and total chlorine.

A free residual, the hypochlorite ion (OCl^-) attained after breakpoint chlorination, may be present in secondary and tertiary effluents with low ammonia concentration. Much more chlorine is required to achieve a free residual than for attaining a residual. After the breakpoint is reached, when all of the chlorine demand has been satisfied, each mg/L of added chlorine dosage adds one mg/L of chlorine residual.

Chlorine demand represents the difference between the concentration of free chlorine applied (hypochlorous acid HOCl and hypochlorite ion OCl^-) and the concentration of chlorine residual remaining at the end of the contact period. Consequently, chlorine demand represents the chlorine quantity that is chemically reduced or converted to less active forms of chlorine by substances in the water, such as ammonia compounds, organic matter, iron, and sulfur compounds. In the case of partial nitrification, the chlorine demand exerted by nitrite may consume significant quantities of chlorine, evidenced by an increase in chlorine demand.

A total residual, also referred as combined residual, is present in any chlorinated effluent. This residual, therefore, is considered as the standard form of chlorine residual. Total chlorine residual includes chloroorganics and chloramines, both of which are less powerful oxidizing agents than free chlorine residual. Nevertheless, the disinfection activity of chloramines persists longer that of free chlorine. The optimum chlorine residual, free or combined, is that sufficient to produce effluent with the required microbiological quality.

7.6.3 Chlorine Dioxide

Chlorine dioxide has some serious advantages as an alternative to chlorine. It has been tested and applied for wastewater disinfection for reuse purposes in China, France, Israel, and the United States.

Many studies define chlorine dioxide as a more effective disinfectant than chlorine for both bacteria and virus removal in a broad range of pH. Spores and cysts may also be effectively inactivated. The required pathogen removal for unrestricted irrigation (up to 4–5 log fecal test germs) has been achieved with ClO_2 doses of 2–5 mg/L and contact times of 5–15 minutes. These design parameters allow removing enterobacteria by 2.4–4 log and *Clostridium perfringens* spores by 1.5–2 log. Higher disinfection efficiency of ClO_2 has been reported in the inactivation of highly resistant viruses such as adenovirus, coxsackievirus, poliovirus, herpes simplex, etc.[28,29] Another important advantage is that the germicidal efficiency of chlorine dioxide is not affected by the presence of ammonia.

7.6.4 UV Disinfection

Ultraviolet irradiation is the most commonly used alternative to chlorination in water reuse systems, with comparable and often more effective disinfection efficiency for virus and bacteria control. Presently, thousands of installations worldwide (>60 in the United Kingdom, > 2000 in the United States) practice UV disinfection for water reuse purposes or protection of sensitive and bathing zones using mostly open-channel facilities equipped with low- or medium-pressure mercury arc lamps.

The success of UV technology in water reuse schemes is largely attributable to low cost, as well as the absence of toxic by-products. Numerous pilot and full-scale studies have indicated that UV disinfection can consistently achieve the water quality objective of <200 FC/100 mL.[19,30] Moreover, many UV systems have met the stringent Title 22 requirement in California for unrestricted irrigation with recycled water of 2.2 TC/100 mL.

The major advantages of UV disinfection compared to chlorination are its simplicity, minimal space requirement, and absence of toxic by-products. As with chlorine, amebic and some protozoan cysts are the most difficult to inactivate. UV systems can be recommended as a competitive solution for disinfection of secondary and tertiary effluents before reuse for unrestricted irrigation.

As mentioned previously (Chapter 4, Table 4.1), the main advantages of UV technology are low cost and absence of any toxic by-products (Table 7.13). It is also easy to use, with low maintenance requirements, and is readily adaptable for automation.

Table 7.13 Main Advantages and Constraints of UV Disinfection

Advantages	Constraints
No by-product generation	Dose difficult to measure
Cost efficient (similar to chlorination)	Lamp aging and fouling difficult to assess
Easy to operate	No disinfectant residue
High efficiency for inactivation of viruses, bacteria, and cysts (*Cryptosporidium*)	Efficiency depends on water quality (removal of suspended solids is required for complete disinfection)
Low footprint	Lamp replacement and disposal concerns

From an operational point of view, the major constraint related to all UV equipment is the lamp cleaning due to fouling of quartz sleeves. Depending on the type of lamp and plant size, automatic wiper systems or manual cleaning with pressurised water or citric acid are implemented. There is as yet no reliable method to predict the coating of quartz sleeves, except for case-by-case pilot studies. A failure of individual UV lamps normally does not affect disinfection performance as long as no more than 5% of the UV lamps are defective, while a failure of an entire UV module leads to a decrease in disinfection efficiency.

Another disadvantage from the point of view of operation is lamp life-time, which is between 6 and 12 months. Lamp disposal is a major concern because of mercury content: for example, low-pressure lamps contain 50–100 mg of mercury (Hg), which cannot be easily disposed of in some countries

Photoreactivation and dark repair have been reported as another disadvantage of UV disinfection. Regrowth depends on the UV dose and reaches a maximum of 1 log increase after irradiation of up to 40 mJ/cm.[1,31] For higher UV doses, photoreactivation is negligible or does not occur. No relationships were observed between repair and suspended solids or UV transmittance in the range of 10–60 mg/L and 10–80%, respectively. No significant repair has been observed for microorganisms with health implications such as fecal streptococci, *Salmonella*, and somatic coliphages. A consensus does not exist among engineers and regulatory agencies regarding the inclusion of repair in UV system design.

With the increasing popularity of UV disinfection, the State of California has developed guidelines for qualification tests for UV equipment in water reuse applications. The latest revision of the California NWRF UV guidelines[32] includes viruses as target microorganisms for qualification tests. Among other requirements, a 5 log inactivation of vaccine-strain poliovirus is required. Other test microorganisms may be used if they are proven to be more resistant to UV than vaccine strain polioviruses. Bacteriophage MS2 is the most commonly used microorganism for UV equipment qualification in the United States and in the State of California.[33]

7.6.4.1 UV Technologies Used in Water Reuse Systems

UV disinfection equipment may be divided into two main categories, depending on the type of lamp featured in the system: low pressure and medium pressure. The major difference in these technologies is the nature of the radiation, monochromatic at 253.7 nm for the low-pressure lamp and polychromatic for the medium-pressure lamp. Because 253.7 nm radiation is near the peak absorption of most target biomolecules, low-pressure lamp systems are more energy efficient than medium-pressure systems. However, low-pressure lamps also produce less intense light, and thus require more lamp and larger installations than medium-pressure systems. Recently developed low-pressure, high-intensity UV lamps (Figure 7.14 and Figure 7.15) offset this disadvantage.

The germicidal effects of UV light are a result of photochemical damage to RNA and DNA, leading to effective cell inactivation. If damaged cells replicate, they produce mutant daughter cells that are unable to replicate.[34] However, photoreactivation of damaged pathogens can occur and has been reported in several studies.

Generally, open channel gravity flow systems have been used for wastewater disinfection. By far the most common type of UV disinfection system is the low-pressure, low-intensity system, which represents more than 90% of the installations in North America and offers the best electrical efficiency of any UV system. Low-pressure lamps have a simple straightforward design. Despite their lower efficiency, medium-pressure lamps are more suited for high through-flow rates, using significantly fewer lamps. For low flow rates ($<4000\,m^3/h$), low-pressure systems are more competitive. For flow rates of 4000–40,000 m^3/h, either technology can be chosen, depending on site-specific criteria. Medium-pressure systems are especially suitable for a combination of high flow rates ($>40{,}000\,m^3/h$) and low power costs. The new generation of low-pressure, high-output lamps represents a promising system.

Figure 7.14 An open-channel UV system with low-pressure, high-intensity horizontal UV lamps (Wedeco system courtesy).

Figure 7.15 An open-channel UV system with low-pressure, high-intensity vertical UV lamps, Aquaray 40 H0 VLS (IDI-Ozonia courtesy).

7.6.4.2 Influence of Water Quality on UV Design

The effectiveness of UV disinfection is greatly affected by the quality of the effluent[30]: higher doses are required with increasing concentrations of suspended solids, UV absorbance, and some inorganic matter that absorbs UV light such as ferric ions (Figure 7.16). Moreover, physical-chemical characteristics of wastewater also greatly influence lamp fouling. No single parameter can be directly related to disinfection performance, and more work is required to fully explain experimental behavior.

Typically, UV transmittance and particle concentration (suspended solids or turbidity) should be considered the major water quality parameters that affect disinfection efficiency. Table 7.14 provides common values of effluent turbidity and transmittance for evaluating UV light treatability.

The qualitative impact of relevant wastewater constituents or bulk parameters on UV and ozone treatment processes is compared in Table 7.15

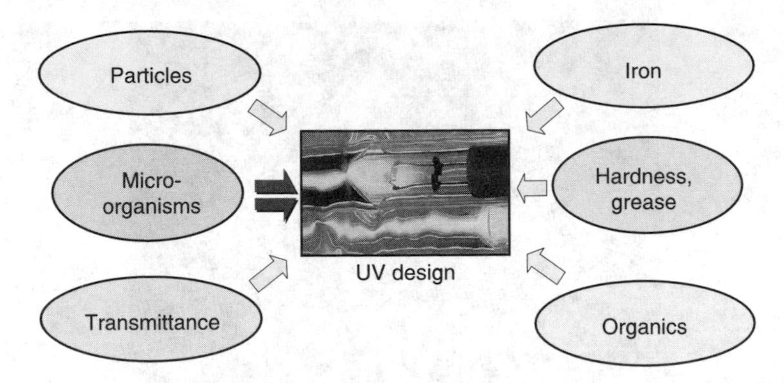

Figure 7.16 Water quality parameters influencing UV disinfection design.

Table 7.14 UV Transmittance and Turbidity of Several Effluents to Consider for UV Design

Parameter	Primary effluents	Secondary effluents	Tertiary effluents	MF/RO effluents
UV transmittance, %	45 (20–55)[a]	60 (35–89)	70 (50–90)	>90
Turbidity, NTU	>15	8 (1.5–20)	1 (0.5–5)	<0.5

[a]mean value (range of variations).

and discussed below. Although these parameters are known to have the greatest impact on disinfection efficiency, efforts to establish any relevant correlations remain unsatisfactory. Empirical or semi-empirical formulas have been used, but they are site-specific and include constants that need to be determined with laboratory or field tests. Therefore, sole water quality data cannot be used to predict process efficiency. When high accuracy is required, pilot testing remains necessary.

Temperature

Low-pressure lamps are effective energy-wise and operate at small wall temperatures (50–200°C for conventional and high-output technologies). This can make them sensitive to variations in effluent temperature in terms of UV-C output. Most systems today feature lamps that are stable with variations in wastewater temperature. The effluent temperature will not impact fouling of these lamps because their operating temperature range remains low.

UV Transmittance

The passage of light through wastewater is affected by the presence of some dissolved compounds and particles. Particulate matter can partly or totally absorb light and/or scatter it. Light availability (irradiance) in the reactor is

Table 7.15 Qualitative Impact of Wastewater-Quality Parameters on UV and Ozone Disinfection Processes

	Impact of wastewater quality		
	UV disinfection		Ozonation,
Quality parameter	Efficiency[a]	Fouling	efficiency
pH	–	–	–
Temperature	+	–	+
UV transmittance	+++	+	+
Particles			
Particle count	+++	–	+++
TSS	+++	–	+++
Turbidity	+++	–	++
Dissolved organic matter			
COD	++	+	++
BOD_5	++	+	++
TOC - DOC	++	+	++
Dissolved inorganic matter			
Iron III	++	+++	–
Iron II	–	–	+
Manganese	+	–	+
Hardness	–	++	–
Alkalinity	–	+	+
Nitrites	+	–	+
Bromide	–	–	+
Chlorine	+	–	+
Grease	+	+++	+

–, no impact; +, ++, +++, small, medium, and strong impact.
[a]With clean lamps.

crucial for UV disinfection in the sense that it governs the dose delivered by the system. Because of the presence of particles that may scatter light out of the detector's reach in a conventional spectrophotometer, it is recommended to measure UV transmittance by spectrophotometry with an integration sphere or by actinometry. These methods require special equipment. In the USEPA UV Guidelines,[33] an empirical formula is given to correct the deviations of a conventional spectrophotometer when the purpose is the estimation of doses in reactors with low-pressure lamps:

$$T = \exp\left[-0.6^{*}\left(\left(-\ln T_{\text{unfiltered}}\right)^{0.64}\right)\right] \tag{7.1}$$

where:

T = corrected transmittance, to be used for calculation of doses

$T_{\text{unfiltered}}$ = transmittance of raw sample

This correction can be used with reasonable error in a large range of transmittance values. For medium-pressure lamps, the estimation of UV dose is more complex because the germicidal dose is delivered over a variety of wavelengths.

Particles

Particles not only consume disinfectant (UV light), they can also embed microorganisms and protect them. In a first approximation, the effect of particles is a function of their quantity, which can be measured with a particle counter. If this apparatus is not available, total suspended solids (TSS) or turbidity can be used, with some restrictions. TSS gives the total mass of particles but does not give a number of particles, so it does not correlate with the number of colonies that will be counted on a culture media. On the other hand, the size of the particles that provoke turbidity might not be in accordance with the particle size range that is critical for disinfection (viruses, bacteria).

In a number of studies[21,31,35] pathogen inactivation has been found to significantly decrease as the TSS concentration of the effluent increases. More precisely, particle size distribution was demonstrated to be an important parameter. It was observed that disinfection performance was impacted adversely as the number of larger particles increased, indicating the critical role of polymer addition upstream of the filters. An investigation of the total number of coliform bacteria-associated particles in secondary and tertiary effluents indicated that 93%, 69%, and 10% were found to occur between 10 and 40 μm, 10 and 20 μm, and 10 and 11 μm average particle diameters, respectively.[35]

It is important to stress that coliform bacteria located within particles are the most resistant to UV disinfection. Great care must be taken to ensure that upstream treatment facilities are properly designed and operated. Undersizing of clarifiers, process interference, (e.g., rising sludge) and lack of appropriate maintenance (filamentous growth) lead to an increase in suspended solids and attached bacteria that will inhibit UV disinfection performance.

Dissolved Organic and Inorganic Matter

The parameters listed in these categories impact disinfection mainly in that they compete with microorganisms for disinfectant. At 254 nm, many dissolved organics absorb radiation, especially those with conjugated rings. Similarly, dissolved inorganics such as ferric ions, nitrites, bromine, manganese, and sulfates are strong absorbers at that wavelength. Textile, food processing, paper, or pharmaceutical industries also produce wastes with low UV transmittance. Other dissolved constituents contribute to fouling, such as hardness and ferric ions.

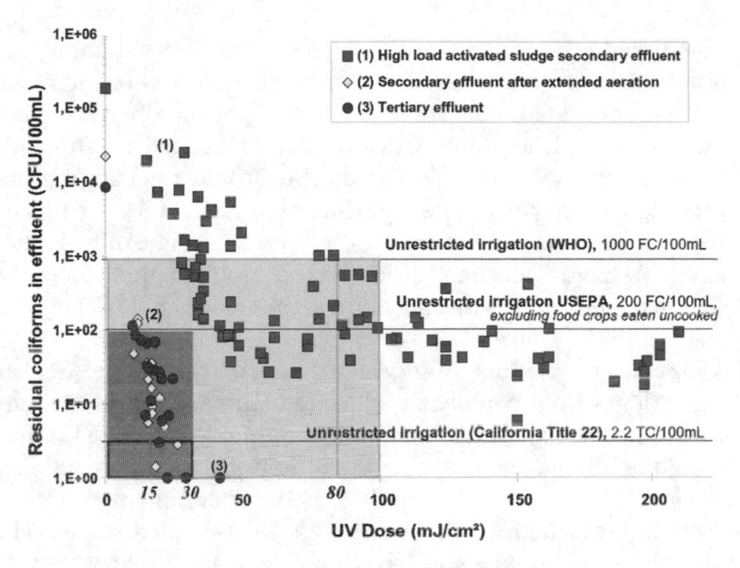

Figure 7.17 Performance of UV irradiation for fecal coliform removal from (1) secondary, (2) high-quality secondary, and (3) tertiary effluents with different requirements for unrestricted irrigation.

7.6.4.3 Influence of Upstream Treatment

The influence of upstream treatment on the design of UV disinfection (open-channel low-pressure UV lamps) is illustrated in Figure 7.17.[19,30] In agreement with literature data, these results indicate that standard UV doses of about 30 mJ/cm^2 are enough for good quality secondary or tertiary effluent (with low concentrations of particles) to comply with both WHO[12] (1000 FC/100 mL) and USEPA Reuse Guidelines[21] (200 FC/100 mL, excluding food crops eaten uncooked) for unrestricted irrigation, removing 3–5 log of coliform bacteria. Moreover, extended disinfection can be achieved in tertiary effluents in compliance with the stringent California Water Recycling Criteria.[22] It is important to underline that much more conservative minimal doses of 140 and 100 mJ/cm^2 at weekly and daily peak flows, respectively, are recommended by California regulations in order to offset wastewater quality fluctuations. In all cases, poor-quality secondary effluents such as those after high load activated sludge require significantly higher UV design doses of 80–100 mJ/cm^2 even for the less stringent WHO disinfection objective. Compliance with more stringent water quality requirements cannot be reached without complementary tertiary treatment such as coagulation/flocculation followed by multimedia filtration or chemically enhanced filtration.

The influence of upstream treatment on UV disinfection design can be summarized as follows[1,35]:

Generally, for efficient UV disinfection, the average value of influent transmittance should be more than 50% and residual concentration of suspended solids less than 15 mg/L.

Activated sludge processes exhibit a wide range in treatability by UV disinfection, associated mainly with particle count and transmittance. Particle-associated coliforms decline exponentially with increasing mean cell residence time. Consequently, extended aeration produces high-quality and easily disinfected effluent that can comply with stringent regulations for extended disinfection (2.2–23 TC or FC/100 mL). High-load systems and those with cell residence time of less than 8 days have problems meeting strict disinfection requirements, but are expected to comply with moderate standards such as WHO for < 200 and <1000 FC/100 mL.

Trickling filters were not observed to be capable of meeting stringent regulations. Even complying with moderate disinfection requirements may be problematic. At a minimum, laboratory collimated beam testing should be conducted on such effluent to evaluate UV treatability.

Lagooning (i.e., non–floc-forming) led to formation of fewer coliform bacteria associated with particles that did activated sludge. However, these effluents are likely to have low UV transmittance. Provided that adequate UV light could be applied to such effluents, extended disinfection can be achieved.

If filtration is to be used to improve UV disinfection, an accurate assessment of the particle size distribution must be obtained. The possibility that filtration may break some large particles that contain numerous coliform bacteria into many small particles should be considered. Alternatively, filtration may result in more small particles being removed than expected due to the effect of autofiltration (i.e., straining of small particles by filter cake). Prudent pilot testing of wastewater of interest is recommended.

7.6.4.4 Influence of Microorganisms

Recently, it was reported that the agents involved in waterborne diseases from water recycling in urban areas are predominantly viral. Considering the inactivation kinetics of several microorganisms, it was concluded that the use of bacterial indicators alone to determine the effect of wastewater disinfection when using chlorination or UV irradiation underestimates human enteric virus inactivation. However, the analysis of enteric viruses is time-consuming, expensive, and requires specialized laboratories. Thus, some regulations recommend the use of viral model organisms for the design or monitoring of wastewater disinfection processes, in the same way that coliform bacteria are used as model organisms for bacterial enteric pathogens.

MS2 bacteriophage is being increasingly used as an indicator for UV system performance and plant design. MS2 are F-specific bacteriophages (or male-specific), i.e., they adsorb on the F pilus of their bacterial host (coliforms) as the first stage of infection. They are RNA phages; their genetic material consists of a single strand of RNA. This bacteriophage is considered one of the best microorganisms to serve as an indicator of viral inactivation, because it

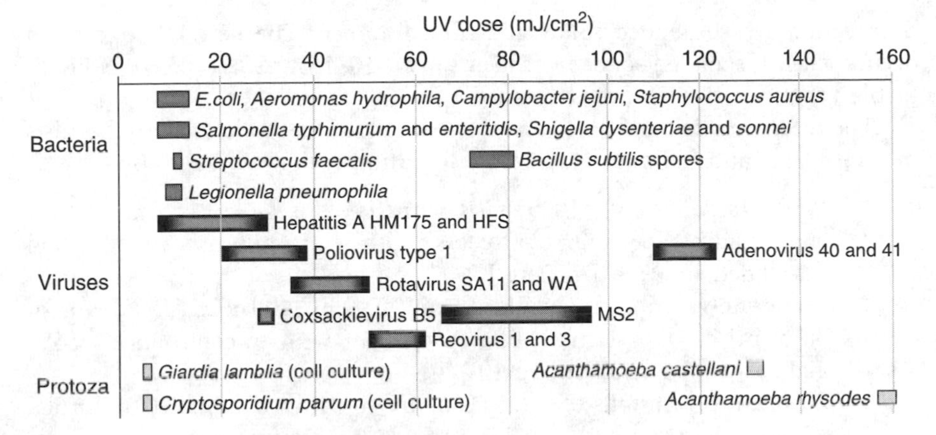

Figure 7.18 UV dose requirements for 4-log inactivation of different microorganisms.

resembles human enteroviruses (polioviruses 1–3, coxsackieviruses, echoviruses, and enteroviruses 68–72) in both their structure and function. On the other hand, analysis of the available literature[36] shows that MS2 bacteriophage is likely to give a conservative estimate of virus inactivation and is extremely resistant to UV light (Figure 7.18).

The choice of microbial culture also influences the measured UV dose. Pure cultures are more sensitive than indigenous viruses or bacteria isolated from sewage effluents.

7.6.4.5 UV Design and Operation

Manufacturers use calculation methods to evaluate average doses delivered by their system according to flow rate. This often results in over- or undersizing. The dose effectively received by the water during its residence time in such a reactor cannot be accurately described by a single average value, as would be the case in laboratory collimated beam experiments where hydrodynamic conditions are totally controlled. The water flow pattern and the difference in intensity field result in a distribution of UV doses over a cross section. When a reduction of more than 2 log units is sought, which is usually the case in wastewater disinfection, the presence of weaker irradiation zones within the reactor (due to low intensity, high velocity, or both) has a strong effect on the global disinfection performance. For this reason, the sizing of full-scale UV plants is usually achieved on the basis of pilot plant studies.

Recent advances in process modeling have allowed for accurate predictions of process behavior on the basis of the UV dose-distribution concept. The combination of dose-response curves from collimated beam tests and of the calculated dose-distribution curve allows predicting more precisely coliform inactivation inside the UV unit.

As indicated previously, wastewater quality has a great influence on UV disinfection performance. The two parameters that can most inhibit UV disinfection are suspended solids and transmittance. The best UV operation occurs when suspended solids in effluents are <10–15 mgSS/L (<5–10 NTU of turbidity) and >55–60% transmittance.

The recommended UV doses under maximum day flow defined on the basis of pilot plants and full-scale studies and literature review are as follows[30,31,36]:

1. WHO guidelines of <1000 FC/100 mL: UV doses of 30–50 mJ/cm^2 for good-quality secondary effluents with transmittance >55% and suspended solids <10–15 mg/L
2. Moderately stringent standards of <200 FC/100 mL: UV doses of 30–50 mJ/cm^2 for tertiary filtered effluent with transmittance >55% and suspended solids <10 mg/L
3. Stringent regulations of <2.2 TC or FC/100 mL or similar (California, Florida, Australia):

   Granular filtration: UV dose should be at least 100 mJ/cm^2, filtered transmittance >55%, suspended solids <5 mg/L, turbidity <2 NTU (24 h average, max 5), 5 log inactivation of polioviruses

   Membrane filtration: UV dose should be at least 80 mJ/cm^2, filtered transmittance >65%, turbidity <0.2 NTU (95% of time, max 0.5), 5 log inactivation of polioviruses

   Reverse osmosis: UV dose should be at least 50 mJ/cm^2, filtered transmittance >90%, turbidity <0.2 NTU (95% of time, max 0.5), 3 log inactivation of polioviruses

The design UV dose must be based on a 50% UV lamp output after an appropriate burn-in period and 80% transmittance through quartz sleeve for manually and automatically cleaned systems.

Reliable operation of UV disinfection units requires proper training and timely maintenance and calibration of system components. The presence of mercury is of concern because it can be detrimental to public health and aquatic life.

UV disinfection systems must be capable of producing disinfected effluent during any component failure prior to distribution. A minimum of two operating reactors per train ensures that some disinfection occurs until a standby reactor is brought on-line. Reliable power supply and back-up power are essential to ensure continuous disinfection.

The most common operational problems of UV disinfection and some solutions are given in Table 7.16.[13,37]

7.6.5 Ozonation

Wastewater ozonation has been used in the United States since the early 1970s as an alternative to chlorine for wastewater disinfection. More recently, municipal wastewater ozonation treatment facilities have been constructed in the Middle East, South Africa, France, and Spain.[19] The treatment of primary effluents is a relatively new application of ozonation under intensive

Table 7.16 Routine Operation and Problems of UV Disinfection and Corrective Actions

Indicator	What to check	Potential problem	Explanation	Corrective actions
Ballast card failure	Surface T°C Proper grounding	Overheating due to poor ventilation Frequent failures	The lack of adequate ventilation and air conditioning during summer, and the associated high temperatures (> 30°C), lead to damage of the electronic compounds of the ballast cards. The life cycle of the ballast cards is at least 3–4 years, and even 10 years according to some suppliers. Besides heating, the ballasts are very sensitive to electromagnetic disturbance due to poor grounding or poor connections.	Add panel ventilation or cooling system Adequate grounding according to supplier recommendations Change ballast cards if necessary
High frequency of lamp replacement	Lamp status Heat build-up GFI indicator O-rings	Burned out Little or no flow Broken quartz sleeve or seal system failure Water leaks Malfunction of level control system	Poor compression of O-rings (poor gasket clamping) and associated water intrusion into the quartz sleeves leads to premature lamp burnout. High frequency of shutdown of UV lamps and short operating cycles lead to vaporization of tungsten and other oxides from the lamp electrodes. Another aging mechanism is associated with excessive temperatures leading to devitrification (crystallization) and chemical structural damage of quartz sleeves. Any blocking of the weir control device due to insufficient lubrication, for example, leads to unacceptable water levels and associated high temperatures and lamp burning.	Replace as necessary Increase water supply Check sleeves for breaks and leaks; dry contacts; replace components Clean level sensors; repair/replace as necessary Check the water level control system and repair if necessary Take appropriate measures for flow rate control to avoid excessive variations of flow rates

(*continued*)

Table 7.16 Continued

Indicator	What to check	Potential problem	Explanation	Corrective actions
Low-intensity signal	UV intensity sensor, its signal and sensor fouling or wearing	Wrong signal in the PLC system associated with parasite interferences, build-up on quartz jackets or on intensity sensor	Because the measurement of low intensity (in particular during low flow operation) and the lack of amplification, the electrical signal is often disturbed. High temperatures and other factors could accelerate the sensor wearing. Because the measurement is carried out on only two lamps (situated as a rule at the top of the module), any chemical and/or organic deposits and build-up on the quartz jackets lead to decrease of the measured intensity (especially after shutdown of UV operation).	Clean routinely UV intensity sensors as necessary Clean routinely quartz sleeves Check and repair if necessary the electronic system: long or overloaded 4–20 mA loop, troubles in the analog-to-digital converter, fault of the protection against electromagnetic disturbance, poor quality of grounding, etc.
Poor lamp cleaning	Wiper compressor	Water leaks	The cleaning systems can be affected by several malfunctions including low pressure in the cleaning loop (compressor fault or too much water in the backwash tank), clogging by rags, or poor adjustment of the wiper-scraping rings.	Tighten the gland nut to compress O-ring or replace Clean lamps as necessary Check the downstream treatment and take measures to improve water quality
Excessive lamp fouling due to iron salts	Check and adjust the Fe dosage downstream	Yellow to brown deposits on lamp sleeves	Excessive iron content (>0.5 mgFe/L) leads to precipitation of brown deposits on the quartz sleeves and decrease of UV intensity, followed by fast development of additional organic fouling.	Check the residual $FeCl_3$ concentration, decrease and control the $FeCl_3$ dose, and/or replace the Fe salts with alum
Failure in water level control and poor disinfection	Weir control device Flow rate	Inadequate level control	Excessive flow rates, after heavy rain for example, or blocking of weir device can lead to unacceptable increase of water level in the UV channel.	Check the water level control system and repair if necessary

investigation. In Montreal, Canada, only physicochemical primary treatment and ozone disinfection have been investigated, using high dry and wet flows of 15 and 45 m^3/s and a target disinfection limit of 5000 FC/100 mL.

It is important to stress that because ozone is generally more expensive to produce and has to be generated on site, ozonation has been considered as less attractive than UV irradiation. Nevertheless, the powerful oxidative and disinfectant power of ozone makes it a good alternative for wastewater disinfection, especially when real pathogens are of concern (e.g., viruses and parasites). In addition, an enhancement of water quality after ozonation has been reported in numerous studies—reduction of COD (up to 20%) and UV absorbance as well as significant color removal.

Ozone is a strong oxidizing agent, effective in destroying bacteria, viruses, and cyst-forming protozoan parasites like *Giardia* and *Cryptosporidium,* which are particularly resistant to most other disinfectants. The germicidal effect of ozone consists of totally or partially destroying the cell wall, resulting in microorganism lysis. In addition, ozone breaks chromosomes, nitrogen-carbon bonds between sugar and bases, DNA hydrogen bonds, as well as phosphate-sugar bonds leading to depolymerization and leakage of cellular constituents and irreversible enzyme inhibition.

Two mechanisms of ozone disinfection occur: a direct oxidation of compounds by the ozone molecule and a reaction involving the radical products of ozone decomposition, principally believed to be the hydroxyl radical. This radical is highly reactive and has a life span only of few microseconds in water. The predominant reaction will depend on the wastewater characteristics.

The available data on generation of toxic compounds by ozone are contradictory.[19] In several studies, no toxic compounds have been found in ozonated secondary effluent. In general, the reaction of ozone with organic molecules leads to destruction, forming more polar biodegradable products with low molecular weight. Moreover, most mutagenicity studies show that ozonation reduces or removes mutagenicity in water. Nevertheless, it has been shown that in some cases, particularly those involving pesticides, more toxic intermediary products may be formed. In the case of bromide-containing effluents, the general concern about bromates in drinking water is not important for wastewater disinfection, because the presence of ammonium prevents bromate generation.

Considerable improvements have been made in ozone generation, focusing on efficiency, reduction of energy output, and control of heat losses inherent in the system (Figure 7.19). In some generators, the energy consumption has been reduced up to 40% since 1993.[38] The production and use of high ozone concentrations (up to 20%) resulted in enhancement of mass transfer and reduction of capital and O&M costs. Other improvements are related to contactor hydraulics.

The main advantage of ozonation is the high disinfection potential to remove not only bacteria, but also viruses. Moreover, this treatment leads to a significant improvement in wastewater quality, in particular color removal and

Figure 7.19 An ozone generator. (Courtesy of Ozonia.)

decrease of COD. Wastewater disinfection by ozone is also characterized by some specific advantages such as the possibility to reuse residual oxygen in other processes of the treatment facility.

7.6.5.1 Influence of Wastewater Quality

As for other disinfection agents, design doses of ozone are highly dependent on wastewater quality (see Table 7.15). Organic matter is the major contribution to ozone demand, which increases with organic content, as well as the ozone doses required for a given log inactivation.[30] The presence of particles is much less critical than for UV irradiation, but particles may also protect microorganisms from ozone action.

Figure 7.20 illustrates the influence of water quality on ozone disinfection of different effluents to water quality levels for unrestricted irrigation in compliance with WHO,[12] USEPA Reuse Guidelines[21] and California Water Recycling Criteria.[22,30,31] Fecal coliform concentration after ozonation is plotted as a function of the transferred ozone dose (TOD), and design doses can be deduced depending on the regulation to be complied with. These results show the feasibility of ozone to treat all effluents to moderate standards such as WHO[12] and USEPA[21] recommendations for unrestricted irrigation. TODs around 15 and 25 mg/L, respectively, would be required for poor quality wastewater such as primary or secondary effluents produced by high load activated sludge treatment. For a good-quality secondary effluent obtained after extended aeration or tertiary effluents, the required ozone dose is reduced

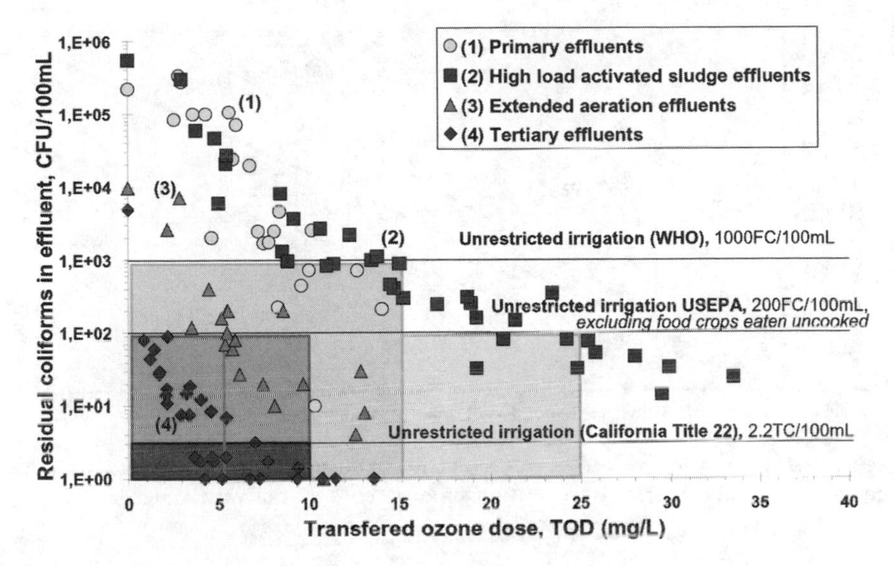

Figure 7.20 Ozonation for fecal coliform inactivation in different effluents with different water quality objectives.

to 5–10 mg/L. If a stringent standard like California Water Recycling Criteria[22] must be complied with, a tertiary filtration step must be implemented before ozonation with a required TOD of around 10 mg/L. Without any filtration step, huge doses of 10–50 mg/L would be necessary to meet the most stringent regulations.

The high oxidative power of ozone also makes it an efficient virucidal agent: viruses are removed more quickly than target bacteria. In general, lower ozone concentrations, in the range of 1–5 mg/L, and lower contact times are necessary for 3–5 log virus removal, including the highly resistant MS2 strain. Ozone is also more effective on certain parasites: *Giardia* and *Cryptosporidium* cysts and *Acanthamoeba* and *Naegleria* amebae.

As shown in Figure 7.21, the transferred ozone doses required for a 2 log reduction of fecal coliforms vary from around 2 mg/L for a tertiary effluent, 6–17 mg/L for secondary effluents, up to 25–30 mg/L for primary effluents.[38] These results clearly show the main trend of less ozone required with each treatment step from primary settling to tertiary filtration.

7.6.5.2 Influence of Hydraulic Residence Time

Recent experimental data showed that very short contact times (only few minutes) were as efficient as the standard 15–20 minutes typically used and derived from drinking water design.[39] This phenomenon can be explained with the fast kinetics of inactivation of viruses and bacteria, which is quite immediate. Because the disinfection objectives for wastewater disinfection are

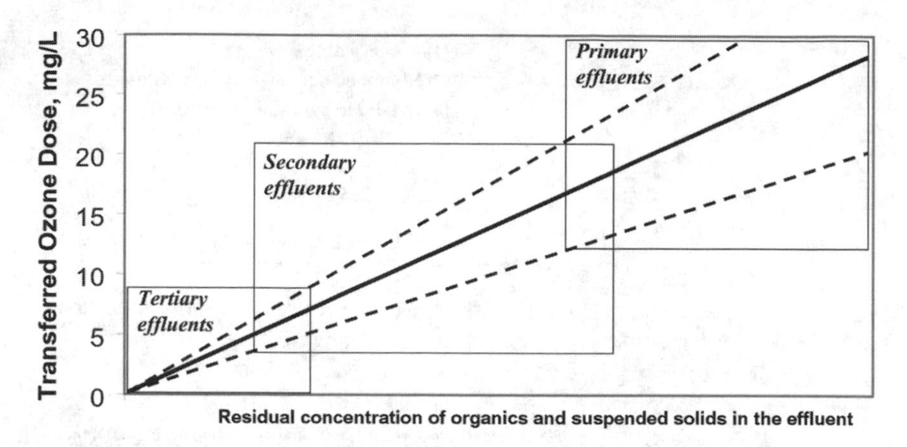

Figure 7.21 Transferred ozone doses (TOD) required for an objective of 2-log reduction in fecal coliform concentrations in different types of wastewater.

usually less severe than for drinking water applications (based on resistant parasite inactivation), this new design approach leads to large reductions in contactor volume and costs. Ozone mass transfer optimization is becoming the critical step for the design of wastewater ozone disinfection.

7.6.5.3 Design Parameters

A new design approach must be applied for wastewater ozonation based on short contact times and enhanced mass transfer because of the fast kinetics between ozone and fecal bacteria. This can lead to a significant reduction in capital and O&M costs, especially when a new disinfection facility has to be built.

Ozone doses are directly determined by the upstream treatment (quality of the effluent to be treated) and the disinfection objectives to be satisfied[30,31,39]:

1. Moderate water quality requirements such as WHO[12] guidelines for irrigation (<1000 FC/100 mL) will involve transferred ozone doses of 2–5, 5–10, and 15–20 mg/L for tertiary, high-quality secondary, and poor-quality secondary or primary effluents, respectively, for hydraulic residence time of 2–4 minutes and an additional beneficial effect of 30% reduction in UV absorbance.
2. Stringent regulations such as the Californian Water Recycling Criteria[22] (<2.2 TC/100 mL) can be reached only after tertiary filtration with transferred ozone dose of 8–10 mg/L for hydraulic residence time of 2–8 minutes.

As mentioned previously for wastewater treatment, ozone transfer is the major element in the selection of appropriate ozone contactors. One method to optimize ozone mass transfer and disinfection could be direct injection of ozone in a static mixer, followed by a mere pipe.[38] Additional cost reduction may be expected for ozonation due to significant improvements made recently in ozone-generation equipment. The extreme adaptability of ozone generators to specific cases must be underlined, i.e., the choice between air or oxygen generation, on-site production of oxygen or liquid oxygen delivery according to local constraints (size, ratio between operating and investment costs).

Finally, site-specific constraints may have a big impact on the choice of the most appropriate disinfection process. Depending on their design and size, existing chlorine chambers can be transformed into ozone contactors or retrofitted for UV disinfection systems. In some cases, the disinfection step can be combined with another step of the treatment line. After being used for ozone generation and as carrier gas, oxygen may be recycled to the activated sludge units in order to enhance the efficiency of the biological oxidation of the carbon and nitrogen, ensuring up to 25% reduction of the required aeration capacity. Likewise, ozonation could be preferable in a plant with decarbonation: air would be simultaneously employed for decarbonation with carbon dioxide, while oxygen would be used for ozone generation. Another combination of processes could be the use of ozone for two purposes: disinfection and sludge reduction in the biological basin.

Eventually, the ability of ozone to treat effluents with poor quality could lead to mixed solutions, where only part of the flow could be ozonated after partial treatment, whereas the other part would undergo a more advanced treatment. In all cases, the cost evaluation must integrate the whole treatment train.

For ozone reactivity and safety reasons, ozone contactors require special covered constructions. Properly calibrated ozone monitor equipment is needed for both work areas and the ozone off-gas destruction units. The established threshold limit values of ozone concentration in air are 0.2 and 0.6 mg/m^3 at 25°C and 100 kPa pressure for work days and maximum limit, respectively, measured at a maximum sampling time of 10 minutes.

7.6.6 Membrane Filtration

Microfiltration (MF), ultrafiltration (UF), and nanofiltration (NF) are being intensively studied in France and the United States for wastewater tertiary treatment, including disinfection and removal of colloids and larger molecular weight organics. These technologies are based on a physical barrier concept. Because of the high quality of the treated water, membrane filtration is used in Australia, Japan, and the United States for specific water reuse applications, such as groundwater recharge, grey water recycling, and industrial wastewater recycling. Moreover, the absence of bacterial regrowth and residual toxicity may give membranes important advantages over other processes for groundwater recharge and potable reuse.

Suspended solids, bacteria, parasites, and viruses are readily removed by NF, UF, and MF, except for viruses that can pass through the more porous MF membranes (0.2 μm cut-off): 1–3 log pfu/100 mL of viruses were found in the microfiltered permeate. Typical design flux rates of MF and UF are between 40 and 80 L/h m^2.

The use of low-pressure membranes for pretreatment to reverse osmosis (RO) is a new, promising technology for the production of high-quality desalinated water. For such applications, both MF and UF could provide SDIs consistently under 1 and decrease the pressure required to operate the downstream RO unit with a given flux. The largest membrane water recycling facility for irrigation, at Sulaibiya, near Kuwait City (Kuwait), will produce 375,000 m^3/d of recycled water by ultrafiltration and reverse osmosis, which, after blending with existing brackish water, will be used mostly for agricultural irrigation.[40]

7.6.7 Membranes Bioreactors

More recently, a new generation of water treatment processes has become available combining membrane-separation techniques and biological reactors: membrane bioreactors (Figure 7.22). This process is a modification of conventional activated sludge, where the clarifier is replaced by a membrane unit for the separation of the mixed liquor from treated effluents. The main advantages of this technology are the high-quality effluent, high flexibility, compactness, and low sludge production.

Two main configurations of membrane bioreactors are used in water reuse applications (Figure 7.23): submerged membranes or membranes with external recirculation. Both UF and MF are used in these MBR technologies with membrane modules of different types, e.g., spiral, plane, or hollow fibers. Either organic or ceramic membranes are used for wastewater treatment. Ceramic membranes are physically more resistant against drastic pH variations than organic membranes. This stability makes ceramic membranes easier to clean with chemicals. Organic MBRs predominate in municipal wastewater reuse schemes because of their lower costs.

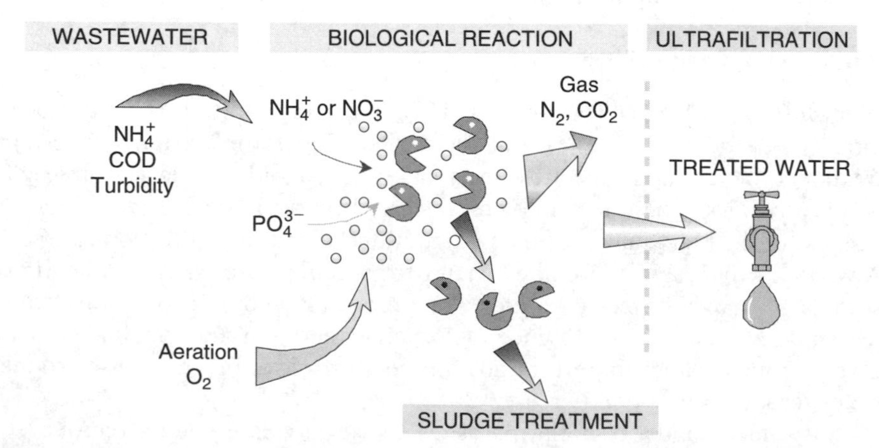

Figure 7.22 Principle of operation of membrane bioreactors. (Courtesy of Degrémont.)

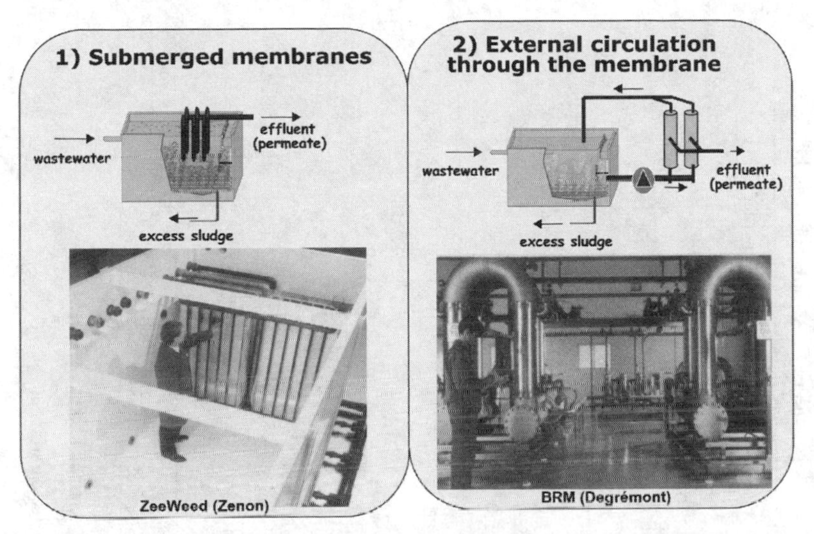

Figure 7.23 Classification of membrane bioreactors: (1) Submerged membrane bioreactors and (2) membrane bioreactors with external recirculation.

Compared to other biotechnologies, MBRs enable the uncoupling of the biomass solid retention time from the hydraulic retention time, facilitating the biodegradation of high molecular weight and slowly degradable organics. Due to high biomass concentrations, high organic loading rates have been achieved with low excess sludge production. MBR disinfection efficiency is high, as a rule, and depends not only on membrane cut-off, but also on operational conditions and other factors.

Thousands of MBRs are under operation worldwide, with many more currently under construction. More than 98% are for aerobic treatment, and approximately 55% of the commercial systems have submerged membranes. MBRs with submerged organic membranes are becoming a very promising treatment process for water reuse systems with very fast growth, in particular in the United States. The main characteristics of the principal MBR technologies used for wastewater treatment and reuse are summarized in Table 7.17.[41–43]

The main design parameter of MBRs is flux rate that ranges from 5 to 300 $L/m^2 h$ (specific flux 20–200 L/m^2 bar) and is typically 25–60 $L/m^2 h$ for submerged systems and 40–120 $L/m^2 h$ for tubular sidestream systems operated in high pressure. It depends on a number of complex interrelated parameters, including membrane characteristics, transmembrane pressure (TMP), cross-flow velocity, and biomass properties.

Volumetric loading rates of aerobic MBRs range typically from 1.2 to 3.2 kg COD/m^3 d and 0.05 to 0.66 kg BOD/m^3 d. Although these values are similar to activated sludge, removal efficiencies are higher than 90% and 97%

Table 7.17 Principal MBR Technologies Used in Wastewater Treatment and Reuse Systems

Parameter	Zenon ZeeWeed[a]	Pall[b] (Asahi)	Memcor[c]	Norit X-Flow[d]	Membra-tek	Koch[e]	Ionics	Mitsubishi[f]	Kubota[g]
Type of membrane	UF immerged Vertical hollow fibers	UF	MF	UF	UF	UF	UF	UF submerged Horizontal hollow fibers	MF submerged Plane vertical
Membrane cut-off, µm	0.035 µm	0.1 µm 13,000 or 80,000 daltons	0.2 µm	0.03 µm	0.1 µm	0.01 µm 10,000–100,000 daltons	0.015 µm	0.04 µm	0.4 µm
Membrane material	Patented	PVDF (polyfluorure of vinylidene)	PP (polypropylene)	PES and PVP (polyvinyl pyrrolidone)	Anaerobic, tubular	PES (polyether-sulfone)	PES	PE	Mineral
Fiber diameter, in/out, mm		0.7/1.3		0.8 or 1.5		0.51 or 0.77			
Pression in max, bar		3	1.4		4.5–5.5	3	1.4		
Transmembrane pression, bar			0, 2–1			0,5			
Flux direction	Out-in	Out-in	Out-in	In-out		In-out			
Chlorine, max	2000 mg/L	< 1000 mg/L	Very sensible 0.01 mg/L	200 mg/L		Tolerant < 5000 mg/L	100 mg/L		
Type of cleaning	Air/water	Air/water	Air/water	Water		Water			
Frequency of chemical cleaning	2–4/year		>1/month			1/month		2–4/year	2/year

[a]Very resistant fibers, pore size 13,000 or 80,000 Daltons, low fouling potential, numerous references in the USA, Canada, Europe, 500 to 19,000 m^3/d
[b]Efficient on-line measure to decrease membrane fouling, numerous references in Australia and the USA, 47–552 m^3/h
[c]High fouling potential, increased by the low chlorine tolerance, references in the USA
[c]References in the Netherlands
[e]Pore size 10,000 to 100,000 Daltons, references in treatment of industrial wastewater and coupling with reverse osmosis
[f]Well developed in Japan
[g]Well developed in Japan, Germany, the Netherlands

for COD and BOD, respectively, with residual concentrations consistently <10 mg BOD/L and <40 mg COD/L. Improved COD removal is due to the complete particulate retention and degradation of slow biodegradable compounds.

Biological performance appears to be relatively insensitive to hydraulic residence times between 2 and 24 hours. Sludge age also appears to have little influence on effluent quality, with reported values between 5 and 3500 days (no sludge removal).

MBRs operate at high mixed-liquor concentrations of between 10 and 20 g/L, which may increase up to 50 g/L without sludge removal. Low sludge-production rates have been reported between 0 and 0.34 kgMLSS/kgCOD removed. To avoid operation problems, it is recommended to maintain mixed liquor concentration below 12 g/L.

Energy consumption rates vary from 0.2–0.4 kWh/m^3 for submerged operation to 2–10 kWh/m^3 for sidestream operation. In the sidestream mode, aeration accounts for 20–50% of the total power requirements, whereas it accounts for 80–100% of the total in submerged systems. The overall power demand of submerged systems was shown to be up to 3 orders of magnitude less than for comparable sidestream systems.

The disinfection efficiency of MBRs is high. Relatively large pore sizes (5 μm) of organic membranes gave 3–5 log rejection of total coliforms compared to 8 log rejection on small pore systems (0.4 μm). Recent reviews of the operational performances of MBRs of about 100 facilities worldwide[44,45] demonstrated excellent effluent disinfection of UF Zenon membranes (membrane cut-off 0.035 μm) with absence of coliform bacteria (fecal, total, or *E. coli*) or residual concentrations under the detection limit of <1 TC/100 mL or <3 *E. coli*/100 mL. Excellent disinfection efficiency (absence of total coliforms in mediane samples and under the detection level for geometric mean) has been demonstrated for the Pall membranes (0.1 μm cut-off) during pilot plant tests in Sacramento, California.[46] In addition to the excellent bacteria removal, MBR technology also provides good virus elimination, e.g., Zenon membranes allowed the removal of more than 4 log of coliphages, which is a good indicator for virus elimination.

7.7 STORAGE AND DISTRIBUTION OF RECYCLED WATER

Storage and distribution of recycled water, as well as the associated evolution in water quality, must be considered as important elements of water-recycling schemes for irrigation. The evolution of water quality during storage of recycled water is often underestimated. Nevertheless, important changes in water quality may take place during storage and distribution and, thus, deteriorate or improve water quality at the point of use. Consequently, storage of recycled water must be considered as part of the overall treatment process.

Depending on the demand for water for irrigation and discharge constraints on the remaining recycled water, two main types of storage exist:

Short-term storage, using predominantly open- or closed-surface storage reservoirs

Long-term seasonal storage, using mainly open-surface reservoirs or aquifer recharge

The main concerns about water quality and management strategies of these two storage processes are completely different and may be taken into account in the design of water reuse schemes for irrigation.

7.7.1 Short-Term Storage

Short-term storage is typically required for almost all landscape projects and is implemented in some agricultural irrigation schemes. In general, landscape irrigation and golf course watering is applied during the night, where wastewater flow is minimal. For this reason, storage reservoirs are considered an integral part of treatment schemes for urban irrigation and are designed for hydraulic residence times from 1–2 days to 1–2 weeks.[2]

Open reservoirs are commonly used for such purposes. The principal problem in the operation of open systems is an increase in fecal coliform concentration. Bacterial regrowth and external contamination can reach relatively high values, up to 2–3 log of coliforms, as reported in numerous irrigation projects (France, Israel, New Caledonia). For this reason, closed reservoirs should be considered. The main advantage of closed reservoirs is to limit bacterial regrowth and external fecal coliform contamination from birds and animals. However, other operational problems can arise, such as odor generation due to water stagnation.

When chlorine residuals must be maintained in distribution networks, short-term storage after chlorination leads to a decrease or complete chlorine residual consumption. The residual chlorine decay is lower in closed reservoirs. In all cases, if residual chlorine is required, it is recommended to implement chlorination after the storage of recycled water.

7.7.2 Long-Term Storage

Long-term storage is applied mostly for seasonal storage of recycled water during the winter in order to satisfy the high irrigation demand during dry seasons or zero discharge requirements in the natural environment. Long-term storage can be performed in surface storage reservoirs or in aquifers.

7.7.2.1 Long-Term Storage in Surface Reservoirs

Recycled water production is fairly constant throughout the year, while irrigation demand depends on the season. Therefore, during the winter and rainy seasons, recycled water is either stored to be used for irrigation in spring and summer or disposed of in streams, lakes, or other receiving bodies.

Consequently, the first function of seasonal storage is to accumulate water during the cold season to satisfy subsequent irrigation requirements.

The mechanisms that control water quality in storage reservoirs are the same as those occurring in lagoons (stabilization ponds). The main difference is the water depth, which entails thermal stratification in deep reservoirs.[47] Stratification is always observed in hot seasons and may disappear in winter and windy periods. Pollutant removal is much more important in the upper water layer, which is aerobic, while anaerobic conditions prevail in deeper zones. The treatment mechanisms occurring in the upper layer are almost the same as those in natural treatment systems, influenced by algae activity, solar radiation, evaporation, wind, and stratification. This specificity of deep storage reservoirs may be taken into account in their design. Surface storage reservoirs are characterized by non–steady-state operation, similar to sequential batch reactors. Therefore, the description of removal kinetics is very complex and requires numerical modeling procedures.

Surface storage reservoirs can operate with continuous or discontinuous input and output.[47] One of the most common configurations is the continuous-flow single reservoir, which is filled throughout the year, empting only during the irrigation season. Deep reservoirs with a small area-to-volume ratio are recommended for semi-arid regions with high evaporation rates. The active depth is 6–8 m on average, with a maximum of 20 m. A dead volume of about 1 m depth must be added to the bottom of the reservoir to avoid pumping of sediments. In such systems, the oxygen supply is limited to the upper layer, reducing the maximum allowed organic loads. The use of aeration or other destratification devices can improve treatment efficiency.

The definition of maximum surface loading of long-term surface storage reservoirs is quite complex and must include simulation of reservoir behavior in time. In fact, the volume and specific area of reservoirs change with time, as well as environmental conditions (temperature, solar radiation, quantity of algae, etc.). An average surface loading of 30–40 kg BOD/ha d is recommended in Israel to maintain fully aerobic or facultative conditions in reservoirs, avoiding sporadic emission of odors.[47] Higher loads are associated with significant odor emission and degradation of water quality, in particular at the end of irrigation season.

Surface storage reservoirs have proved to be efficient means of organic matter, nitrogen, phosphorus, parasites and bacteria removal.[1,47] COD and BOD removal percentages are 70–85% and 50–85%, respectively, in the Kishon reservoir (Figure 7.24), the biggest reservoir in Israel. Mean N and P removals are 75 and 60%, respectively. Coliform removal can reach 3–4 log units as orders of magnitude and sometimes as much as 1000 FC/100 mL. However, removal efficiency has been observed to depend greatly on hydraulic residence time distribution and particularly on the percentage of water during short detention times, i.e., less than a few days. Drastically reducing short detention times and, better, closing the input into the reservoir several weeks before using the stored water (batch operation) will result in a dramatic improvement in water quality.

Figure 7.24 Long-term seasonal storage reservoir, Upper Kishon reservoir, Israel. (Courtesy of Mekorot.)

Studies conducted on reclaimed water stored in interseasonal ponds emphasized the treatment capacity of this step. In Sicily[48], the storage of raw wastewater is recommended as a physical-chemical and microbiological treatment, delivering water that can be used for unrestricted irrigation. The study of interseasonal storage ponds in real conditions, i.e., with associated reuse, showed the importance of hydraulic operation on the ponds' treatment capacity. Proper management of the operational regime of the ponds allows optimal microbial abatements for the whole irrigation period.[49–51]

The main recommendations for design and operation of surface storage reservoirs can be summarized as follows:

Irrigation water quality can be improved by sequential batch operation of storage reservoirs. The feed of continuous-flow reservoirs must be stopped before the use of stored water for irrigation. Implementation of three to six independent storage units is recommended to allow optimal reservoir operation without interruption of wastewater input.

Low organic loads make it possible to improve irrigation water quality. An average value of 30 kg BOD/ha d could be used as indicative value to maintain good oxygen transfer and limit odor generation. However, it is important to stress that maximum organic loads depend on reservoir configuration, environmental conditions, wastewater quality, and other specific factors.

Specific considerations are needed for the location of the inlet and outlet of the storage reservoir. The inlet must be located as far as possible from the outlet and at the bottom of the reservoir. The outlet must be located at the leeward side of the reservoir to avoid an adverse influence of wind on recycled water quality. For the same reason, the inlet-outlet axis must be perpendicular to the direction of predominant winds.

The optimum active depth of storage reservoirs is 6–8 m. Deeper reservoirs up to 20 m with low area-to-volume ratio can be used in semi-arid regions with high evaporation rates. In such cases, measures for destratification and aeration could be necessary to improve treatment efficiency and water quality.

7.7.2.2 Long-Term Storage in Aquifers

Groundwater recharge is an important step prior to reuse for irrigation in many regions. Some recharge processes incorporate both the objectives of wastewater treatment and groundwater recharge, while others can be considered as recharge techniques only, the performance of each method being highly site-specific.

There are several advantages to storing water underground[1]:

1. Economic and construction advantages:

   The cost of artificial recharge may be less than the cost of equivalent surface reservoirs, as no construction is needed. However, the cost of treatment required prior to recharge, in particular before direct aquifer recharge, may reverse this advantage.

   Suitable sites for large-surface reservoirs may not be available or environmentally acceptable.

2. Water quality improvement and water conservation:

   Water stored in surface reservoirs is subject to evaporation, potential taste and odor problems due to algae and other aquatic productivity, and pollution. These may be avoided by underground storage.

3. Enhancement of public acceptance:

   The addition of groundwater recharge in a water-reuse project may also provide psychological and aesthetic secondary benefits as a result of the apparent transition from recycled water to groundwater.

   This aspect is particularly significant when a possibility exists in the water-reuse planning to augment substantially potable water supplies.

A variety of methods have been developed to recharge groundwater that can be classified as surface techniques and direct injection.

Surface techniques or recharge by means of soil infiltration is the most common practice, which includes the following methods:

Slow-rate infiltration
Over irrigation

Infiltration-percolation
Stream-channel modification
Stream augmentation
Flooding

Direct surface spreading is the simplest, oldest, and most widely applied method of artificial recharge. Recharge waters such as reclaimed municipal wastewater percolate from spreading basins to the aquifer through the vadose zone. This method is also called soil aquifer treatment (SAT) (see also §7.5.4).

Annual application rates range from 0.5 to 6 m for the slow-rate infiltration process to 6–125 m of water for SAT and infiltration-percolation.[1] Project design and operation include not only the management of water loading but also the control of the mixing zone and residence time in the unconfined aquifer.

The main advantages of groundwater recharge by surface techniques are as follows:

Surface spreading provides the added benefits of the treatment during infiltration through the unsaturated zone (such as the SAT process).
Aquifer storage ensures transporting facilities and, thus, leads to significant cost savings.

Direct subsurface recharge is achieved when water is conveyed and injected directly into an aquifer. As a rule in direct injection, highly treated wastewater is pumped directly into the aquifer layer of a confined or phreatic aquifer. Groundwater recharge by direct injection is practiced where the topography or existing land use makes surface spreading impractical or too expensive and with confined aquifer in which surface infiltration is obviously impossible. Specific treatment of recycled water before injection is the most important requirement, since it allows one to comply with the expected aquifer water quality as well as avoid clogging problems and precipitation.

7.7.3 Management of Recycled Water Storage Reservoirs

Adequate reservoir management is important to improve water quality. In general, the management strategies currently used in freshwater reservoirs may be applied to recycled water storage. Natural ponds require a significantly different management approach. Compared to storage of freshwater in both open and closed reservoirs, recycled water requires more frequent maintenance. For example, reservoir cleaning must be planned every 3–5 years to avoid excessive sludge accumulation on the bottom of the reservoir. This sludge is usually removed by pumping and flushing. In this respect, appropriate access to the reservoirs as well as means for sludge disposal and/or transportation must be provided during the planning and design.

Depending on site-specific conditions, the following management strategies can be applied for improvement of stored recycled-water quality.[52]

1. Destratification: To improve oxygen supply in deeper zones and avoid anoxic conditions, reservoirs can be destratified by addition of compressed air and recirculation using a wide range of oxygen systems. This method is very effective for improving oxygen concentrations and reducing odors during storage. However, this method disrupts the thermocline and could affect the fish population.
2. Hypolimnetic aeration: If it is necessary to prevent mixing of hypo- and epilimnion, oxygen can be added to the bottom layers only without vertical circulation. Similarly to destratification, this method provides good oxygenation and odor reduction.
3. Chemical treatment: The addition of algicide chemicals such as copper sulfate is used as temporary measure to limit algae growth. However, chemical dosage greatly depends on site-specific conditions, and overdosing has adverse impacts on the reservoir ecosystem. Another chemical treatment like alum addition could be used to reduce internal recycling of phosphorus.
4. Wetland treatment: The addition of macrophytes in constructed wetlands makes it possible to remove nutrients. However, phosphorus removal is rather small compared to nitrogen removal, and consequently the efficiency of algae control could be low.
5. Dredging: Dredging is applied to remove sediments and nuisance macrophytes, reducing the input of phosphorus under anoxic conditions. Mechanical or hydraulic dredge can be used, as well as reservoir empting with removal of sludge by bulldozer. This method is efficient in the long term but is expensive and requires sludge disposal.
6. Biomanipulation: In many cases, biomanipulation has been successfully applied to control algae growth in shallow open reservoirs. The effective application of this strategy requires good understanding and control of the entire ecosystem. For example, zooplankton can effectively control algae growth in many species. However, to increase its quantity, it is necessary to replace planktivorous fish with piscivorous fish. The main constraint for the application of this strategy is the high phosphorus level in recycled water.

7.7.4 Control of Water Quality in Distribution Systems

Generally, water-quality requirements for irrigation are applied at the outlet of the water-recycling plant. However, in some new regulations, water-quality monitoring is also recommended at the point of use. In this case, significant change in water quality can occur, in particular in specific conditions such as long length of distribution network, hot climate, inhomogeneous water withdrawal, etc. The main problems related to degradation of water quality in distribution systems are odor, color, and bacterial growth (Figure 7.25). As mentioned in Chapter 4, the most important action to prevent biofilm growth

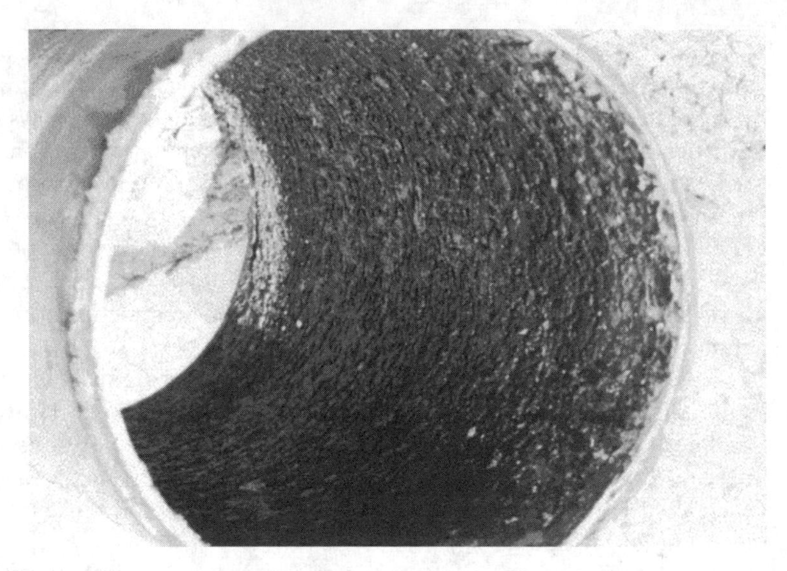

Figure 7.25 Biofilm growth in pipelines with recycled secondary effluent. (Courtesy of Mekorot.)

and associated degradation of water quality is flushing pipes before irrigation and chlorination.

A recent study of water quality in a large distribution network for unrestricted irrigation (about 100 km) in southern California demonstrated rapid chlorine decay and associated bacteria regrowth, especially during winter, when residence time may increase greatly. To prevent strong bacteria regrowth, two strategies may be recommended: periodic purging of distribution network to lower residence time and additional in-pipe rechlorination at some critical points of irrigation distribution systems.

7.8 CRITERIA FOR SELECTION OF APPROPRIATE POLISHING PROCESS BEFORE IRRIGATION

The choice of the most appropriate additional treatment in water-reuse projects for irrigation is driven by a number of criteria, such as costs, regulatory requirements, ease of implementation and operation, potential adverse environmental and health impacts, as well as specific local conditions.

7.8.1 Cost of Additional Treatment and Reuse

The distribution of capital and O&M costs of additional treatment and reuse varies from one project to another and depends on the type of the applied treatment processes. These costs are also highly influenced by local constraints: price of the building site, distance between the production site and the consumers, and need to install a dual distribution system or retrofitting. The latter two constraints are important, as in many projects the main capital investment concerns the distribution system and can reach 70–200% of the overall costs depending on site-specific conditions. Storage, mainly seasonal storage, represents significant part of investment. New systems cost less than retrofitting of existing networks.

Cost evaluation is based on operational experience and unit prices in Western Europe and the Mediterranean region, with a life cycle of 20 years and return rate of 8%. It is important to stress that the costs reported here only illustrate the influence of plant size on treatment costs. The costs' values cannot be extrapolated to other case studies or countries because the unit cost of reclaimed water depends not only on the plant size and treatment chain, but also on wastewater composition, water-quality requirements, and other local conditions (energy costs, labor, etc.). Moreover, the main components of recycled water costs are not the same from one plant or country to another.

Although it represents a relatively small part of the costs of most water-recycling projects, the attention of practitioners often focuses on the cost of tertiary treatment and disinfection. Filtration step construction results in a two- to threefold increase in the capital and operating costs of disinfection processes.

Among the tertiary treatments, polishing pond treatment is the most rustic but has proven to be a competitive, efficient solution for small communities. This technology is the cheapest solution for flows under 3000 m^3/d (15,000 p.e.) with average total annualized cost of about US$ 5–7 cents/m^3. In this range, polishing pond treatment might be a practical and efficient solution for water-reuse purposes such as irrigation. In this case, final in-line chlorination might be used to protect against microbial regrowth in the distribution network (commonly used chlorine dose: 1 mg/L).

As the project size increases, polishing pond treatment becomes less and less competitive compared to other solutions, not taking its storage function into account. There are two main reasons for this. First, the capital expenses for polishing pond treatment do not greatly benefit from scale economy. Second, the operational expenses per cubic meter do not decrease because they are largely dictated by the cost of sludge evacuation and disposal. This cost is typically a fixed cost per cubic meter, as agreed upon with local farmers. Additional difficulties arise when the project increases. The surface area of the ponds (300,000 m^2 for 30,000 m^3/d) as well as the need for sludge evacuation (at least once every 10 years) may become prohibitive.

For project sizes more than 7500 m^3/d (50,000 p.e.), the cost for UV or chlorine treatment becomes comparable to maturation ponds within the error margin of the cost estimation. Knowing the critical issue of chlorination by-products, it is widely recommended to use UV disinfection, at least when the cost is comparable to chlorine disinfection. The addition of chlorine to treated wastewater has been shown to produce carcinogenous compounds. However, concerns related to potable water might not be extrapolated to all reuse applications. After years of debate, this issue is now a major concern for the regulator and for environmental associations, which makes the future of wastewater chlorine disinfection uncertain.

For small and medium-size water-recycling works (< 50,000 p.e.), chlorination and UV irradiation are more competitive than ozonation, with average total annualized cost of about US$ 2.2–8 cents/m^3. The cost difference between UV and ozone decreases with plant size. The competitiveness of ozonation appears clear for large recycling plants (>100,000 p.e.), where total costs are in the typical range of US$ 0.8–2.5 cents/m^3, and in some cases could be less than UV irradiation. Given that ozone also improves the visual aspect of the recycled water and sometimes lessens its odor, this process should be considered as a viable option for large plants.

The costs of membrane filtration (MF and UF) are significantly higher compared to the other disinfection processes and typically reach US$ 0.40–0.70/m^3 for plant capacity of 20,000–500,000 p.e. The cost difference decreases when compared with combined sand filtration and UV or ozone disinfection. In all cases, recycled water quality is significantly higher, as well as operational reliability, which is often the decisive criteria for the choice of treatment train for some urban applications for unrestricted landscape irrigation.

The high cost of membranes is also the main constraint for the widespread application of membrane bioreactors (MBRs), despite all the process advantages. Compared to activated sludge, the overall costs remain up to 20% and 50% higher than activated sludge, depending on plant size. Reported MBR costs typically vary from US$ 0.095 to 0.20/m^3 for treatment plant size up to 50,000 p.e.

The operating costs are about 45–50% of the total annual costs for UV irradiation and increase up to 50–70% for chlorination and ozonation, respectively, for small to large water-recycling plants. Operation and maintenance costs incurred by chlorination and ozonation are primarily those associated with chemical costs. Operating costs for UV systems consist mostly of lamp replacement and cleaning. Energy costs are about 2–5% of the operating costs for chlorination. Energy costs for UV irradiation and ozonation are between 15 and 35% respectively, depending on plant size. Higher reagent costs are characteristic for chlorination—up to 60% of the operating costs.

7.8.2 Main Criteria for Selection of Disinfection Process

The choice of a given disinfection process is generally driven by several criteria, such as regulatory requirements, cost-effectiveness, safety, practicality, environmental impact, and public health-related issues (Table 7.18). Systematic procedures cannot be used to ease the choice, because site-specific constraints prevail in many cases (permit or safety regulations, existing treatment chain, etc.). However, some general trends, advantages, and disadvantages can provide helpful guidelines for disinfection-process selection. A good definition of the appropriate criteria for process selection is especially important nowadays, when new indicators or pathogens other than bacteria

Table 7.18 Criteria for Choice of Disinfection Technology

Characteristics/ Criteria	Chlorination	UV	Ozone	UF/MF	MBR	Maturation ponds
Size of plant	All sizes	Small to medium	Medium to large	Small to medium	Small to medium	Small
Pretreatment level	All levels	Secondary	At least primary, in preference secondary	Secondary	Primary	Secondary
Reliability	Good	Good	Good	Very good	Very good	Fair to good
Relative complexity	Simple to moderate	Simple to moderate	Complex	Complex	Moderate to complex	Simple
Safety concerns	Yes	No	Yes	No	No	No
Transportation on site	Substantial	Minimal	Minimal	Minimal	Minimal	No
Bacteria removal	Good	Good	Good	Very good	Very good	Poor to good
Virus removal	Poor	Good	Very good	Very good	Very good	Poor to good
Protozoa and cyst removal	Poor	Good	Good	Very good	Very good	Good
Residual toxicity (fishes)	Toxic	Non toxic	None expected	None	None	None

(Continued)

Table 7.18 Continued

Characteristics/ Criteria	Chlorination	UV	Ozone	UF/MF	MBR	Maturation ponds
Bacteria regrowth	Low	Moderate	Low	None	None	Possible "wild" contamination
Hazardous by-products	Yes	No	No; yes in some conditions	No	No	No
Persistant residue	Long	None	None	None	None	None
Contributes dissolved oxygen	No	No	Yes	No	No	No
Reacts with NH_4	Yes	No	No	No	No	No
Color removal	No	No	Yes	Low effect expected	Low effect expected	No
Algal growth	No	No	No	No	No	Substantial
Increased dissolved solids	Yes	No	No	No	No	No
pH-dependent	Yes	No	Slight at high pH	No	No	Yes
O&M sensitive	Minimal	Moderate	High	High	High	Minimal
Corrosive effect	Yes	No	Yes	No	No	No
O&M costs	Low	Low	Moderate	High	High	Very low
Investment costs	Moderate	Moderate	High	Very high	High	Low to moderate

are often issued in wastewater discharge and reuse permits (enteroviruses, *E. coli*, *Giardia*, *Cryptosporidium*, etc.).

Chlorination has proved to be a reliable means of removing bacteria and respecting conventional permits for disinfection of primary, secondary, and tertiary effluents. However, toxic by-products may present a risk to public health. In most cases, the presence of residual chlorine represents a threat to the environment (discharge in water streams, groundwater recharge), and dechlorination must be implemented, increasing disinfection costs. Also, the effectiveness of chlorine on some viruses is questionable. Protozoa are not affected by the commonly applied chlorine doses and residence times. Chlorination is not longer authorized for wastewater disinfection in France or the United Kingdom.

The reliability of UV disinfection is also well established for disinfection of secondary and tertiary effluents. Its main advantage over chlorination is the absence of toxicity and by-product formation and comparable costs. Moreover, UV systems require no specific safety control or equipment. These advantages make UV irradiation particularly suitable for wastewater disinfection and various reuse applications, including unrestricted irrigation.

With slightly higher costs, ozonation may be recommended for large plants when viruses and/or protozoan parasites are targeted. This might become increasingly the case because of the increasing concern about some epidemic microorganisms such as viruses and *Giardia* and *Cryptosporidium*, in particular for unrestricted landscape irrigation or irrigation of crops eaten uncooked. Ozone also removes odor and color, which would be desirable for some reuse applications. Recent studies reported high efficiency of ozonation for the oxidation of some emerging organic pollutants such as endocrine disruptors and pharmaceutical products that are of increasing concern in new regulations. Ozonation also leads to an increase in bulk oxygen concentration, as well as to an enhancement of the biodegradability of residual organic matter. After its first contact with the effluent, the carrier gas can be recycled to the activated sludge to reduce secondary treatment size requirements.

UV and ozone are both considered as safe processes, but safety measures with UV are more straightforward. The maintenance of an ozone system usually requires more skill than a UV system. Finally, most studies show that UV and ozone will not increase effluent toxicity.

Membrane filtration is a highly efficient process for wastewater disinfection. The excellent water quality of the effluent makes it appropriate for unrestricted landscape irrigation. Its main disadvantage is still relatively high cost, but it is the only technology that guarantees reliability, absence of toxicity, and almost complete disinfection.

REFERENCES

1. Lazarova, V., Role of water reuse in enhancing integrated water resource management, Final Report of the EU project CatchWater, EU Commission, 2001.

2. Metcalf & Eddy, *Wastewater Engineering: Treatment and Reuse*, 4rd ed., Mc-Graw-Hill Inc., New York, 2003.
3. Petty grove, S.G., and Asano, T., *Irrigation with Reclaimed Municipal Wastewater—A Guidance Manual*, Lewis Publishers Inc., Chelsea, 1985.
4. Metcalf & Eddy, *Wastewater Engineering. Treatment, Disposal, Reuse*, 3rd ed., Mc-Graw-Hill Inc., New York, 1991.
5. WEF Manual of Practices 8, *Design of Municipal Wastewater Treatment Plants*, 4th ed., ASCE Manuals and Reports on Engineering Practices N°76, WEF, Alexandria, 1998.
6. Degremont, *Memento Technique de l'Eau*, Degremont, Paris, 1989.
7. Lazarova, V., et al., Production of high quality water for reuse purposes: the West Basin experience, *Water Supply*, 3, 3, 167, 2003.
8. Lazarova, V., and Manem, J., Innovative biofilm treatment technologies for water and wastewater treatment, in *Biofilms II: Process Analysis and Applications*, Bryers, J.D., ed., New York, Wiley-Liss Inc., 2000, 159.
9. Lazarova, V., Perera, J., Bowen, M., and Sheilds, P., Application of aerated biofilters for production of high quality water for industrial water reuse in West Basin, *Wat. Sci. Techn.*, 41, 4/5, 417–424, 2000.
10. Lazarova, V., Guidelines for irrigation with recycled water, Internal Report Suez-Environment-CIRSEE, France, 2003.
11. Berland et al., *Procédés extensifs d'épuration des eaux usées, adaptés aux petites et moyennes collectivités (500-1500 eq-hab)*, Guide, Office International de l'Eau, Office des publications officielles des Communautés européennes, Luxembourg, 2001.
12. WHO, World Health Organization, *Health Guidelines for the Use of Wastewater in Agriculture and Aquaculture*, Report of a WHO Scientific Group, Technical Report Series 778, World Health Organization, Geneva, Switzerland, 1989.
13. WEF Manual of Practices 11, *Operation of Municipal Wastewater Treatment Plants*, Technical Practice Committee Control Group, WEF, Alexandria, 1990.
14. Lazarova, V., Levine, B., and Renaud, P., Wastewater reclamation in Africa: assessment of the reuse applications and available technologies, in *Actes IXème Congrès de l'Union Africaine des Distributeurs d'Eau*, Casablanca, February 16–20, 1998.
15. Aguilera, G.S., Audic, J.M., and Geneys, C., Rhizopur: a green solution for the treatment of wastewaters of small communities, in *Proc. 9th International IWA Specialist Group Conference on Wetlands Systems*, Avignon, France, September 27–30, 2004.
16. Brissaud, F., and Lesavre, J., Infiltration percolation in France: 10 years experience, *Wat. Sci. Techn.*, 28, 10, 73–81, 1993.
17. Icekson-Tal, N., Abraham, O., Sack, J., and Cikurel, H., Water reuse in Israel–the Dan Region Project: evaluation of water quality and reliability of plant's operation, *Wat. Sci. Techn.: Water Supply*, 3, 4, 231–237, 2003.
18. Idelovitch, E., Icekson-Tal, N., Abraham, O., and Michall, M., The long-term performance of soil aquifer treatment (SAT) for effluent reuse, *Wat. Sci. Techn.: Water Supply*, 3, 4, 239–246, 2003.
19. Lazarova, V., Wastewater disinfection: assessment of the available technologies for water reclamation, in *Water Conservation* Vol. 3. *Water Management, Purification and Conservation in Arid Climates*, Goosen, M.F.A. and Shayya, W.H., eds., Technomic Publishing Co. Inc., 2000, 171.
20. Connell, G.F., *The Chlorination Dechlorination Handbook*, 2002. Water Environment Federation (WEF), Alexandria, VA, 2002.

21. USEPA, *Guidelines for Water Reuse*, manual, EPA (United States Environmental Protection Agency) & USAID (United States Agency for International Development), Cincinnati, OH, 1992.
22. State of California, *Water recycling criteria*, California Code of Regulations, Title 22, Division 4, Chapter 3, California Department of Health Services, Sacramento, CA, 2000.
23. King, C.H., Shotts, E.B., Wooley, R.E., and Porter, K.G., Survival of coliforms and bacterial pathogens within protozoa during chlorination. *Appl. Environ. Microbiol.*, 54, 12, 3023–3033, 1988.
24. Venczel, L.V., Arrowood, M., Hurd, M., and Sobsey, M.D., Inactivation of Cryptosporidium parvum oocysts and *Clostridium perfringens* spores by a mixed-oxidant disinfectant and by free chlorine, *Appl. Environ. Microbiol.*, 63, 4, 1598–1601, 1997.
25. Abarnou, A., and Miossec, L., Chlorinated waters discharged to the marine environment chemistry and environmental impact. An overview, *Sci. Total Environ.*, 126, 173–197, 1992.
26. Szal, G.M., Nola, P.M., Kennedy, L.E., Barr, C.P., and Bilger, M.D., The toxicity of chlorinated wastewater: instream and laboratory case studies, *Res. J. WPCF*, 63, 6, 910 920, 1991.
27. Heltz, G.R., and Nweke, A.C., Incompleteness of wastewater dechlorination, *Environ. Sci. Tech.*, 29, 4, 1018–1022, 1995.
28. Huang, J.L., Wang, L., Ren, N.Q., Liu, X.L., Sun, R.F., and Yang, G.L., Disinfection effects of chlorine dioxide on viruses, algae and animal planktons in water, *Wat. Res.*, 31, 3, 455–460, 1997.
29. Junli, H., Li, W., Nenqi, L., Li, L.X., Fun, S.R., and Guanle, Y., Disinfection effect of chlorine dioxide on viruses, algae and animal planktons in water, *Wat. Res.*, 31, 3, 455–460, 1997.
30. Savoye, P., Janex, M.L., and Lazarova, V., Wastewater disinfection by low-pressure UV and ozone: a design approach based on water quality, *Wat. Sci. Techn.*, 43, 10, 163–171, 2001.
31. Lazarova, V., Savoye, P., Janex, M.L., Blatchley, III, E.R., and Pommepuy, M., Advanced wastewater disinfection technonogies: state of the art and perspective, *Wat. Sci. Techn.*, 40, 4/5, 203, 1999.
32. National Water Research Institute, NWRI and American Water Works Association Research Foundation, AWWARF, *Ultraviolet Disinfection Guidelines for Drinking Water and Water Reuse*, NWRI-00 03, Fountain Valley, CA, 2000.
33. Lazarova, V., et al., Wastewater disinfection by UV: evaluation of the MS2 phages as a biodosimeter for plant design, in *Proc. WaterReuse Asssos. Symposium XV*, Sept 12–15, Napa, CA, 2000.
34. United States Environment Protection Agency (USEPA), *Ultraviolet Disinfection Guidance Manual*, United States Environment Protection Agency, Office of Water (4601), EPA 815-D-03-007, Draft, Washington DC, June 2003.
35. Darby, J., Emerick, R., Loge, F., and Tchobanoglous, G., Effect of Upstream Treatment Processes on UV Disinfection Performance, Water Environment Research Foundation (WERF), Alexandria, VA, 1999.
36. Savoye, P., et al., Evaluation of UV disinfection for wastewater recycling at West Basin, California: technical and sanitary aspects, in *Proc.73rd Annual Conference & Exposition on Water Quality and Wastewater Treatment, WEFTEC 2000*, Anaheim, CA, October, 10–14, 2000, Session 15.

37. Lazarova, V., and Savoye, Ph., *Guidelines for UV Wastewater Disinfection,* Internal Report Suez Environment—CIRSEE, 2004.
38. Lazarova, V., Technical and economic evaluation of UV and ozone for wastewater disinfection, in *Proc. IWSA World Water Congress, Symp. IWA/AIDIS*, Reuse of treated sewage effluents, September 18, Buenos Aires, 1999.
39. Xu, P., Janex, M.L., Savoye, P., Cockx, A., and Lazarova, V., Wastewater disinfection by ozone: main parameters for process design, *Wat. Res.*, 36, 4, 1043–1055, 2002.
40. Gottberg, A., and Vaccaro, G., Kuwait's giant membrane plant starts to take shape, *Desalin. Wat. Reuse*, 13, 2, 30, 2003.
41. Manem, J., and Sanderson, R., Membrane bioreactors, in *Water Treatment Membrane Processes*, Mallevialle J., Odendaal P., and Wiesner M.R., eds., McGraw-Hill, New York, 1996, chapter 17.
42. Stephenson, T., Judd, S., Jefferson, B., and Brindle, K., *Membrane Bioreactors for Wastewater Treatment*, IWA Publishing, London, 2000.
43. Wallis-Lage, C., MBR systems: similarities and differences between manufacturers, in *Proc.WEFTEC*, Los Angeles, CA, WEF, 2003.
44. Foussereau, X., Roderick, P., and Sudhanva, P., The current status of the use of membranes for wastewater treatment, in *Proc. WEFTEC*, Los Angeles, CA, WEF, 2003.
45. Thompson, D., Clinghab, P., Scheinder, C., and Thibault, N., Municipal membrane biorector in the Northeast, in *Proc. WEFTEC*, Los Angeles, CA, WEF, 2003.
46. Sethi, S., Juby, G., Schuler, P., and Holmes, L., Evaluation of MF for microbial removal in reuse applications: performance assessment from three pilot studies, in *Proc. AWWA Annual Conf.*, Nashville, TN, AWWA, 2001.
47. Juanicó, M., Process design and operation, in *Reservoirs for Wastewater Storage and Reuse*, Juanicó, M. and Dor, I., eds., Springer, New York, 2002.
48. Indelicato, S., Barbagallo, S., and Cirelli, G., Change in wastewater quality during seasonal storage, in *Proc. Conf. Natural and Constructed Wetlands Wastewater Treatment and Reuse*, Perugia, Italy, 195, 1995.
49. Juanicó, M., The performance of batch stabilization reservoirs for wastewater treatment, storage and reuse in Israel, *Wat. Res.*, 33, 10, 149, 1996.
50. Juanicó, M., and Shelef, G., The performance of stabilization reservoirs as a function of design and operation parameters, *Wat. Sci. Techn.*, 23, 7, 1991.
51. Liran, A., Juanicó, M., and Shelef, G., Bacterial removal in a stabilization reservoir for wastewater irrigation in Israel, *Wat. Res.*, 28, 6, 1305, 1994.
52. Miller, G., et al., *Impact of surface storage on reclaimed water: seasonal and long term,* Report 99-PUM-4, Water Environment Research Foundation, Alexandria, VA, and IWA Publishing, London, 2003.

8 Adverse Effects of Sewage Irrigation on Plants, Crops, Soil, and Groundwater

Herman Bouwer

CONTENTS

8.1 Toward a healthy environment and sustainable development 236
8.2 Compounds with potential adverse effects on recycled water for irrigation 237
8.3 Behavior of some compounds during irrigation with sewage effluent 238
8.3.1 Salt and water relations in irrigation soils 239
8.3.2 Behavior and potential adverse effects of nutrients in irrigation soils 244
8.3.3 Effects of disinfection by-products on groundwater 247
8.3.4 Effects of pharmaceuticals and other organic contaminants 248
8.4 Salt and groundwater water-table management for sustainable irrigation 253
8.4.1 Salt loadings 254
8.4.2 Salt tolerance of plants 255
8.4.3 Management of salty water 256
8.4.4 Future aspects for salinity management in south-central Arizona 259
References 260

1-56670-649-1/05/$0.00 + $1.50

8.1 TOWARD A HEALTHY ENVIRONMENT AND SUSTAINABLE DEVELOPMENT

There was a time when human and other wastes were simply thrown out of the window. The only environmental concern then was a direct hit on the people in the street. Thus, as a courtesy to passers-by, the thrower would yell "gardez l'eau" (watch out for the water). This term was anglicized to gardyloo, which is the name of a British ship used for ocean dumping of municipal sludge. From the streets, the waste could readily run into streams or other surface water, along with raw wastewater from the early sewers. The discovery that sewage contamination of drinking water was the main cause of outbreaks of diseases such as dysentery, cholera, and typhoid made it necessary to keep wastewater out of surface water, since adequate treatment and disinfection technology for drinking water had not yet been developed. This led to the establishment of "sewage farms" around many cities. Applying sewage to land rather than discharging it into surface water was an early form of zero discharge.

In the early part of the previous century, disinfection of drinking water by chlorination was discovered and put into use. This allowed the discharge of sewage into surface water to be resumed because now the surface water could be treated and disinfected for drinking, and microbiological contamination was no longer a health problem.

As a result, the average human life expectancy increased dramatically. Also, the cities were growing and needed the sewage farms around them for more streets and houses. As the sewage farms disappeared, better sewage-treatment processes were developed and applied, primarily to prevent undue oxygen "sags" in the streams and to not exceed the "assimilative" and "self-purification" power of the receiving water. Removal of biochemical oxygen demand (BOD) and suspended solids (SS) were the main objectives. Then came the era of "better living through chemistry," causing more and more chemicals to enter sewage through discharges from households, industries, hospitals, etc.

This era was followed by increasing environmental awareness and a realization that pollution of surface water should be drastically minimized or avoided. Now environmental concerns are calling for increasingly stringent standards for discharging sewage or other wastewater into surface water to prevent eutrophication and to protect aquatic life, recreational areas, and water reuse opportunities. For example, the U.S. Clean Water Act of 1972[1] called for the elimination of all pollutant discharges into the nation's waters. Until recently this law was not strongly enforced, but that is changing now with the introduction of the total maximum daily load (TMDL) principle, which also applies to nonpoint (watershed) sources of surface water pollution.[2]

The trend towards more stringent regulations will undoubtedly continue until wastewater treatment becomes so expensive that municipalities may want to stop discharging their sewage effluent into the surface water and use it themselves. When that happens, we will have come full circle to zero discharge. Reuse and zero discharge of wastewater are the ultimate forms of prevention of point-source pollution of surface water. Combined with stricter regulations on non–point-source pollution, such as watershed runoff, this should maintain drinkable, swimmable, fishable, and optimum recreational conditions in our rivers and streams.

8.2 COMPOUNDS WITH POTENTIAL ADVERSE EFFECTS ON RECYCLED WATER FOR IRRIGATION

In principle, sewage effluent or other wastewater can be used for any purpose, provided it is treated to meet the quality requirements of the intended use. As mentioned in Chapter 1, due to treatment costs and economic feasibility, wastewater will most commonly be used for nonpotable purposes that do not require water of a very high quality, as does drinking water. These nonpotable purposes include industrial use (power plant cooling, processing, construction, aggregate washing, dust control, etc.) and agricultural and urban irrigation.

Since sewage treatment for irrigation use is primarily aimed at removing pathogens (primary and secondary treatment followed by granular media filtration and/or disinfection), the effluent may still contain a wide variety of other chemicals that can have adverse effects on plants and groundwater quality and its use for drinking and irrigation. This concern is especially acute for efficient irrigation in dry climates, where chemical concentrations in the drainage water or deep-percolation can be many times those in the wastewater itself (typically by factors of 2–10, depending on rainfall and irrigation efficiency).

The chemicals of concern include salts, pesticide residues, nitrogen (mostly as nitrates in the drainage water), disinfection by-products, pharmaceutically active chemicals, other chemicals, and precursors of disinfection by-products, like humic substances and other dissolved organic matter, that form a new group of disinfection by-products when the groundwater is pumped up again and chlorinated or otherwise disinfected for drinking.

Most domestic effluents meet the normal chemical requirements for crop irrigation. Industrial discharges into the sewer system can cause excessively high concentrations of heavy metals and other trace elements.

In the case of irrigation, the most important factor in terms of potential agronomic adverse effects is salinity. All natural waters contain some salt, which is expressed as total dissolved solids (TDS). Rainfall and other atmospheric precipitation have the lowest TDS content, averaging about 10 mg/L.[3] Surface waters in streams and lakes have higher TDS contents because the water has been in contact with soil and rocks, from which it picks up dissolved minerals and other constituents. Also, water evaporates from the watershed, which increases TDS in the remaining water. TDS contents of surface water typically are on the order of a few tens to a few hundred mg/L. As mentioned in Chapter 2, urban uses add about 300 ± 100 mg/L of dissolved salts. This is due to the addition of salts and other chemicals in homes and industries and the removal of distilled or very pure water by evaporation (evaporative coolers or cooling towers) or membrane filtration (reverse osmosis) by industries needing ultra-pure water and putting the reject brine into the sewers. If all the reject brines are returned to the sewer and all the "good" water is not returned because of, for example, outdoor use and evaporation, sewage effluent has a higher salt content than the input water.

The TDS increase from one cycle of municipal use also depends on the in-house water use, which may vary from 400 L per person per day in water-rich countries to 60 L per person per day or less in water-poor countries. A great deal of water also evaporates during the agricultural and urban irrigation of crops, plants, and turf, leaving salts behind in the soil which must be leached out of the root zone by applying more irrigation water than needed to meet the evaporative needs of the plants. The salty "deep-percolation" water created by this leaching moves down to underlying groundwater, where it increases the salt content of the groundwater and causes groundwater levels to rise where there is not enough drainage or groundwater pumping.

Of course, the biggest evaporators are the oceans themselves, the salt content of which is now about 35,000 mg/L. Oceans contain about 97% of the global water.[3] Of the remaining 3%, about 2% is in the form of snow and ice in our polar regions and mountain ranges. This leaves only about 1% as liquid fresh water, almost all of which occurs as groundwater and very little as surface water, which often is fed by groundwater. This shows the importance of groundwater and the need for proper management of that resource to prevent depletion and quality degradation.

Sometimes toxic chemicals are leached from the soil. Serious environmental problems are resulting from such leaching.[4] The major concern in this case study was the selenium leaching in drainage water from irrigated land in California's Central Valley that was discharged into Lake Kesterson.

8.3 BEHAVIOR OF SOME COMPOUNDS DURING IRRIGATION WITH SEWAGE EFFLUENT

Increasingly, sewage effluent will be used for urban and agricultural irrigation. The main concern, as discussed in Chapter 3, is the potential for infectious diseases in farm workers and city dwellers exposed to the effluent, as well as in people who consume crops irrigated with effluent, especially when those crops are eaten raw or brought raw into the kitchen.

Prevention of diseases requires adequate disinfection of the effluent to meet the strict California Water Recycling Criteria (Table 8.1)[5] or the less strict WHO guidelines[6] that are more achievable in developing countries. The effluent also must meet normal irrigation water requirements for parameters such as salt content, boron, sodium adsorption ratio, and trace elements,[7,8] as presented in detail in Chapter 5 (see Tables 5.6, 5.11 and 5.13–5.17).

Table 8.1 California Title 22 Regulations for Fully Treated Effluent

Reuse conditions	Unrestricted irrigation, including spray and surface irrigation of food crops consumed raw and high-exposure landscape irrigation as in parks, playgrounds, and residential yards
Total coliforms	Median not to exceed 2.2/100 mL, single sample not to exceed 23/100 mL in 30 days
Wastewater treatment requirements	Secondary treatment followed by tertiary treatment consisting of filtration and disinfections

Source: Ref. 5.

Unfortunately, little or no attention is paid to the long-term effects of sewage irrigation on underlying groundwater. Since most of the water applied for irrigation in dry climates evaporates, the concentrations of nonbiodegradable chemicals in the drainage or deep-percolation water going down to the groundwater can be much higher than in the effluent itself (about five times higher for an irrigation efficiency of 80%). These chemicals comprise not only the salts, nitrates, and possibly pesticide residues normally expected in irrigated agriculture, but also "sewage chemicals" like synthetic organic compounds, disinfection by-products, including the very carcinogenic compound nitrosodimethyl-amine (NDMA),[9] pharmaceuticals, and pharmaceutically active chemicals such as the endocrine disrupters fulvic and humic acids.

The humic acids, for example, are known precursors of disinfection by-products formed when drainage water ends up in drinking water supplies that are then chlorinated. Thus, groundwater below sewage irrigated areas may become unfit for drinking, which in some cases may raise questions of liability. More research on the long-term effects of sewage irrigation on groundwater is urgently needed. Long-range planning is necessary to ensure sustainability of such irrigation with minimum adverse environmental and health effects.

Two parameters of special concern in irrigation with sewage effluent are the salinity or salt content of the effluent and the nitrogen concentration.

8.3.1 Salt and Water Relations in Irrigation Soils

To avoid salt accumulation in the root zone of irrigated land, more irrigation water needs to be applied with irrigation than is needed for crop evapotranspiration (see also Chapter 5, § 5.5.1). This causes excess water to move down through the root zone and then out of the root zone to underlying groundwater with the salts that had accumulated in the root zone after the previous irrigation. This process is described in its simplest form by Equation (8.1):

$$D_i C_i = D_d C_d \tag{8.1}$$

where:

D_i = amount of irrigation water applied

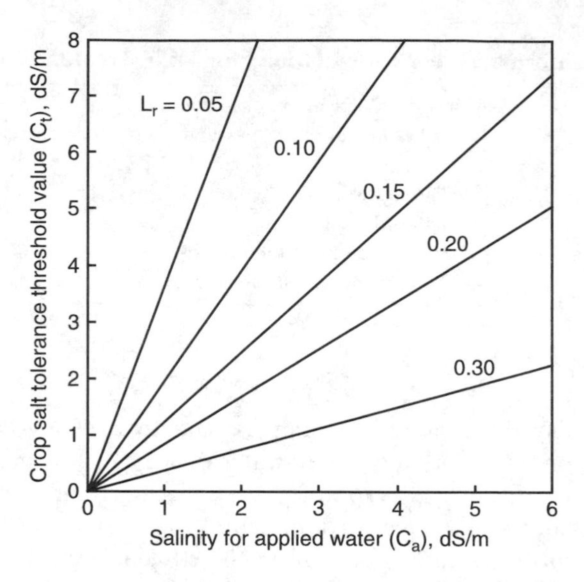

Figure 8.1 Leaching requirement (L_r) as a function of the salinity of the applied water and salt-tolerant threshold value. (From Ref. 11, with permission.)

C_i = salt concentration of irrigation water
D_d = amount of drainage water leaving the root zone
C_c = salt concentration of drainage water

D_i is equal to crop evaporation D_{ET} plus D_d. Where rainfall is significant, it should be included in the equation. The ratio D_d/D_i is the leaching requirement. The leaching requirement increases with increasing C_i if crop yield reductions are to be minimized. Also, the lower the salt tolerance of the crop, the higher the leaching requirement becomes if normal crop yields are to be obtained, as shown in Figure 8.1.[10,11] Thus, for a crop with an annual D_{ET} of 1.5 m/year and a leaching requirement of 0.3 m/year and irrigation with water that has a salt concentration of 500 mg/L, the salts in 1.8 m are leached out in 0.3 m, so that the salt concentration C_d of the drainage water at salt balance is as follows:

$$1.8 \times 500 = 0.3 \times C_d \text{ or } C_d = 900/0.3 = 3000 \text{ mg/L} \quad (8.2)$$

This is six times the concentration C_i of the irrigation water, making the water unsuitable for general irrigation and drinking. Other conservative constituents in the irrigation water also are concentrated six times in the drainage water.

The minimum leaching requirement for salt balance and acceptable salt levels in the root zone for normal crop yield depends on crop salinity tolerance and salt content of irrigation water, as shown in Figure 8.2.[11–13] Typically, a leaching ratio of 10% is suitable for most cases, giving a maximum irrigation efficiency of 90%. Most farm irrigation systems have efficiencies well below 90% (see Chapter 5, Table 5.3). Well-designed and well-managed irrigation

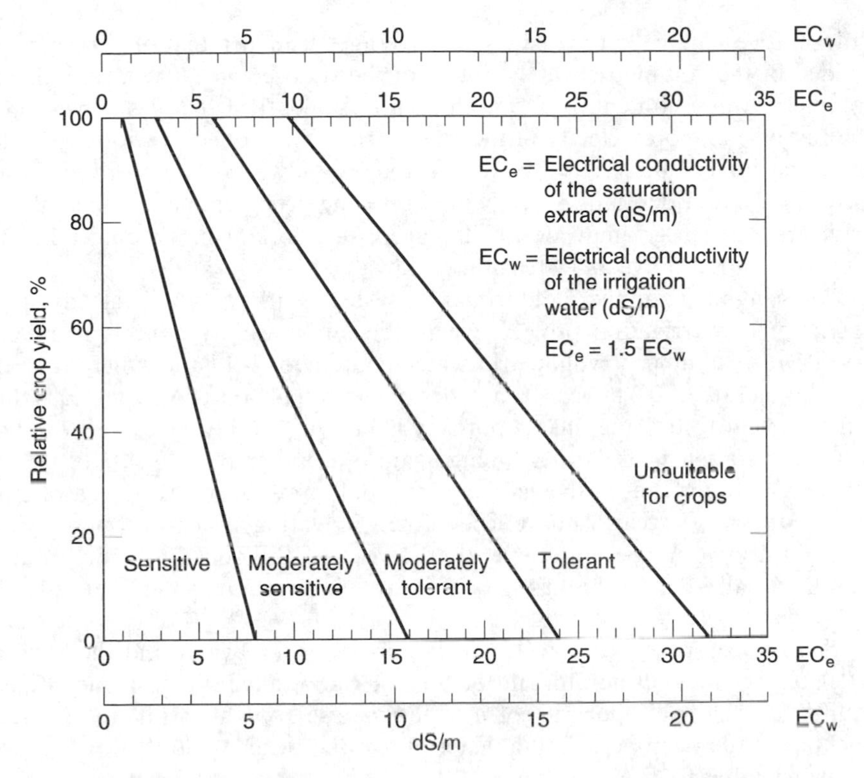

Figure 8.2 Relative salt tolerance ratings of agricultural crops. (From Ref. 13, with permission.)

systems may have an efficiency of about 80%. Many surface irrigation systems have much lower efficiencies, for example, 60% or less.

The higher the TDS of the irrigation water, the larger the amounts and frequencies of leaching need to be. Thus, normal inefficiencies of irrigation systems often are more than sufficient for adequate leaching of salts and other chemicals out of the root zone. This leaching avoids build-up of salts and other chemicals in the soil and maintains a salt or chemical balance for the root zone. Eventually, however, these chemicals will show up in underlying groundwater and from there in surface water via natural drainage of groundwater into surface water, via discharge from ditch or tile drains or from pumped drainage wells, or via sewage effluent discharges in areas where the affected groundwater is first used for the municipal water supply. A sustained irrigation efficiency of 100%, as advocated by some, is only possible if distilled water or other water with a TDS content of zero is used for irrigation.

While downward flow of deep percolation water below the root zone is unsteady and occurs in pulses after each irrigation, the pulses flatten out with depth so that actual downward water velocities or pore velocities deeper in the vadose zone can be estimated as the average deep percolation Darcy flux divided by the volumetric water content of the vadose zone. Thus, assuming a volumetric water content of 0.15 in the vadose zone, the 0.3 m/year downward macroscopic or pore velocity in the vadose zone mentioned previously would be on the order of 0.3/0.15 = 2 m/year. If the groundwater is relatively shallow, for example, at a depth of 3 m, it would take the deep percolation then about 1.5 years to reach groundwater. If it is deep, such as 100 m, it would take the deep percolation 50 years to reach groundwater.

These numbers apply to old irrigation systems with essentially steady-state flow. For new irrigation projects where the initial vadose zone is relatively dry at, for example, a volumetric water content of 0.1, and where a deep percolation rate of 0.3 m/year from a new irrigation project would increase this water content to 0.3, the fillable porosity would be the difference between the two water contents, or 0.2. Assuming again a deep percolation rate of 0.3 m/year, the wetting front in the vadose zone would move downward at a rate of 0.3/0.2 = 1.5 m per year. Thus, effects of new irrigation systems on underlying groundwater would be noticeable after 2 years if the depth to groundwater is rather small—3 m in this case—or 67 years if the groundwater is at a depth of 100 m.

As the deep-percolation water arrives at the groundwater and the aquifer is unconfined, it will accumulate on top of the aquifer. For an unconfined aquifer and a fillable porosity of n in the vadose zone above it, the vertical stacking of the deep-percolation water above the water table then will cause the water table to rise a distance of D_d/n, where D_d is the deep-percolation flow. Thus, if D_d is 0.3 m/year and $n = 0.25$, the groundwater table would rise 1.2 m/year, assuming that there are no other additions or reductions in groundwater due to, for example, artificial or natural recharge, pumping from wells, or lateral flow in the aquifer. This rise has been observed in practice in the southeastern part of the Salt River Valley of Phoenix, Arizona, which still has much irrigation. When groundwater pumping for irrigation was stopped and surface water was used more often for irrigation, groundwater levels in the area rose about 0.6 m/year, while the TDS and nitrate contents of the well water rose 500–1300 mg/L and 5–15 mg/L, respectively.[14] Eventually, where groundwater is pumped for irrigation, the increase in TDS can decrease crop yields (see Figure 8.2) and restrict the choice of crops to the more salt-tolerant types (see Chapter 5, Table 5.8).

The pore velocity in the vadose zone of 2 m/year and the water table rise of 1.2 m/year in the previous example are based on year-round irrigation. For more seasonal irrigation, with only one crop per year and fallowing between crops, these values will be less, closer to about 1 m/year for the pore velocity in the vadose zone and about 0.6 m/year for the rise of the groundwater table. For mixed irrigated agriculture with a combination of seasonal and year-round irrigation, downward pore velocities in the vadose zone thus may range between 0.6 and 1.5 m/year, and groundwater rises may be between 1 and 2 m/year. Thus, the long-term effects of irrigation on underlying groundwater are water-quality degradation and rising groundwater levels.

On the other hand, where overpumping occurs and groundwater levels are dropping, arrival rates of deep-percolation water at the groundwater are reduced and can even reach zero if groundwater levels are dropping faster than the pore velocity of the deep-percolation water in the vadose zone. Of course, groundwater pumping and depletion cannot go on forever, so when groundwater level declines are reduced to where the deep-percolation water can "catch up" with the water table, slower declines and even rising groundwater levels and significant groundwater-quality reductions can be expected.

Urban irrigation can also cause groundwater levels to rise. For example, groundwater levels rose from a depth of about 36 m to a depth of about 15 m in a few decades below an old residential area with flood irrigated yards in north central Phoenix, Arizona. This rise was mainly in response to shutting down several large-capacity irrigation and water supply wells in the area. The rate of rise of the groundwater level in the affected area then became about 0.3–0.6 m per year. At one area (Camelback and Central), rising groundwater levels flooded the lowest level of a five-level underground parking garage below an office building. Initially, groundwater levels were adequately controlled by draining the ABC layer (mostly sand and gravel) below the concrete floor slab. Eventually, however, wells had to be installed around the building to lower groundwater levels. The discharge water from the wells was contaminated by local leaking underground fuel tanks. This required expensive treatment of the water before it could be discharged into a storm drain.

The effect of deep-percolation water moving to the aquifer and entering pumping wells will still be gradual, as the drainage water stacks up on top of the natural groundwater and only slowly moves deeper into the aquifer and finally into the wells. Wells that are perforated or screened deeper into the natural groundwater at first will not show TDS increases in the water pumped from that well. Only when enough agricultural drainage water is stacked up above the natural groundwater will some of it be drawn into the well when it is pumping. More and more of the salty drainage water will then enter the well as it continues to be pumped. Consequently, the cone of groundwater depression around the well produces vertical gradients, which will cause salty upper groundwater to move deeper into the aquifer and the well.

The portion of salty deep-percolation water in the well discharge is a function of time of pumping. Since the contaminated water will remain mostly in the upper part of the aquifer according to the vertical stacking principle, wells with their screens or perforated sections near the water table will show the quality degradation first. Wells in unconfined aquifers with deeper screens will be affected later, as pumping produces vertical flow components in the aquifer and upper groundwater is drawn deeper into the aquifer and into the well, even if the deeper aquifers are semi-confined. Eventually, wells may produce mostly deep-percolation water from the irrigation practices. Such water will not meet drinking water standards and may be too salty for general agricultural use. Options then include blending the well water with better quality water, drilling the wells deeper or sealing off upper portions of screens to buy more time before the well water gets saltier, and treatment of the well water with, for example, reverse osmosis which, of course, is expensive and produces a reject brine that may present disposal problems.

Where deep percolation rates are very small, as with very efficient irrigation systems, deficit irrigation, or low-water-use landscaping (xeriscapes), evaporation of water deeper in the vadose zone may become significant. Deep-percolation rates will then decrease with depth to the point where TDS concentrations become so high that salts precipitate in the vadose zone and maybe even in the root zone itself, which would have adverse effects on the plants. Low deep-percolation rates would cause water contents in the soil of the vadose zone to be low, which would increase the permeability of the soil to air. Evaporation of water in the vadose zone could then be caused by diurnal barometric pressure variations that typically occur in desert environments in the absence of major weather systems moving through. Barometric pressures then increase during the night when the air cools down and becomes heavier and decrease during the day as the air warms up again and becomes lighter. This could cause the vadose zone to "breathe," "inhaling" dry atmospheric air during the night that causes vadose zone water to evaporate into the soil air, and "exhaling" this damp vadose zone air into the atmosphere during the day. This "deep" evaporation could cause significant amounts of salt to precipitate and, hence, to be stored in the vadose zone, which reduces the salt and water loads on the underlying groundwater. More research on this phenomenon is necessary, especially on long-term effects to determine if salts could build up to the point where they form caliche-like layers that impede downward movement of water and could cause water logging of the upper soil, evaporation from the soil surface, and formation of salt flats where nothing will grow.

The main physical effect on soil of an increase in TDS of water applied to that soil is a change in soil structure and resulting change in hydraulic conductivity, K. A decrease in K may be undesirable where the effluent is used for irrigation because it will reduce infiltration rates and will adversely affect the structure of the soil. For sewage effluent, the TDS increase from about 200 to 400 mg/L per cycle of municipal use is often accompanied by a significant increase in sodium chloride (NaCl) concentration. This also increases the sodium adsorption ratio (SAR) of the water, which may decrease K of the soil. On the other hand, an increase in TDS may increase K of the soil. The combined effect of these two parameters on K is shown in Figure 8.3.[11] In reality, the relation between K, SAR, and TDS is more scattered than indicated by the curves in Figure 8.3.

8.3.2 Behavior and Potential Adverse Effects of Nutrients in Irrigation Soils

Another concern where sewage effluent is used for irrigation is nitrogen, which may be present as organic, ammonia, or nitrate nitrogen. All of these nitrogen compounds tend to be converted to nitrate in the soil and vadose zone.

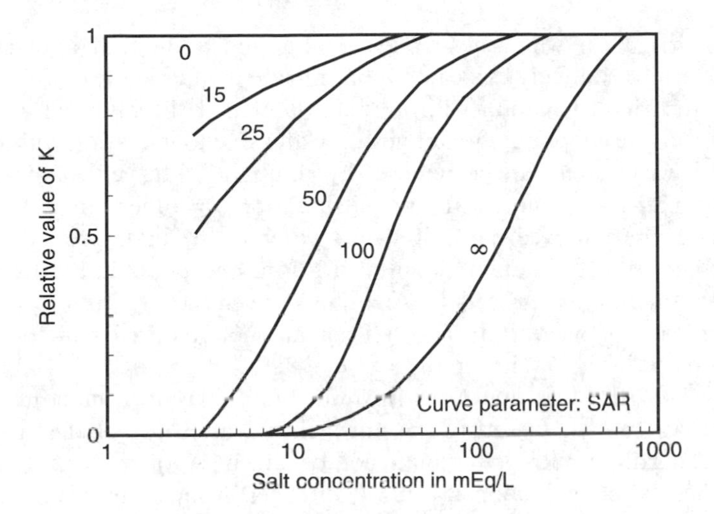

Figure 8.3 Effect of SAR and salt concentration of soil solution on hydraulic conductivity of Pachappa sandy loam.

The main losses of nitrogen from the soil system would occur by denitrification and possibly by the anammox process.[15] This is a recently discovered process that is autotrophic and takes place under anaerobic conditions. The process requires the presence of approximately equal amounts of nitrate and ammonium nitrogen, of which about 90% can be removed, leaving a 10% residue that is mostly in the nitrate form.[15] While crop uptake of nitrogen is very significant, it is not enough to prevent nitrates from being leached out of the root zone and move downward with the deep-percolation water to underlying groundwater.

Nitrogen fertilizer requirements are about 50–500 kg/ha per crop.[7,16] As a rule-of-thumb, half of this nitrogen is absorbed by the crop, one fourth is lost by denitrification and returns to the atmosphere as nitrogen gas and oxides of nitrogen, and one fourth or up to 67 kg/ha is leached out of the root zone as nitrate in the deep percolation water.[17] Assuming 0.3 m/year of deep-percolation water and 400 kgN/ha of nitrogen applied as fertilizer, of which 100 kgN/ha is leached out of the root zone, this would give a nitrate nitrogen concentration in the deep percolation water of 33 mg/L. This is well above the maximum limit of 10 mg/L for drinking water.

Conventionally treated secondary sewage effluent (activated sludge) may contain about 30 mg/L total nitrogen, mostly as ammonium.[18] If this effluent was used for irrigation with a total application of 1.8 m per year or growing season, the amount of nitrogen applied with the water would be about 540 kg/ha per year or growing season, more than twice the average requirement. Assuming no luxury uptake of nitrogen by the crop, about one fourth of this nitrogen is leached out as nitrate with the deep percolation water and, assuming also that the irrigation efficiency again is about 80%, would then give a nitrate nitrogen concentration in the drainage water of about 37 mg/L.

Thus, irrigation with sewage effluent and no additional application of nitrogen fertilizer already can cause more nitrate contamination of underlying groundwater than irrigation with normal water and the nitrogen applied as fertilizer. Nitrate contamination of groundwater due to irrigation with recycled municipal wastewater can be reduced by removing nitrogen in the sewage-treatment plant with nitrification-denitrification or other processes. Also, nitrogen can be removed naturally from water in the underground environment by denitrification, ammonium adsorption, and possibly by the recently discovered anammox process.[15] Anammox bacteria are autotrophic and anaerobic, so organic carbon and oxygen are not required and the process can take place in the aquifer itself.

If sewage effluent is used for irrigation, the nitrogen in the effluent often is more than enough to satisfy the nitrogen requirements of the crops, and additional fertilizer nitrogen should not be given. As a matter of fact, the effluent may already contain too much nitrogen, which can adversely affect not only the underlying groundwater, but also the crop itself. Adverse crop effects due to excess nitrogen include delay of harvest, too much vegetative growth and not enough reproductive growth (seeds), impaired quality of crop (reduced sugar contents in beets and cane, reduced starch content in potatoes), reduced yield of marketable fruit, and nitrate toxicity in people and animals consuming the crop.[16]

The nitrogen problem where crops are irrigated with sewage effluent mainly stems from the inability of farmers to properly schedule irrigations with sewage effluent and to get both the desired amount of water into the ground at the desired time from an irrigation standpoint and the desired amount of nitrogen therein from an agronomic standpoint. Often this means that the crops get too much nitrogen at the wrong times in the growing season. Giving the crops too much nitrogen usually produces too much vegetative growth (stems and leaves) and not enough reproductive growth (flowers, fruit, and seed). For example, nitrogen applications to sugar beets and sugar cane should be stopped toward the end of the growing season to get more sugar stored in the roots (sugar beets) and stems (sugar cane), thus increasing the sugar content and also the percentage of the sugar that readily crystallizes into "sugar" in the refinery.

For fruit crops, the yield of marketable produce can be reduced if extensive amounts of nitrogen are applied to the crop, especially if the crop is a perennial like apples, pears, peaches, oranges, etc. Sometimes the total yield (in kg/ha) may not be reduced, but the maturation will be delayed, fruit sizes may be decreased, or the quality of the fruit (e.g., texture and taste) may be adversely affected. Too much nitrogen on potatoes may produce fewer and smaller tubers with lower sugar content, probably due to the excess vegetative growth resulting from too much nitrogen fertilizer application. Navel and Valencia oranges have produced grainy and pulpy fruit with less juice than trees receiving normal nitrogen applications. Also, overfertilized Valencia oranges show regreening of the rind, reducing their marketability. Delays in fruit maturation due to excessive nitrogen application were also observed in apricots and peaches, so that they missed the early markets and the higher prices that they command. Excessive vegetative growth also gives production problems like fruit rot in melons and grapes due to more leaves and shade that increases moisture levels inside the plant canopy. Increased lodging of grain crops like wheat, barley, and oats caused by increased vegetative growth due to excessive nitrogen in the soil can create harvesting problems.[16]

High nitrate levels in food can also produce problems in persons and animals consuming those foods, as nitrate is reduced to nitrite during digestion.

Phosphates in sewage effluent are essential for plant growth and may not be toxic even at high concentrations. However, excessive applications of phosphates may cause deficiencies of mobile copper and zinc in the soil, which are important micronutrients. In high-pH soils, phosphates precipitate out, mostly as Ca and Mg phosphates. As a rule, phosphate content in recycled urban wastewater is lower than a crop's needs, and consequently, the addition of phosphate fertilizers is necessary during irrigation with recycled water.

8.3.3 Effects of Disinfection By-Products on Groundwater

Unfortunately, the normal water-quality requirements for irrigation with municipal wastewater do not address disinfection by-products, pharmaceutically active chemicals, humic substances, and other potential contaminants (see Chapters 3 and 5). Disinfection by-products may already be present in the wastewater entering the treatment plant due to chlorination of the drinking water. By-products can also be formed by chlorination of effluent from the wastewater treatment plant, particularly with the high chlorine doses and long contact times used for Title 22–type tertiary treatment.

Disinfection with ultraviolet (UV) irradiation after granular media filtration would give lower disinfection by-product levels in the tertiary effluent. Some disinfection by-products, like trihalomethanes and haloacetic acids, have been found to be biodegradable in aquifers near aquifer storage and recovery wells.[19] Still, disinfection by-products comprise a whole suite of halogenated organic compounds with yet-to-be-discovered identities, fates in the underground environment and health effects. Trihalomethanes can include chloroform and bromodichloromethane. The latter is of concern because it may increase miscarriages in women.

As with the dissolved salts, concentrations of refractory and nonvolatile disinfection by-products in the deep-percolation or drainage water would also be about five times higher than those in the effluent used for irrigation, again assuming an irrigation efficiency of 80%. Because of their potential toxicity and carcinogenicity, disinfection by-product levels in drinking water are continually scrutinized, and maximum contaminant levels (MCLs) may be lowered in the future (see Chapter 2, Table 2.4). For example, the U.S. EPA[20] has lowered the drinking water MCL for trihalomethanes from 100 to 80 μg/L. This does not bode well for potable use of groundwater affected by deep-percolation water from effluent-irrigated areas.

The challenge for the drinking-water industry of balancing disinfection by-product formation against microbial control[21] also applies to water reuse issues. The choice is between the possibility of immediate acute illness caused by pathogens (diarrhea and worse) and much more serious diseases like cancer caused by chemicals after years of ingestion.

The high nutrient and organic carbon levels in effluent can be expected to enhance plant growth and bioactivity in the soil. Decaying roots and other plant parts and biomass can then form humic and fulvic acids as stable end-products. These are nonbiodegradable, and they are known disinfection by-product precursors in water that is to be chlorinated. Thus, when groundwater from below effluent-irrigated areas is pumped and chlorinated for potable use, a new suite of disinfection by-products can be formed.

8.3.4 Effects of Pharmaceuticals and Other Organic Contaminants

Concern also is rising about pharmaceutically active chemicals that have entered the sewers with domestic, industrial, pharmaceutical, and hospital waste discharges.[22]

8.3.4.1 Hormones

Sewage effluent usually contains a variety of hormones, which, when the effluent is used for irrigation, increases the endogenous production of hormones (phyto-hormones) in legumes like alfalfa. These phyto-hormones can then cause fertility problems in sheep and cattle that eat the forage.[23,24,29]

8.3.4.2 Endocrine disruptors

Other substances are not hormones themselves, but they disrupt the hormone (endocrine) system in the body. These endocrine disruptors (EDCs) can be hormones themselves or chemicals that interfere with the hormone system in the body. EDCs can function as hormones where they mimic and, hence, increase the normal hormone activity, or they may block hormone-binding sites where they decrease hormonal activity. Endocrine disruption can be caused not only by normal hormones, but also by other chemicals, such as PCBs and many others. It was indicated[25] that possibly more than 70,000 chemicals have endocrine disruptive potential, many of which have been detected in sewage effluent. However, hormones tend to be several thousand times more potent than industrial chemicals, pesticides, and metals.[26,27]

8.3.4.3 Behavior of Pharmaceuticals in Water Reuse Schemes and Natural Environment

Pharmaceutically active chemicals seem to survive wastewater treatment and may not adsorb well to soil particles, so they may be rather refractory in the underground environment. While these chemicals may not be directly toxic, they can produce adverse health effects by affecting the immune and hormone systems of animals and humans, i.e., they can act as endocrine disruptors.[28]

At least 45 chemicals have been identified as potential endocrine-disrupting contaminants, including industrial contaminants like dioxins and PCBs, insecticides like carbaryl and DDT, and herbicides like 2,4-D and atrazine.[29] More research is needed on the occurrence and fate of pharmaceutically active chemicals in the underground environment and about synergistic effects when a whole spectrum of pharmaceutically active chemicals and other contaminants occurs and is ingested.

To get some idea of the pharmaceuticals that can be expected nationwide in surface water, water samples from 139 U.S. streams were selected for monitoring with known sewage effluent discharges.[30] These samples were then analyzed in laboratories with equipment and procedures that could identify and determine concentrations of 95 different pharmaceuticals, from which 82 have been found. Thus, surface water into which sewage effluent is discharged is likely to contain a large assortment of pharmaceuticals.

Many of these chemicals are not removed by passage through soil, so that groundwater below losing streams, below land irrigated with effluent or effluent contaminated water, below septic tanks, and below systems for artificial recharge can contain pharmaceuticals or pharmaceutically active compounds.[31] This will be particularly true for agricultural soils with shallow groundwater (less than a few m deep, for example) where root activity, tillage, decomposing stems of plants and roots that have been plowed under, worm holes, etc., and spatial variability have created a system of macropores through which water with its dissolved chemicals can move rapidly to greater depth without interaction with the soil and the chemical and microbiological processes that otherwise would adsorb, accumulate, transform, or degrade undesirable chemicals.[17,32,33] This so-called preferential flow basically gives the water and chemicals a short-cut to greater depths, where the water could join groundwater. Thus, groundwater below sewage-irrigated fields can be expected to contain effluent chemicals and agricultural chemicals.

Concerns are rapidly rising and spreading about residues of pharmaceutical and personal care products (PPCPs) in sewage, their persistence in sewage-treatment plants and during percolation in soil which puts them into surface water and groundwater, and their health effects, which include interference with the hormone system, like endocrine disruption and feminization of male fish.[27,34] Unknown and unspecified toxic effects can also be expected, as would development of antibiotic-resistant bacteria by repeated exposure of the pathogens to antibiotic levels in wastewater and contaminated streams.[35] Some industrial wastes in sewage effluent like PCBs and other organic wastes may also have some of these health effects.

The fate of PPCPs in groundwater recharge systems has been studied[31] where secondary sewage effluent after filtration through tertiary filters was put into shallow basins for infiltration and recharge of underlying groundwater. The soils were predominantly alluvial sands and gravels, and the groundwater table was about 13–15 m below the bottom of the infiltration basins. Groundwater-monitoring wells had screened intervals from 23 to 56 m depth and from 14 to 25 m depth. One well was screened over its entire depth of 56 m in the aquifer. While DOC was reduced from 15 mg/L in the effluent to 2 mg/L in the aquifer, some pharmaceuticals like caffeine and analgesic/anti-inflammatory drugs such as diclofenac, ibuprofen, ketoprofen, naproxen, and fenoprofen and blood lipid regulators such as gemfibrozil were efficiently removed to concentrations near or below detection limits. The antiepileptics carbamazepine and primidone were not removed during groundwater recharge under either anoxic saturated or unsaturated or aerobic unsaturated flow conditions during underground travel times of up to 8 years.

Another application of improving water quality by underground movement of water is riverbank filtration, where water in rivers and streams is induced by groundwater pumping to infiltrate into the banks and bottoms and then to move laterally away from the stream to pumped wells that are installed a distance away from the stream bank. While improving the river water quality, the bank filtration process does not remove all undesirable organic chemicals.[36,37]

8.3.4.4 Management of Adverse Effects of Pharmaceutical Products in Irrigation Systems

Residuals of PPCPs in sewage effluent are entering the surface water environment through direct discharge of variously treated sewage effluents into streams and lakes. PPCPs can then enter groundwater, where the groundwater level is lower than the water level in affected streams or lakes, causing them to lose water to the underground environment. Other pathways to groundwater are via irrigation or artificial groundwater recharge systems that use sewage effluent or sewage contaminated water and via septic tank leach fields. Pharmaceuticals used in animal production and present in animal waste could enter surface water via surface runoff and groundwater via infiltration and deep percolation from farms and manured fields.

Health effects so far have been detected mainly in aquatic life (fish, amphibians) and animals up the food chain, but not positively in humans, although there are significant indications of potential adverse effects.[29] Even if the effects were known, eliminating PPCPs may be difficult, and some form of source control and treatment may be a first step to minimize their presence and concentrations in surface water and groundwater.

On the one hand, it may be argued that since the amounts of PPCPs ingested with drinking water are so small compared with the medical doses at which they are prescribed that significant adverse human health effects may be of no concern. Then again, there is little information on long-term and synergistic effects. There may still be biological effects, and some researchers[22] stated that PPCPs in water "should be avoided in principle."

For aquatic organisms, the exposure is maximum at complete immersion for 24 hours a day. Also, concentrations of the chemicals in organisms and animals increase up the food chain. Thus, there are real concerns about wildlife. PPCPs typically occur at ng/l levels, which seem completely insignificant considering that 1 ng/L is equivalent to 1 second in about 31,000 years. On the other hand, 1 ng/L of a compound with a molecular weight of 300 still contains 2×10^{12} molecules per liter, which for endocrine disruptors could still bind to a lot of hormone receptors.

Public opinion, as fostered through the media, must also be considered. Catchy headlines with words like "ecological disaster," "drugs in drinking water," or "AIDS-like symptoms" can easily stir up serious public concerns, to which scientists may be unable to respond adequately in the absence of adequate and reliable data. Insufficient information breeds concern, and concern leads to fear.

Historically, the United States has pursued a "straightforward and simple policy that no risk can be tolerated in the nation's food supply" and "that all food additives be proved safe before marketing and explicitly prohibits any food additive found to induce cancer in test animals",[38] and, by implication, also in drinking water. Despite public objections on the grounds that this inhibited freedom of choice, this policy was rigorously enforced until the late 1970s, when saccharine was discovered to cause cancer in rats. However, plans for taking saccharine off the market caused a serious public outcry against the federal government dictating what people could and could not eat. In response, the policy shifted to one of informing the public about risks and letting the people decide for themselves what they want to eat and drink. Thus, an informed public making its own decisions was the new policy, an acknowledgment that there is no such thing as a risk-free society.

This also led to more studies of carcinogens naturally occurring in food and of the risks associated with recreational activities and sports, risks in common human activities and environmental effects, occupational risks, and various cancer risks to show that life as a whole is not risk-free.[38] The matter of choice also is very important. Often people are willing to accept higher risks in eating habits, sports, recreational and other activities which they chose to do than in the quality of the food they buy or of the water that comes out of the tap, over which they have no control.[31,38]

The fear about carcinogens naturally occurring in food and drink or added artificially with food processing and in polluted water was somewhat attenuated by some studies,[39] which showed that chemicals caused cancer in rodent bioassays not because they were carcinogenic, but because they were administered in such high doses that they caused cell damage in the test animals. Subsequent cell division to heal the damaged tissue then could produce mutations that caused malignant tumors. Thus, the linear response theory to extrapolate positive responses from high doses in the bioassays to low doses that are more realistic in real life is flawed because it does not recognize threshold concentrations below which positive responses do not occur.[40] About half the chemicals found carcinogenic in such tests in actuality may not have been carcinogenic, but rather produced cancer because of the high doses administered to the test animals,[39] thus confirming Paracelsus' conclusion that it is the dose that makes the poison.

For hormonally active compounds, however, dose-response relations may not be linear and may even be odd-shaped like an inverted "U".[29] This makes extrapolation to different doses and exposures very problematic. Rodent bioassays also do not recognize genetic effects, as some chemicals caused cancer in mice but not in rats.[41] Thus, if rodent bioassay results cannot be transferred to different rodent species, how can they be transferred to humans? However, this may not be true for hormones, which may function basically the same in all mammals.[29]

Large systems such as the ecosystems we live in are inherently chaotic, even though they may appear to be in equilibrium. Therefore any technological fix of one variable will change all the other variables in a totally unpredictable manner. It is, therefore, inadvisable to make large-scale changes to correct pollution without numerous small-scale studies. This requires long-term database collection. Solutions based on small databases and extrapolated models are, therefore, not recommended. What is needed is ecological "common sense," e.g., if the fish are dying in the rivers used for drinking water, action must be taken immediately. If minor or localized disturbances are seen in wildlife ecology, the best course of action would be long-term data collection.

The true significance of PPCPs in the aquatic environment and in water supplies is still a big question. Adverse effects on aquatic life and microorganisms observed so far are serious enough to warrant more research, including effects on humans. Only if these effects are better understood can the public be sufficiently informed to make its own decisions. Because PPCPs play such an important role in the well-being of people and animals, some adverse effects on aquatic organisms living in affected water and on people and animals drinking that water may have to be accepted, just as side effects of medical drugs are accepted. The question then is, what is acceptable and what can be done about it?

Ideally, dose-response relationships are developed on which regulators can base appropriate maximum contaminant levels. Hormone-disrupting chemicals, however, may not follow classical dose-response relations.[29] Developing more biodegradable PCPPs is another avenue toward reducing their harmful impact. Phasing out a compound is, of course, an action of last resort. However, if there are serious enough concerns about a certain product, action may have to be taken before there is absolute scientific proof of harm. This is where eco-toxicology and eco-epidemiology become important.

Lessons can also be obtained from the accepted and established practice of potable use of municipal wastewater after it has had the usual in-plant treatments followed by rapid infiltration for recharge of groundwater and soil-aquifer-treatment (SAT) and to break the dreaded toilet-to-tap connection of potable water reuse and make it indirect.[42,43] While some pharmaceuticals and other organic wastes seem to survive SAT via recharge and irrigation, so far two major health effect studies in California have failed to show any adverse health effects in people drinking the water after SAT.[44,45] To protect the public health, California has set an upper limit of 1 mg/L for the total organic carbon (TOC) content of the water after recharge and SAT that is due to the sewage effluent. This is achieved by placing recovery wells at least 150 m from the irrigation or infiltration system to allow adequate underground detention times (at least 6 months) and mixing with native groundwater so that the latter comprises at least 50% of the water pumped from the well to ensure a sewage TOC in the well water of less than 1 mg/L. Another advantage of letting the sewage effluent move underground after recharge or irrigation before pumping it up for reuse (including potable) is that it makes water reuse more acceptable in countries that have a religious taboo against the use of "unclean" water.[46,47]

In view of all the experiences obtained with SAT and the rules developed for potable reuse of sewage effluent after SAT, irrigation with sewage effluent does not seem to present significant or unacceptable risks to potable use of local groundwater that has received drainage water from effluent-irrigated areas above it and where there is sufficient blending with native groundwater before extraction from the aquifer.

The role of scientists in all this was already defined almost 400 years ago by Francis Bacon,[48] who wrote: "And we do also declare natural divinations (forecasting by natural observation) of diseases, plagues, swarms of hurtful creatures, scarcity, tempests, earthquakes, great inundations, comets, temperature of the years, and diverse other things; and we give counsel thereupon, what the people shall do for the prevention and remedy of them." Of course the list of "diverse other things" has greatly expanded over the years, and now definitely includes PPCPs. However, information about PPCPs in general and endocrine disruptors in particular is "limited, with sparse data, few answers, great uncertainties and a definite need for further research".[49] Indeed, much more research needs to be done before scientists can give, in Bacon's words, counsel thereupon and what the people shall do.

Meanwhile, irrigation with sewage effluent may be an acceptable practice if done properly and with the right precautions to protect human health and the environment and to ensure sustainability of the practice. Ultimately, sewage irrigation may well be done because rather than a good solution, it may be the least undesirable solution to a wastewater problem.

8.4 SALT AND GROUNDWATER WATER-TABLE MANAGEMENT FOR SUSTAINABLE IRRIGATION

Although sewage chemicals like nitrate, pesticides, pharmaceuticals, and other industrial chemicals can cause problems when the effluent is used for irrigation, the most insidious and serious contaminants in sewage effluent or any other irrigation water may well be the salts dissolved in that water, as expressed by

TDS. These salts become concentrated in the drainage water that moves from the root zone to the underlying groundwater after most of the irrigation water is returned to the atmosphere via evapotraspiration, leaving the salts behind in the soil and groundwater. This increases the TDS of the groundwater and causes groundwater levels to rise until they eventually can damage underground pipes, basements, cemeteries, landfills, old trees, etc. Continually rising groundwater levels may eventually waterlog the surface soil from which the water directly evaporates. This leaves behind a salt crust where nothing will grow anymore.

Inadequate management of salt and rises in groundwater below irrigated areas have forced local people to leave and have caused the demise of old civilizations like Mesopotamia and possibly the Hohokams in Arizona, that had an irrigated agriculture that was abandoned in the fifteenth century, possibly because of high groundwater levels and salt accumulation in the upper soil. Waterlogging and salinization of surface soil are still a threat to many irrigated areas in the world.[50] An example of the latter is the roughly 150×300 km Phoenix–Tucson region with intensively irrigated areas and a rapidly expanding urban population, which now is about 4 million. The area also has both surface water and groundwater resources for water supplies and little or no export of water and salts.

8.4.1 Salt Loadings

The main renewable water resources, i.e., surface water, for south-central Arizona are the Salt River system (about 1000 million m^3/year with a TDS of about 500 mg/L) and the Central Arizona Project Aqueduct (about 1500 million m^3/year of Colorado River water with a TDS of about 650 mg/L). Groundwater in this area is essentially a nonrenewable resource, because natural recharge in a dry climate is very small, on the order of a few mm per year, and there is essentially no "new" groundwater being formed.[43,51] Almost all of the recharge in the Phoenix–Tucson area is deep-percolation water from irrigated areas, which is a return flow and does not represent "new" water. Groundwater pumping for irrigation in the Phoenix, Tucson, and Pinal Active Management Areas is about 1100 million m^3/year with an estimated average TDS of about 1000 mg/L[52]. This represents a total salt load in delivered surface water and groundwater of about 2.6 million tons per year.[53] For the present population of about 4 million people in the area, this amounts to 0.65 ton or 650 kg per person per year or about 1.8 kg per person per day. This is much more than the amount of salt ingested with food and drink and excreted again into the sewers or, for that matter, the salt added by other sources like water softeners. Eventually, these salts end up in the deep-percolation water in the vadose zone. If groundwater pumping and resulting groundwater level declines then are reduced to where this deep-percolation water can "catch up" with the water table, rising groundwater levels and significant groundwater quality reductions can be expected. The amount of salt leaving south-central Arizona probably is very small. There may be some salty groundwater leaving the area at the west side, but no surface water. Thus, south central Arizona is an area of major salt imports with little or no exports.

The calculated increases in groundwater TDS, nitrate levels, and groundwater levels themselves agree with observed values in a study conducted by the Salt River Project in the southeastern part of the Salt River Valley where groundwater pumping was greatly reduced and irrigation was mostly done with surface water starting in the late 1970s and continuing throughout the 1980s.[14] For example, nitrate levels in groundwater pumped from the aquifer below the affected area increased from a range of 2–7 mg/L as nitrogen to a range of 10–20 mg/L. TDS increased from about 500 mg/L to about 1000 mg/L for some wells, and from 500 to 1800 and from 700 to 1500 mg/L for others, while groundwater levels rose about 0.6 m/year. The TDS values are significantly lower than the TDS contents of the deep-percolation water, which for efficient irrigation systems would be about 2500 mg/L. Also, nitrate levels were lower than expected in the deep-percolation water. This is because the wells are perforated or screened for a significant depth interval, whereas the deep percolation water accumulates at the top of the aquifer.

Thus, the well water consists of a mixture of salty deep percolation water from the upper part of the aquifer and much less salty natural groundwater from deeper in the aquifer. Simple calculations can be made to predict the TDS and nitrate increases of the well water as a function of time after the arrival of deep percolation water, for example, by calculating the TDS of the well water as a mixture of natural groundwater overlain by a layer of high TDS from the screen lengths in each and the corresponding transmissivities of each. If the situation is more complicated, like well screens only in the deeper portion of the aquifer and/or presence of a middle fine-grained unit or other layers of low permeability, modeling techniques can be used to predict TDS increases in well water as a function of time of pumping.

As discussed previously, the vadose zone may breathe, inhaling dry atmospheric air when barometric pressures are increasing and exhaling moist soil air when barometric pressures are decreasing. The resulting "deep" evaporation could cause significant amounts of salt to be stored in the vadose zone, which reduces the salt load on the underlying groundwater.

8.4.2 Salt Tolerance of Plants

Increasing TDS contents of well water, or, for that matter, of any water, are undesirable because for health and aesthetic reasons they should be below 500 mg/L for potable water. TDS increases are also undesirable because they shorten the useful life of pipes, water heaters, etc., and make water treatment more expensive for industrial uses where high water qualities, including ultrapure water, are needed.

TDS increases are also undesirable for urban and agricultural irrigation of plants and crops (see Chapter 5, § 5.5.1). As a rule, water with a TDS content of less than 500 mg/L can be used to irrigate any plants, including salt-sensitive plants (See Chapter 5, Table 5.6). Between 500 and 2000 mg/L TDS, there can be slight restrictions on its uses, and above 2000 mg/L there can be moderate to severe restrictions like growing salt-tolerant crops only and adequate leaching of salts out of the root zone.[11,13] For agricultural purposes, salt contents of irrigation water and water in soils and aquifers are often measured as electrical conductivity, EC, expressed in dS/m. For most natural waters 1 dS/m is equivalent to a TDS content of about 640 mg/L.

The basic relationships between the EC of irrigation water and relative crop yields are shown in Figure 8.2. Typically, such relations show no decrease in crop yield with increases in the salt content of the irrigation water, as expressed by EC_w, as long as EC_w is small. Then, as EC_w of the irrigation water is increased, a threshold value is reached where crop yields start to decrease linearly with further increases in EC_w. This threshold value is about 0.7 dS/m (450 mg/L) for salt-sensitive crops, 1.8 dS/m (1150 mg/L) for moderately salt-sensitive crops, 4.0 dS/m (2600 mg/L) for moderately salt-tolerant crops, and 6.5 dS/m (4200 mg/L) for salt-tolerant crops. Examples of crops in these categories are shown in Table 5.8 (Chapter 5).

The lines in Figure 8.2 show that if the EC_w of the irrigation water increases beyond the threshold value, farmers have to accept a reduction in crop yield or switch to a more salt-tolerant crop. Considerable research is being done to increase the salt tolerance of crops.[13,54]

8.4.3 Management of Salty Water

The first reaction to a decreasing quality of well water often is to shut the well down and use other sources of water, if available. However, where groundwater is not pumped at adequate rates, water tables will then continue to rise due to continued arrival of deep-percolation water from the irrigated areas until they become so high that they flood basements, damage underground pipelines, come too close to landfills or cemeteries, kill trees, reduce crop yields, and eventually waterlog the surface soil so that water can evaporate directly from the soil, leaving the salts behind and creating salt flats.

Failure to control groundwater levels in irrigated areas and resulting salinization of the soil has caused the demise of old civilizations and is still causing irrigated land to go out of production at alarming rates.[50] In addition to developed and developing countries, there now are also deteriorating countries where salts and groundwater levels are not adequately controlled. Where there are rises of salty groundwater, groundwater must eventually be pumped again to keep groundwater levels at a safe depth. For agricultural areas where higher groundwater levels can be tolerated than in urban areas, water tables can also be controlled by tile or ditch drainage.

Great care must be taken that the poor-quality salty drainage water that comes out of these wells and drainage systems is discharged into the surface environment in an ecologically responsible manner. Options include discharge into oceans or big rivers, where dilution is the solution to pollution, or into dedicated "salt" lakes for accumulation and storage of salts in perpetuity. Where the salty water needs to be transported over long distances to proper disposal areas, concentrating the salts into smaller volumes of water may be necessary to reduce the cost of pipelines, aqueducts, or other conveyance systems and to reduce the volume of water that leaves the area. Ideally, the salts are concentrated with appropriate treatment technologies to help defray the cost of net disposal.

One way to concentrate the salts into smaller water volumes while making economic use of the desalted water is membrane filtration. The desalted water could then be used for potable or industrial purposes. As a matter of fact, mildly brackish groundwater could be an important reserve water resource in periods of drought since desalting this water is relatively inexpensive compared to desalting much saltier water like seawater.

Concentration of salts into smaller water volumes can also be achieved with sequential irrigation of increasingly salt-tolerant crops where the deep-percolation water from one crop is used to irrigate a more salt-tolerant crop, etc., starting with salt-sensitive crops and ending with halophytes.[55] This can increase the salt concentrations of the drainage water to seawater levels (about 30,000 mg/L) and in volumes that are a small fraction of the original irrigation water volumes, as illustrated in Table 8.2. Depending on local conditions, sequential irrigation to halophytes may not be needed, and the sequence may be stopped if the salt content of the deep-percolation water has become high enough to achieve sustainable disposal at acceptable costs. The wells for pumping salty deep-percolation water from the aquifer in sequential irrigation projects should be rather shallow so that they pump primarily deep-percolation water from the top of the aquifer and a minimum of deeper native and less salty groundwater. Also, sequential irrigation is best carried out by growing increasingly salt tolerant crops in relatively large blocs so that there is not much lateral flow in the aquifer that could interfere with proper control of the deep-percolation water from the different crops.

A third option for concentrating salts into smaller volumes is via evaporation ponds.[53] For the Salt River Valley, evaporation rates of free water surfaces are about 1.8 m/year. Thus, if flows of drainage water are significant, large land areas will be required for such ponds. The ponds may also become environmental hazards. For example, if the irrigation amount is 1.5 m/year and the irrigation efficiency is 75%, evaporation ponds with a surface area of about 20% of the irrigated area would be required if all the deep percolation water must be evaporated. This will eventually increase salt concentrations in the ponds to values well in excess of those for seawater, as happened in the Salton Sea in California with about 45,000 mg/L and the Dead Sea between Jordan and Israel with about 340,000 mg/L. However, complete evaporation may not be necessary if the main purpose of the pond is to concentrate the salts into more manageable smaller volumes of water that can then be more economically exported to an ocean or designated inland salt lake. In that case, pond areas will be less.

Another possibility for concentrating the salts into smaller volumes of water by evaporation is to use the salty well water for power plant cooling. For example, the 3810 MW nuclear power plant west of Phoenix is cooled with

Table 8.2 Sequential Irrigation of Increasingly Salt-Tolerant Crops with Drainage Water from Less Tolerant Crops

	Type of crop				
Parameters	**Sensitive (peas, beans, strawberries, stone, pome, and citrus fruits)**	**Moderately sensitive (lettuce, kale, broccoli, celery, potato)**	**Tolerant (wheat, sorghum, rye, beet)**	**Very tolerant (barley, cotton, sugar beet, bermuda grass, salt cedar, eucalyptus, poplar)**	**Halophyte (salicornia)**
Irrigation volume[a]	100	25	10	5	2
TDS, mg/L	*200*	*800*	*2,000*	*4,000*	*10,000*
Efficiency, %	*75*	*60*	*50*	*60*	*67*
Drainage volume[a]	25	10	5	2	0.67
TDS, mg/L	*800*	*2,000*	*4,000*	*10,000*	*30,000*

[a]Volumes expressed in arbitrary units.

Source: Adapted from Ref. 55.

about 2800 million m^3/year of treated sewage effluent. The effluent is recycled 15–20 times through the plant and is then discharged into 200 ha of evaporation ponds, where it completely evaporates. At an annual evaporation of about 1.8 m, the evaporation from the ponds is about 3.7 million m^3/year. Thus, the salts in 80 million m^3 of effluent are concentrated into 3.7 million m^3 of cooling tower outflow, giving a volume reduction of about 95% and a 20-fold increase in salt concentration.

Perhaps the evaporation ponds can be constructed as solar ponds, which can be used to generate hot water and/or electricity. For example, in an experimental solar pond project in El Paso, Texas, the pond was 3 m deep with a 1 m layer of low-salinity water on top, a 1 m layer of medium-salinity water in the middle, and a 1 m layer of high-salinity (brine) water at the bottom.[56] This created a density gradient so that sun energy was trapped as heat in the bottom layer, while the lighter top layers prevented thermal convection currents and acted as insulators. The hot brine from the bottom layer was pumped to a heat exchanger, where a working fluid like isobutane or freon was vaporized, which then went through a turbine to generate power. The working fluid was condensed in another heat exchanger that was cooled with normal water, which was recirculated through a cooling tower. The working fluid then returned to the brine heat exchanger, where it was preheated by the brine return flow from the heat exchanger to the pond before it was vaporized again. The El Paso pond had a surface area of 0.3 ha and generated 60–70 kW. At this rate, a solar pond system of about 5000 ha or an area of about 7 × 7 km could generate about 1000 MW of electricity, which is typical of a good-sized power plant. There was enough heat stored in the hot brine layer to also generate power at night. The El Paso studies have demonstrated the principles of solar power generation. Considerable research is still necessary to see how a large-scale system should be designed and managed.

Concentrating the salts into smaller and smaller volumes with revenue-producing techniques will be of special benefit to inland or other areas where salts need to be transported over long distances to reach suitable (or least objectionable) places for final disposal, such as, an ocean or a dedicated lake. Concentrating the salts into small volumes of water will then minimize the cost of pipelines and other conveyance structures. The ultimate concentration of salt is, of course, achieved by complete evaporation of the water, so that the salts crystallize and can be stored in perpetuity in landfills or used commercially if beneficial uses can be developed.

8.4.4 Future Aspects for Salinity Management in South-Central Arizona

As the population in south-central Arizona continues to increase and the Phoenix–Tucson corridor expands into a Prescott-Nogales corridor, more and more water will be needed for municipal water supply and more and more sewage effluent will be produced. If all the main renewable water resources, i.e., the Salt River and Colorado River, were solely used for municipal water supply, the 2500 million m^3/year brought in by these rivers could support a population of about 9 million, assuming a use of 760 L/inhab.d (inhabitant per day), which is between the present of 950 L/inhab.d for Phoenix and 570 L/inhab.d for Tucson.[53] At a sewage flow of 380 L/inhab.d, the 9 million people would produce 1200 million m^3 of effluent per year. At an application rate of 1.5 m/year, this could irrigate almost 80,000 ha, urban as well as agricultural. The salt content of the effluent would be below 1000 mg/L, which is the center of the range where moderately salt-sensitive plants or crops can be grown (see Figure 8.2). The effluent could also be used for potable water reuse via reverse osmosis followed by artificial recharge of groundwater, for cooling water for power plants, and for environmental purposes like restoration of stream flow and riparian habitats. Potable reuse of the effluent could add another 2 or 3 million people to the sustainable population. Such reuse would require more membrane filtration, which produces reject brine that adds to the salt burden. Groundwater will be used where still available and of good quality. However, without incidental recharge from irrigation or without artificial recharge in engineered projects, natural recharge rates in dry climates are so low that groundwater basically is a nonrenewable resource.[43]

As described earlier, the salts in the water used for irrigation are concentrated in the deep-percolation water that moves from the root zone to underlying groundwater, where it will increase the salt content of the groundwater. It will also cause groundwater to rise where there is no serious overpumping of groundwater. Eventually, groundwater must then be pumped to prevent groundwater levels from rising too high. The salty water from the pumped wells then should be reduced in volume so that the salt in this water can be exported in relatively small amounts of water. Such concentration of salt into smaller volumes can be achieved with revenue-producing processes, including membrane filtration that also produces drinking water, sequential irrigation of increasingly salt-tolerant crops, and evaporation ponds that may be used as solar ponds for power generation. Final disposal of salt to obtain regional salt balances may then be via export to an ocean or storage in perpetuity in inland salt lakes or landfills.

REFERENCES

1. U.S. Clean Water Act, 1972.
2. Bouwer, H., Integrated water management for the 21st century: problems and solutions. *Am. Soc. Civ. Engrs*, 150 year Jubilee paper, *J. Irrig. Drain. Eng.*, 128, 4, 193, 2002.
3. Bouwer, H., *Groundwater Hydrology*, McGraw-Hill, New York, 1978.
4. Lemly, A.D., Subsurface agricultural irrigation drainage: the need for regulation, *Reg. Toxicol. Pharmacol.* 17, 157, 1993.
5. State of California, Water Recycling Criteria. California Code of Regulations, Title 22, Division 4, Chapter 3. California Department of Health Services, Sacramento, CA, 2000.

6. World Health Organization (WHO), *Health Guidelines for the Use of Wastewater in Agriculture and Aquaculture*, Report of a WHO Scientific Group, Technical Report Series 778, World Health Organization, Geneva, Switzerland, 1989.
7. Bouwer, H., and Idelovitch, E., Quality requirements for irrigation with sewage effluent. *J. Irrig. Drain. Div.*, Am. Soc. Civil Engrs., 113, 4, 516, 1987.
8. Pettygrove, S.G., and Asano, T., *Irrigation with Reclaimed Municipal Wastewater—A Guidance Manual*, Lewis Publishers Inc., Chelsea, 1985.
9. California State Department of Health Services (DHS), NDMA in drinking water, CSDHS, Sacramento, CA, 1998.
10. Hoffman, G.J., and van Genuchten, M.T., Soil properties and efficient water use; water management for salinity control, in *Limitations to Efficient Water Use in Crop Production*, Taylor, H.M., Jordon, W., and Sinclair, Y., eds, Am. Soc. Agronomy, Madison, WI, 73, 1983.
11. Tanji, K.K., Agricultural salinity assessment and management, Manual of Engineering Practice No. 71, *Am. Soc. Civil Engrs.*, Reston, VA, 1990.
12. Maas, E.V., *Salt Tolerance of Plants. The Handbook of Plant Science in Agriculture*, Christie, B.R., ed., CRC Press, Bora Raton, FL, 1984.
13. Food and Agriculture Organization of the United Nations (FAO), Water quality for agriculture, FAO Irrigation and Drainage Paper 29, Ayers, R.S., and Westcott, D.W, eds., Rome, 1985.
14. Wolf, K.O., Salt River Project, personal communication, 2002.
15. Van de Graaf, A., et al., Anaerobic oxidation of ammonium is a biologically mediated process, *Appli. Environ. Microbiol.*, 61, 4, 1256, 1995.
16. Baier, D.C., and Fryer, W.B., Undesirable plant responses with sewage irrigation, *J. Irrig, Drain. Div.*, Proc. Am. Soc. Civil Engrs. 99-IR2, 133,1973.
17. Bouwer, H., Agricultural chemicals and groundwater quality, *J. Soil Water Conserv.*, 45, 184, 1990.
18. Bouwer, H., Lance, J.C., and Riggs, M.S., High-rate land treatment. II. Water quality and economic aspects of the Flushing Meadows project, *J. Water Pollut. Contr. Fed.*, 46, 5, 844, 1974.
19. Singer, P.C., Pyne, R.D.G., Miller, C.T., and Mojoonnier, C., Examining the impact of aquifer storage and recovery on DBPs, *J. Am. Wat. Works Assoc.*, Nov., 85, 1993.
20. U.S. Environmental Protection Agency, *National Interim Primary Drinking Water Regulations*, EPA 570/9-76-003, Washington, DC, 1976.
21. Ozekin, K. and Westerhoff, P., Bromate formation under cryptosporidium inactivation conditions, *Water Quality Int.*, May/June, 16, 1998.
22. Zullei-Seiber, N., Your daily drugs in drinking water? State of the art for artificial groundwater recharge, in *Third Internat. Symp. on Artificial Recharge of Groundwater*, Peters, J.M. and Balkema, A.A., eds., Rotterdam, The Netherlands, 405, 1998.
23. Shore, L.S., et al., Induction of phytoestrogen production in *Medico sativa* leaves by irrigation with sewage water, *Environ. Exp. Botany*, 35, 363, 1995.
24. Guan, M., and Roddick, J.G., Comparison of the effects of epibrassinolide and steroidal estrogens on adventitious root growth and early shoot development in mung bean cuttings, *Physiol. Plant.*, 73, 246 1998.
25. Bradley, E.G., and Zacharewski, T.R., Exoestrogens: mechanisms of action and strategies for identification and assessment, *Environ. Toxicol. Contam.*, 17, 49, 1998.
26. Khan, S., and Ongerth, J., Hormonally active agents in domestic wastewater treatment —a literature review, Report prepared for CGE Australia, 2000.

27. Lim, R., et al., Endocrine disrupting compounds in sewage treatment plants (STP) effluent reused in agriculture. Water Recycling Australia—is there a concern? in *Proc. of the First Symp. Water Recycling in Australia 2000*, Adelaide, October 19–20, Peter J. Dillon, ed., 2000.
28. Goodbred, S.L., Gilliom, R.J., Gross, T.S., Denslow, N.P., Bryant, W.L., and Schoeb, T.R., Reconnaissance of 17β-estradiol, 11-ketotestosterone, vitellogenin, and gonad histopathology in common carp of United States streams: potential for contaminant-induced endocrine disruption, U.S. Geol. Survey Open-File Report 96–627, Sacramento, CA, 1997.
29. Colborn, T., Vom Saal, F.S., and Soto, A.M., Developmental effects of endocrine disrupting chemicals in wildlife and humans, *Environ. Health Perspect.* 101, 378, 1993.
30. Kolpin, D.W., Furlong, E., Meyer, M.T., Thurman, E.M., Zautt, S.D., Baiber, L.B., and Buxton, H.T., Pharmaceuticals, hormones, and other organic wastewater contaminants in U.S. streams, 1999–2000: a national reconnaissance, pp. 1202–1211. *Environ. Sci. Technol.*, 36, 1202, 2002.
31. Drewes, J.E., Heberer, T., Rauch, T., and Reddersen, K., Fate of pharmaceuticals during ground water recharge, *Ground Water Monitoring and Remediation*, 23, 64, 2003.
32. Bouwer, H., Simple derivation of the retardation equation and application to preferential flow and macrodispersion, *Ground Water J.*, 29, 1, 41, 1991.
33. Am. Soc. Agr. Engrs., Preferential flow, in *Proc. 2nd International Symp*, Bosch, D., ed., Honolulu 2001.
34. McGovern, P., and McDonald, M.S., Endocrine disruptors, *Water Environ. Technol.*, 35, 2003.
35. Henry, C.M., Antibiotic resistance, *Chem. Eng. News*, 78, 10, 41, 2000.
36. Ray, C., Melin, G., and Linsky, R.B., Riverbank filtration, in *Proc. International Riverbank Filtration Conf.*, Louisville, KY, Nov. 1999, National Water Research Inst. And USEPA, and Cincinnati Water Works, Kluwer Academic Publishers, Dordrecht, The Netherlands, 2002.
37. Melin, G., in *Proc. Second International Riverbank Filtration Conf.*, Cincinnati, OH, National Water Research Institute, Fountain Valley, CA, 2003.
38. Hutt, P.B., Unresolved issues in the conflict between individual freedom and government control of food safety, *Ecotoxic. Environ. Safety*, 2, 447, 1987.
39. Ames, B.N., and L. Swirsky, G., Too many rodent carcinogens: mitogenesis increases mutagenesis, *Science*, 249, 940, 1990.
40. Calabrese, E.J., Animal extrapolation—a look inside the toxicologists' black box, *J. Environ. Sci. Technol.*, 21, 618, 1987.
41. Lave, L.B., Ennever, F. K., Rosenkranz, H.S., and Omenn, G.S., Information value of the rodent bioassay, *Nature*, 336, 631, 1988.
42. Crook, J., MacDonald, J.A., and Trussel, R.R., Potable use of reclaimed water, *J. AWWA*, August, 40, 1999.
43. Bouwer, H., Artificial recharge of groundwater: hydrogeology and engineering, *Hydrogeol. J.*, 10, 121, 2002.
44. Nellor, M.H., Baird, R.B., and Smith, J.R., Summary of health effects study: final report, County Sanitation Districts of Los Angeles County, Whittier, CA, 1984.
45. Sloss, E.M., Geschwind, S.A., McCaffrey, D.F., and Ritz, B.R., *Ground-water Recharge with Reclaimed Water: An Epidemiologic Assessment in Los Angeles County, 1987–1991*, RAND Corp., Santa Monica, CA, 1996.

46. Ishaq, A.M., and Khan, A.A., Recharge of aquifers with reclaimed wastewater: a case for Saudi Arabia, *Arabian J. Sci. Eng.*, 22, 16, 133, 1997.
47. Warner, W.S., The influence of religion on wastewater treatment, *Water 21*, 11–13, 2000.
48. Bacon, F., see Encyclopedia Brittanica, Vol. 2. Encyclopedia Brittanica, Inc. William Benton, Publishers, Chicago, London, Toronto, 1958, for more information about Francis Bacon.
49. Maczka, C., Pong, S., Policansky, D., and Wedge, R., Evaluating impacts of hormonally active agents in the environment, *Environ. Sci. Technol. News*, 136A, 2000.
50. Postel, S., Last oasis, Worldwatch Institute, Washington, DC, 1992.
51. Bouwer, H., Estimating and enhancing groundwater recharge, in *Groundwater Recharge*, M.L. Sharma, ed., Balkema Publishers, Rotterdam, The Netherlands, p. 1–10, 1989.
52. Swieczkowski, D., Arizona Department of Water Resources, personal communication, December 2001.
53. Bouwer, H., Accumulation and management of salts in South Central Arizona, Technical Appendix P, *Central Arizona Salinity Study (CASS) Phase I Report*, T. Poulson, ed. U.S. Bureau of Reclamation, Phoenix, AZ, and Brown and Caldwell, and 14 municipalities and water companies in Salt River Valley, 2003.
54. Apse, M.P., Akaron, G.S., Sneeden, W.A., and Blumwald, E., Salt tolerance conferred by over-expression of a vacuolar Na^+/H^+ antiport in Aribidopsis, *Science*, 285, 31, 1256, 1999.
55. Shannon, M., Cervinka, V., Daniel, D.A., Drainage water reuse, in *Management of Agricultural Drainage Water Quality*, Water Reports No. 13, Madromootoo, C.A., Johnston, W.R., Wallardson, L.S., eds., Food and Agricultural Organization of the United Nations, Rome, Italy, 1997, Chapter 4.
56. Xu, H., *Salinity Gradient Solar Ponds—A Practical Manual*, Vol. 1 (Solar Pond Design & Construction) and Vol. 2 (Solar Pond Operation and Maintenance), Dept. of Industrial and Mechanical Engineering, Univ. of Texas, El Paso, TX, 1993.

9 Economics of Water Recycling for Irrigation

Joe Morris, Valentina Lazarova, and Sean Tyrrel

CONTENTS

9.1 General principles 266
9.2 Financial analysis 266
9.3 Economic analysis 267
9.4 Benefits of recycled water for irrigation 267
9.5 Factors influencing irrigation benefits 268
9.6 Components of recycling systems for irrigation 269
9.7 Irrigation water supply options 269
9.8 Water-recycling options 271
9.9 Costs of water-recycling options 272
9.10 Prices for recycled water 275
9.11 Function of water prices 275
9.12 Criteria for setting prices for recycled water 277
9.13 Pricing instruments 278
9.14 Examples of recycled water prices 279
9.15 Conclusions 282
References 282

1-56670-649-1/05/$0.00 + $1.50

This section considers the financial and economic aspects of wastewater reuse for irrigation. Following an explanation of the distinction between financial and economic perspectives, the main benefits and costs of water reuse for irrigation are discussed, as well as the factors influencing them. The main components of recycling systems are identified together with the options for wastewater recovery and supply. The characteristics of the costs of water-recovery systems and the criteria that might be applied to determine appropriate water-pricing strategies are disscussed. The overall message is that the type and scale of benefits and costs of water recycling and reuse are very location-specific, such that generalizations are difficult and can be misleading. Emphasis is given here to matters of principle that can be used to guide the economic and financial appraisal of water-recycling projects.

9.1 GENERAL PRINCIPLES

Financial and economic analyses are concerned with the identification, valuation, and comparison of costs and benefits with a view to judging whether a proposed activity is worthwhile or not. In the case of wastewater recycling for irrigation, the analyses consider whether a particular wastewater-recovery solution is absolutely worthwhile in itself, that is, whether the benefits are greater than the costs. Furthermore, analyses consider whether the activity is relatively worthwhile compared to other possible solutions, including doing nothing.

9.2 FINANCIAL ANALYSIS

Financial analysis adopts the perspective of the private individual or organization considering whether or not to engage in a specific activity. It considers estimated revenues and costs accruing to such private parties based on actual receipts and payments, inclusive of any subsidies received and taxes paid. In this respect, financial appraisal makes an assessment of the profitability of an activity from the viewpoint of a particular agent. For example, a private water company is likely to decide in favor of a wastewater-recovery project if the revenues from water sales will recover the capital and operating costs of the venture by a satisfactory margin within an appropriate time frame. It will make the judgment on the basis of return on investment of this activity compared to some other one, including leaving the money in the bank.

Financial analysis is also concerned with identifying the need for funding, recognizing the patterns of cash flow over the project life, and the need for borrowing. Expensive projects involving large up-front investments with payback over the longer term are notoriously difficult to finance and often involve public support in the form of grant, subsidies, or guarantees.

9.3 ECONOMIC ANALYSIS

By comparison, economic analysis takes a broader perspective, viewing a particular activity from the point of view of society as a whole. It applies the principle of profitability but from a public-interest viewpoint. Economic analysis usually begins with a financial-type analysis, but then:

- Removes the effect of subsidies and taxes, such as capital grants for investment in water treatment works (which are really mechanisms for transferring money to and from government rather than real costs or benefits)
- Modifies estimates of benefits and costs where the existing market prices on which they are based are considered not to represent social values, such as where protection of local markets from external competition for agricultural commodities keep prices artificially high
- Identifies and places values on hidden, external effects, which are borne by third parties or society as a whole without compensation, such as those associated with increased or reduced environmental pollution
- Adjusts for nonincremental effects by identifying and excluding from the benefit assessment, for example, that part of a new investment in a wastewater-recycling facility treatment works, which displaces an existing provision rather than adding to it
- Allows for consumer surplus, which recognizes, for example, that when water prices fall because of enhanced supply, existing (and some new) users derive benefits over and above the new price they actually pay (denoted by the fact that they would have been willing to pay higher prices than they did before)

These and other adjustments are made to produce an economic assessment of water-recycling options. Thus, compared to financial assessment, the economic analysis extends the boundary of the analysis to include external impacts, which have financial consequences for third parties, such as income loss to fishermen because of water pollution, and for society as a whole in the case of damage to valued habitats. Given the nature and context of investment in water recycling for irrigation, it is important that these economic dimensions be considered carefully.

9.4 BENEFITS OF RECYCLED WATER FOR IRRIGATION

As demonstrated in Chapter 1 (Table 1.2), irrigation with recycled water is associated with numerous benefits, of which the most important for agriculture are:

- Increased and/or less variable crop yield
- Increased or less variable quality of crops
- Opportunities for new crop and livestock systems and products

The benefits of water recycling for irrigated agriculture are thus the enhancement or maintenance of these types of benefits, assuming of course that recycled water is fit for the purpose intended. For example, a case study[1] from India demonstrated that the yields of vegetables and fish are significantly increased as a result of wastewater irrigation. These irrigation benefits will vary according to context. Thus, the financial benefits to users of recycled water depend very much on the overall financial value of water used for the purpose of irrigation.

9.5 FACTORS INFLUENCING IRRIGATION BENEFITS

Generally, irrigation benefits and the value of water are highest where the available natural water supply, in terms of quantity, reliability, and quality, acts as a major constraint on crop production and associated yields and quality. Benefits from irrigation can be high where cropping is not feasible at all in the absence of irrigation, as in many semi-arid areas during the dry season. However, under these circumstances, irrigation depths applied (mm) are often high and financial benefits per unit of water applied ($\$/m^3$) are often relatively low. This is often the case for cereals (including rice) and grass/forage crops, unless these crops receive high levels of subsidy. It is true that such crops are commonly irrigated in tropical climates, but usually in situations where water is relatively plentiful or relatively cheap to deliver to the field. In many cases, however, water users do not pay the full cost of water services such that, although irrigation may be financially attractive to them, the economic (as defined earlier) value of irrigation is sometimes questionable.

Where irrigation is supplementary to rainfall, as it is in many temperate climates such as northern Europe, application depths are relatively low and returns per unit of water are often relatively high. Financial benefits are particularly high for produce where quality assurance more than yield is the key determinant of value, such as in the case of potatoes and fresh vegetables. Such high potential benefits (sometimes exceeding US$ $2/m^3$) can justify investment in expensive irrigation water-supply systems, including the use of potable water. It is noted, however, that the water-quality requirements for such fresh produce are often much more stringent than for crops for processing or animal feed; the World Health Organization's guidelines[2] for the use of wastewater in agriculture are an example of this. In comparison with high-value produce, the irrigation of other crops such as cereals, grass, and sugar beet adds limited benefits to supplemental irrigation and will not justify high-cost water-supply solutions. The reality is sometimes distorted by subsidies given to these crops (for example, under the European Common Agricultural Policy) and through public sector subsidies for water supply, which do not require full cost recovery from users. Once again it is important to take an economic perspective.

The key point here as far as the economics of water recycling for irrigation is concerned is that benefits depend on the value of water used in irrigation. From a financial viewpoint, this depends on the incentives available to farmers to add value through irrigation, inclusive of any taxes and subsidies. From an economic viewpoint, it is important, however, that the effects of the latter are stripped away to get a better estimate of the real value of water used in the irrigation sector, absolutely and relative to values of water use elsewhere.

This last point is also important. Recycled water delivered to irrigators could take the pressure off water resources elsewhere or release water for other uses. However, this does not in itself increase the value of water beyond that of use in the irrigation sector because redirection of existing water from irrigation to other uses could be carried out in the absence of recycling, if this was deemed more valuable. On a general point, recycled water should be valued in terms of the use to which it is put, rather than higher value uses that might be relieved of pressure.

9.6 COMPONENTS OF RECYCLING SYSTEMS FOR IRRIGATION

Figure 9.1 shows the options for water supply for agricultural irrigation and the various ways in which water treatment can be applied to deliver water of suitable quality. The components of this scheme are shown in Table 9.1.

9.7 IRRIGATION WATER SUPPLY OPTIONS

The conventional water-supply options for irrigation are to abstract water from surface or groundwater sources (i and m, respectively, in Figure 9.1) and

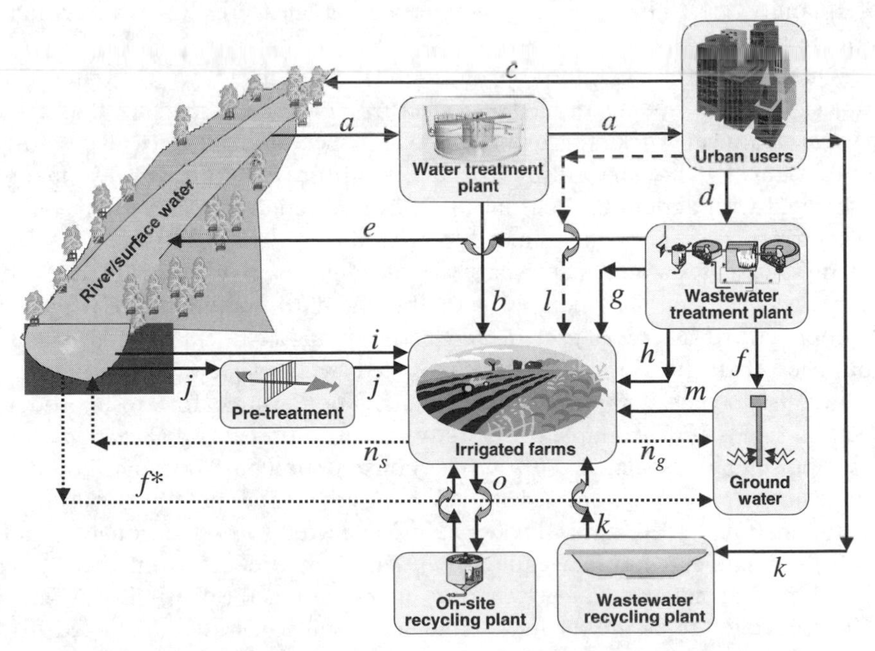

Figure 9.1 Water-recycling components for irrigated agriculture (see Table 9.1 for symbols).

Table 9.1 Components of Recycling Systems for Irrigation[a]

a	Water abstracted from surface or groundwater and treated to potable water standard
b	Potable water delivered to farms for use in irrigation; practiced where irrigation benefits are high or very high water quality standards are required to meet crop specifications
c	Untreated urban wastewater discharged to surface waters
d	Urban sewage delivered to wastewater treatment works
e	Treated urban wastewater discharged to surface waters
f	Treated urban wastewater to recharge groundwater (f^* via surface waters)
g	Treated urban wastewater delivered to farms for irrigation
h	Treated urban wastewater receiving secondary treatment before delivery to farms for irrigation
i	Untreated river/surface waters delivered to farm for irrigation
j	River/surface waters pretreated before delivered to farm for irrigation
k	Urban wastewater treated in dedicated water recycling works to meet irrigation requirements
l	Untreated urban wastewater applied to crops
m	Groundwater delivered to farm for irrigation (treatment may be needed)
n_s, n_g	Diffuse pollution from land: n_s to surface, n_g to ground
o	On-farm dedicated recycling plant

[a]Letter codes refer to Figure 9.1.

to deliver it to the irrigation area. Water quantity and quality aspects are determined by the characteristics of the water sources themselves, in the case of quantity, possibly modified by intervention measures such as artificial storage facilities.

Quality in source waters depends on the extent to which waters are subject to point source discharges, whether controlled, such as that from sewage treatment works (*e*), or uncontrolled, such as from untreated sewage outfalls (*c*). Quality may also depend on diffuse pollution (n_s, n_g), mainly associated with agricultural chemicals from farm land or stormwater run-off from urban areas and contaminated industrial sites.

In some cases water in the natural environment may be polluted to the point where it does not comply with the standards required for irrigation of food crops. There is growing concern among retailers in high-income countries that surface waters contaminated with wastewater and diffuse pollutants may be inappropriate as an irrigation source for fresh produce such as salads. For example, in a study[3] of an outbreak of *Escherichia coli* food poisoning associated with lettuce consumption in Montana, contaminated irrigation water was identified as one of the possible sources of contamination. Where natural waters are considered to pose an unacceptable risk, pre-treatment may be required (*j*). There is evidence from the United Kingdom that major salad growers are investigating the feasibility of using UV disinfection to assure the microbiological quality of water abstracted from surface waters.[4] Even waters from rivers of relatively high ecological status may be deemed not to meet the required standards for fresh produce demanded by the food industry. This, together with issues of reliability of water supply, has encouraged some growers of salad crops in the United Kingdom to use potable water supplies at water charges of US$ 1/m^3 or more. A recent survey suggests that 5% of the irrigated salad area in the United Kingdom receives water from the public main supply.[4]

In some parts of the developing world, there are examples of direct use of poor-quality, partially treated effluent and, more significantly, irrigation with highly polluted rural and urban waters that receive untreated wastewater discharges (*c*). For example, in many urban centers in Africa, peri-urban agriculture is flourishing. Farmers who abstract water from watercourses with urban catchments are practicing indirect reuse of urban wastewater. This water is often highly polluted from direct discharge of sewage and runoff from settlements with inadequate sanitation. Surveys have shown that the microbiological contamination of water used for irrigation may exceed by orders of magnitude the WHO guidelines for unrestricted irrigation.[2] The high demand for fresh produce and lack of regulation of production practices mean that this practice is likely to continue unchecked in many poor communities, discouraging improved wastewater treatment.

The foregoing circumstances describe the context for the financial and economic appraisal of water recycling for irrigation. The control or "counterfactual" against which recycling must be assessed is the use of untreated water from surface or groundwater (*i* and *m*). Thus, the assessment must determine the extra benefits and costs associated with recycling option over and above those options using fresh water. Generally, water recycling does not increase water supply. There is no increment in water as such. Rather it changes its condition and makes it fit for purpose. The key issue is whether the value of the uses to which the recycled water is put will recover the extra costs involved in cleaning it up.

9.8 WATER-RECYCLING OPTIONS

The main wastewater-recycling options for irrigation are associated with the following treatment schemes:

1. Treatment to a standard suited for return to either the water environment or for use as irrigation water (*d* through *e*, *f*, or *g* in Figure 9.1)
2. Treatment to a standard suited for irrigation where this is higher than that for return to the water environment (*g*, where $Quality_g > Quality_e$, or *h*, or *j*)
3. Designated treatment of raw source water for irrigation (*k*)

The critical determinants are, therefore, the quality of the source and receiving waters and the quality of water required for irrigation. If the quality demanded by irrigation is higher than that required in the river system, then some form of treatment for irrigation will be required beyond that which would be needed to permit discharge into the river and water environment.

Table 9.2 Classification of Irrigated Crops by Type, Value, and Water-Quality Requirements

Category	*Type*	Unit value of crops, $/t equivalent	Water-quality standards and likely cost of meeting these
A	Industrial nonfood crops, such as timber, textile, and energy crops	Low	Low
B	Processed food crops, such as sugar beet, oil seeds, protein crops, and cereals	Medium	Medium
	Unprocessed nonfood crops associated with public contact, such as flowers		
C	Fresh food crops or those associated with moderate processing (excluding heating), such as fruits and vegetables	High	High

Thus, the quality standards for irrigation water are critical. These were reviewed in earlier chapters in terms of health (Chapter 3) and agronomic impact (Chapters 2 and 5). Broadly, the standards required reflect the type, purpose, and, for the most part, the value of the crops produced. Table 9.2 classifies crops by type, value, and water-quality standards. Generally, higher-value crops tend to require higher water quality standards, except where lack of regulation in some countries allows otherwise. By implication, high-value crops are more likely to be able to carry the higher costs associated with meeting higher water quality standards. So, the greater the sensitivity of the market price of a commodity to its own quality aspects, the greater will be the justification for expenditure on water recycling for irrigation. Fresh produce delivered to quality-oriented markets offers the greatest scope for investment in water-recycling facilities, and industrial crops the least.

More precisely, as mentioned above, it is the value of water of a given quality, expressed in $\$/m^3$, that captures the benefit of irrigation. This will reflect application rates (m^3/ha) and the added value from irrigation after deducting additional costs associated with crop production (but excluding irrigation costs). It is not simply a matter of extra crop value, but of margin after deducting crop expenses. Some high-value crops are particularly expensive to grow and deliver to market, especially in terms of labor costs. They also tend to exhibit bigger variation in commodity prices than lower-value crops and are therefore more risky. These factors need to be considered.

9.9 COSTS OF WATER-RECYCLING OPTIONS

The definition and valuation of the costs of a recycling installation depend on the purpose of the analysis—whether, for example, the purpose is to determine overall financial feasibility, to determine charges to water users, to determine the need for borrowing to finance the project, or whether the purpose is to assess the wider economic performance of the investment, including the value of any environmental impact. From the outset it is important to be clear on the purpose of the cost estimation. It is conventional to distinguish the following costs: initial capital costs, annual operating costs, annual fixed costs, and total average annual costs.

Initial capital costs include the cost of constructing the wastewater treatment and delivery system. These will vary according to the characteristics shown in Figure 9.1. They include the cost of land, design, supervision, and works including earthworks, civil engineering (canals, pipes, and buildings), electricity supply, treatment and pumping installations, control gear, workshop and office equipment, communications, and vehicles. In some cases grants may be available to reduce capital costs for the investor.

Capital items provide services over the life of the project, although some, such as pump equipment will require replacement at regular intervals. These *replacement costs* need to be identified over the project life, where the latter is taken to be the economic life of the major investment items such as the civil and mechanical engineering works, often 20–30 years for civil works and 10–15 years for mechanical equipment.

Annual operating costs include the costs of repairs and maintenance, labor and management, fuel and energy, licenses or charges for water supply or abstraction, and sludge or bio-solid waste disposal. There may also be annual charges for interest payment on loans, as well as subsidies and taxes, which affect annual costs.

Operating costs by definition vary in total according to the throughput of the plant, although they may be reasonably constant per unit of output (expressed in $\$/m^3$). Major operating costs are likely to be energy for water lifting, repairs to treatment equipment, and labor.

It is usual to express capital investment costs as an *annual equivalent fixed cost*. This is derived by calculating an amortization cost for each capital item according to Equation 9.1:

$$A = P\left(1 \Big/ \sum_{n=20}^{n=1} (1/(1+r)^n)\right) \tag{9.1}$$

where:

A = the annual amortization payment (which includes both depreciation and charges for interest on capital)
P = the capital investment
r = the annual rate of interest as a decimal paid on borrowed funds
n = the life of the particular capital item (20 in the example)

Aggregating these items provides an estimate of average annual fixed costs ($/year). It is probably best to assume that capital items are depreciated and therefore have zero remaining value (see Table 9.3).

Table 9.3 Example of a Costing Method for a Water-Treatment Project

Parameter	Capital costs ($'000)	Life (years)	Amortization factor at 10%	Annual costs[a] ($'000)
Reservoir and civil engineering	5000	20	0.1175	587.5
Treatment equipment	1800	10	0.1627	292.9
Subtotal				880.4
Operation and maintenance				500
Total annual cost				1380.4
Water output, million m^3/year				4.0
Average cost[a], $/$m^3$				0.35

[a]Constant 2004 values.

By definition, fixed costs are fixed in the short term, within which the basic infrastructure and management regime of the treatment plant cannot be changed. Thus, amortization costs are unavoidable in the short term: they have to be paid even if the plant is not used. There may be some other unavoidable costs that are not directly linked to the degree of use. These include the cost of routine site maintenance, licenses, inspection, security, and insurance.

Having derived an estimate of fixed cost per year for a given treatment plant, these can be expressed per unit of output of treated water in $/$m^3$. For a given plant size, fixed costs ($/$m^3$) will be lowest when the plant is operating at full capacity.

Total average annual costs are the sum of annual fixed and total annual operating costs. Water services are generally characterized by relatively high capital investment costs, which means that fixed costs account for a relatively high proportion of total average costs ($/$m^3$). This is an important feature. It means that generally investments of this kind usually have a relatively economic long life (20 years or so) over which capital costs need to be recovered. They are relatively inflexible in their cost structure in that a large proportion of cost is unavoidable and they need relatively long-term investment funding. For these reasons, without guaranteed demand for treated water or assistance with funding, such investments may be regarded as risky for many private investors.

The structure of costs does mean, however, that once a treatment plant has been constructed, a large part of the costs are "sunk" and nonrecoverable whether or not a treatment plant continues to operate. In the short term this might justify operation of existing plant charges, which at least recover operating costs, even though they fail to recover full average total costs ($/$m^3$). However, failure to recover full costs in the longer term, in the absence of subsidies to make good the deficit, will lead to plant closure.

Table 9.3 contains a simple example of the derivation of an average cost ($/$m^3$) of treated water. Depending on treatment level, plant size, and equipment, typical wastewater treatment costs[6] are in the range of 0.05–0.6 $\$/m^3$.

9.10 PRICES FOR RECYCLED WATER

Setting prices for water services is a critical issue with implications for water users, water service providers, and, given the characteristics of water as an essential ingredient of human and natural systems, for environmental quality and public welfare. The public good aspects of water require that decisions about investments in water-resource management and charging for water services usually incorporate a much broad range of economic, environmental, and social considerations. These will, however, be very context specific, for example, drawing distinctions between water to support rural development through irrigation in semi-arid areas and water supply for amenity use in relatively prosperous urban communities. The context will determine the nature of benefits, the distribution of these benefits, and the ability to pay for water on the part of users.

For the most part, the scope for investment in water recovery and reuse technologies will be greater where demand for water exceeds available fresh water supplies, where recycling and reuse can provide water fit for the purpose, and where it is cost-effective to do so. This implies that benefits exceed costs, absolutely and relative to other options for water supply. In situations of water deficit, water can only be put to new uses by reducing that given to existing uses. As mentioned earlier, this opportunity cost of water at the margin of existing use determines the value of recycled water, as shown by the extra value of production generated by the last, say, thousand m^3 applied in agriculture. This value sets the limit for the price to be paid to acquire water of a given quality and the costs to be incurred in its supply.

9.11 FUNCTION OF WATER PRICES

For water services that are exchanged between buyers and sellers, price denotes two things. First, it shows a willingness to pay on the part of the buyer, and hence a measure of benefit derived in use relative to spending an equal amount on some other good or service. Second, it shows a willingness to supply on the part of the service provider and an indication that the price received is sufficient to cover all (or at least a sufficient part) of the costs involved in making the water service available.

Thus, the upper bound for water prices that irrigators will pay is the value of water of a given quality and reliability of supply in irrigation. This is determined by the incremental benefit of irrigation compared to rain-fed cropping (or amenity services such as gardens and sports grounds). This will be the major determinant of willingness to pay where there is no alternative supply of irrigation water. In situations where recycled water substitutes for either abstracted water or potable water, the cost to the user of recycled water must be less than or equal to that of existing supplies, unless of course additional quantities of the latter are no longer available. This assumes that recycled water and water from other sources are perfectly substitutable. Where there are quality differences, these will be reflected in price differentials. For example, where treated recycled water has a higher quality and/or a more reliable supply than river water, it can command a higher price. Of course, where users can freely abstract water from open-access surface and ground sources, they may face relatively low (but nevertheless positive) supply costs. Where these resources are under pressure, increased abstractions are likely to negatively affect water qualities and quantities for other users and for the water environment. Although water recycling can take pressure off these natural sources, this is not a substitute for addressing the failure to manage natural sources properly, including definition of water rights. Indeed, such market failures are likely to make water-recycling projects appear relatively expensive, when in fact the costs of uncontrolled abstraction are high, very real, but not readily apparent.

A similar and common situation arises where existing managed potable water supplies are heavily subsidized by municipal or regional authorities. Water prices are commonly set at levels below full cost recovery in many parts of the world, often for political reasons, even though the world's poorest people usually do not have access to this subsidized water. Subsidized prices not only tend to discourage wise use of water among those who often could afford to pay more, but may also reduce the incentive for investment in water treatment and reuse.

Where recycled water is provided for nonpotable uses, especially irrigation, it is often offered at a lower price than potable water to encourage its use. This price discounting, combined with underpricing of potable water, has meant that few recycling projects attain financial sustainability with full cost recovery. For example, a recent survey of 79 projects[7] found that only 5 in the United States and 7 elsewhere recovered full costs. For the other U.S. projects, operating revenues covered between 0 and 80% of the full cost, implying a high level of subsidy.

Of course, following the earlier discussion, it does not mean that failure to recover costs directly from users implies that reuse schemes are uneconomic: the costs of water-reuse schemes may be justified in terms of broad economic, social, and environmental objectives where the overall target is wise use of available water supplies in support of local, regional, or national development objectives. For example, the wider contribution of water-reuse schemes to regional development may exceed the direct benefits accruing to individual users and their ability to pay for the full cost of a scheme. Here water reuse is serving important strategic objectives. Nevertheless, it is important that full costs are identified, that subsidies if they are provided are transparent and justified against objectives, and that the implications of providing water to users at less than full costs are understood and deemed appropriate to those who fund the subsidies.

For these reasons it is important that water recycling for irrigation is regarded in the context of an overall water resource management strategy.

This involves segmenting the market for water according to different uses, with water quality and security of supply defined accordingly and recycled water targeted at specific uses where benefits exceed costs. For a large water service provider, there may in practice be a degree of cross-subsidy, where different willingness to pay between potable and "grey" water markets can be exploited, where revenue from potable water users may be used to fund lower-quality grey water systems as a preferred option to investment in more expensive potable water projects.

9.12 CRITERIA FOR SETTING PRICES FOR RECYCLED WATER

Pricing decisions vary according to circumstances, as discussed above. Much depends on whether prices are set by a commercial organization with a view to obtaining a return on investment or, in the case of public investment, whether the purpose is to achieve nonprofit objectives such as economic development or wise water use.

Where the investment is private, pricing will be set to recover full costs and provide a satisfactory rate of return on investment. The private operator will predict market demand and willingness to pay for water quality and security of supply and the sensitivity of demand to changes in prices in each segment of the market where the latter might be distinguished in terms of water quality or geographical zone. Given the potential for sole, monopolistic water suppliers to exploit customers in noncontested markets, prices may be subject to external regulation.

Where the supply agency is a not-for-profit organization, water pricing will reflect the objectives to be achieved. These could include cost recovery (part or all), revenue raising beyond cost recovery, wise and efficient use of water, or management objectives such as assistance to a particular user group, regional development, or environmental protection. Thus, actual prices could vary according to the objectives concerned.

Overall, however, the ceiling for price setting will be the willingness and ability of users to pay. Where development objectives are important, users are relatively poor, or where water supply provides benefits beyond users themselves, there may be a case for subsidies. Here user charges can be set to ensure proper operation and maintenance of the scheme at its design specification. All or part of capital investment costs might be funded through the public purse and written off as a development cost in the same way that investments are made in public infrastructure such as roads. In some instances, part of the cost of recycled water supply might be recovered through other taxes on water users, such as through controlled prices for industrial crops such as cotton or sugar cane or government land development or income taxes.

In situations where water is short in supply and the benefits of irrigation are high, users will express a strong willingness to pay for water. They may, in economic terminology, demonstrate price inelasticity of demand whereby water consumption is not very sensitive to change in prices (at existing price and consumption levels). An increase in price brings about a less than proportionate decrease in demand (and vice versa). This characteristic of demand is important, especially when the supply organization or regulatory body is setting prices to users.

A monopolistic supplier will want to charge higher prices because any (small) reduction in demand will be more than made up by increased revenues. A regulatory body may step in to stop these restrictive practices by setting price limits. Indeed, the regulatory organization may want to achieve wise use of water at the same time as supporting the welfare of water users. Where demand is strong, increasing prices to reduce water use is unlikely to prove effective, at least in the short term: farmers will absorb higher prices and suffer income losses before they change their water-use habits. In this way, neither water resource objectives nor development objectives are served. Of course, moneys will transfer to the regulatory organization, and this may or may not be desirable depending on what is done with it. It may be used, for example, to fund other water-resource programs. In cases where price inelasticity of demand is high and welfare issues are important, the regulation of supply supported by advice on wise water use might be a preferred option. Again, there is a need to determine management solutions that are locally relevant.

Where recycled water substitutes for potable water for amenity or agriculture, users expect that prices for recycled water will be lower. But much depends on perceptions of differences in quality and security of supply. The latter is particularly important where nonessential, out-of-house uses are given lower priority of supply from potable sources during drought periods, for example, where restrictions on hoses and sprinklers are issued. In such situations, where reliability of supply is a critical aspect of water service, price discounting due to other aspects of perceived quality may be reduced. Evidence in Florida and California appears to support this. Use of recycled water may require additional capital costs for service connections and on-site distribution. Where these are the responsibility of users, account must be taken of this in the determination of prices, which will encourage take-up of recycled water.

Furthermore, the process of water recycling is often associated with increased storage facilities, and this serves to enhance security of supply. Water storage is itself an expensive undertaking. Combining storage with wastewater recovery offers the benefit of shared costs and economies of scale between the two functions.

9.13 PRICING INSTRUMENTS

A large number of price rate structures are used in water reuse projects, either alone or in combination, including:

Flat monthly charge (e.g., for irrigation, \$/ha/month)
Flat charge per unit volume (\$/m^3)

Base fee plus volume charge
Seasonally adjusted rate (winter/summer)
Ascending block rate
Declining block rate
Time of day–based rate (peak/off peak)
Take or pay–based contracts
Customer-specific negotiated rate

Rates may also vary according to reliability of supply, being highest where supply is supported by a reservoir. They may also vary according to whether use is consumptive (as with irrigation), whether water is quickly returned to the water system (as with mineral washing), and whether its quality is significantly changed (as with discharges from industrial processes).

In the case of irrigation, charge systems mainly include fixed charges per ha per season (sometimes weighted by the type of crop grown according to its water requirements) and volumetric charges per unit of water delivered. In developing countries, where surface irrigation systems are commonly practiced, charges are mostly per ha of irrigation. Where overhead sprinkler or surface drip systems are practiced, volumetric methods are used. As mentioned earlier, water prices may be hidden in a complex of subsidies or taxes to farmers, recovered through land charges, as in Egypt and Ghana, or levies charged for services on industrial crops delivered to processing factories such as sugar cane, oil palm, tea, and cotton in many parts of the world.

Conventional wisdom suggests that volumetric charges for water encourage wise use, with users balancing the price paid against the benefit obtained. Conversely, fixed charges are perceived to be inefficient because there is no charge for additional use. However, where water is scarce in supply and therefore valuable, factors other than price encourage wise use (the Bedouins understand this very well). Furthermore, volumetric pricing requires a relatively sophisticated administrative infrastructure, which may not prove cost-effective in developing country situations.

Different prices may be charged according to volume of consumption. Declining rate structure might be used to attract the larger water-reuse customers and achieve economies in delivery costs. An increasing rate structure may be used to discourage high consumption and encourage wise use. Much depends on cost structures and the overall objectives to be met.

Special rate structures are often designed for different types of water reuse reflecting the different levels of treatment and volumes of water required. Agreements on prices with high-usage customers are often made on a case-by-case basis. In most cases, recycled water is charged for independently of potable water or sewage services where these apply.

9.14 EXAMPLES OF RECYCLED WATER PRICES

A wide variation in recycled water unit pricing exists depending on the type of reuse, flow rates, and local conditions, ranging from 0 to 0.52 US$/m^3.

Almost 50% of 34 reuse projects recently assessed by WERF[7] ranged from 0.15 to 0.52 US$/m^3. Among existing water-reuse projects, the prices of recycled water appear consistently lower than those of potable water, ranging from zero to near potable water rates.

Consequently, recycled water revenue appears to recover operating costs but in most cases tends to rely on some degree of subsidy to recover full costs. Water pricing is driven by the need to discount recycled water either to encourage its use or make it competitive against other sources, many of which are also subsidized. As a rule, the end-users expect to pay no more for recycled water than for alternative water supply of at least the same quality and security of supply. Table 9.4 shows the price of recycled water as a function of the price of potable water supply in some large water-reuse projects in California.[8] The price differentials are apparent.

Table 9.5 presents some examples of recycled water prices mainly for non-potable reuse applications. The reported costs are given only for illustration; they cannot be transferred to other projects because of the strong influence of local factors and the use of different cost-estimation methodologies. This supports the argument that cost estimation and determination of prices is very context-specific.

Table 9.4 Examples of Recycled Water Sale Prices in California

Water agency	Type of reuse	Recycled water price as percentage of potable water rates (%)
City of Long Beach	*Irrigation*	53
Marin Municipal Water District	*Landscape and agricultural irrigation*	56
City of Milpitas	*Landscape irrigation*	80
Orange County Water District	*Indirect potable reuse*	80
San Jose Water Company	*Agricultural and landscape irrigation, industry*	85
Irvine Ranch Water District	*Agricultural and landscape irrigation (90% of uses), toilet flushing, industry*	90
North Coty, San Diego	*Landscape irrigation*	90
Carlsbad Municipal Water District		100
East Bay Municipal Utility District	*Landscape irrigation, industrial uses*	100
Otay Water District		100
West Basin Municipal Water District	*Urban uses and irrigation, aquifer recharge, industrial uses*	80 (53–90%)

Source: Adapted from Ref. 8.

Table 9.5 Examples of User Fees for Recycled Water in the United States

Type of rate	Number of utilities	Range of recycled water charges for end-users
Monthly flat residential charge	3	$7.00 (limited to 0.4 ha), unlimited use: $7.50 or 8.00
		St. Petersburg, Florida: (not metered) $10.35 for the first acre (0.4 ha), $5.92 for each additional 0.4 ha
		Cocoa Beach, Florida: residential (not metered) $8.00 per 0.4 ha
Commodity-based rate generally for commercial and industrial uses landscape irrigation	8	0.08 $/m^3 to 0.45 $/m^3
		St. Petersburg, Florida: few metered large users: 0.08 $/m^3 ($10.36 per month minimum)
		Cocoa Beach, Florida: commercial (metered) 0.07 $/m^3
		Henderson, Nevada: 0.19 $/m^3
		Wheaton, Illinois: 0.05 $/m^3
		County of Maui, Hawaii: major agriculture 0.026 $/m^3, agriculture, golf courses 0.05 $/m^3, other 0.15 $/m^3
		South Bay (California): 0.04 $/m^3 for agricultural irrigation; 0.4 $/m^3 for urban irrigation
		San Diego—north city, California (90% of drinking water): 0.51 $/m^3
Base charge plus volume charge	1	3.25 $ + 0.03 $/m^3
Seasonal rate	2	low: 0.27 $/m^3–0.43 $/m^3; medium 0.32 $/m^3; high: 0.42 $/m^3–0.53 $/m^3
Declining block rate encourage, large industrial users such as industry, water supply augmentation	2	First block: 0.13 $/m^3; second block: 0.05 $/m^3; third block: 0.03 $/m^3
		South Bay (California): 0.23 $/m^3 (up to 31,000 m^3/month; 0.21 $/m^3 (31,000–62,000 m^3/month) 0.196 $/m^3 (62,000–123,000 m^3/month); 0.13 $/m^3 (123,000–246,000 m^3/month); 0.16 $/m^3 (over 246,000 m^3/month)
		West Basin, California (encourage large industrial users):
		a) Title 22 effluent — 0.23 $/m^3 (up to 31,000 m^3/month); 0.21 $/m^3 (31,000–62,000 m^3/month); 0.19 $/m^3 (62,000–123,000 m^3/month); 0.18 $/m^3 (123,000–246,000 m^3/month); 0.16 $/m^3 (over 246,000 m^3/month)
		b) 0.35 $/m^3 with declining block pricing
Inverted block rate encourages conservation: landscape irrigation	2	First block: 0.16 $/m^3; second block: 0.21 $/m^3; third block: 0.42 $/m^3; fourth block: 0.84 $/m^3; fifth block: 1.67 $/m^3
		Irvine Ranch, California (90% of potable water rates): 0.20 $/m^3 (0–100% of base volume); 0.40 $/m^3 (100–150%); 0.81 $/m^3 (150–200%), 1.78 $/m^3 (over 200%)
		San Rafael, California: 0.71 $/m^3 (0–100% of water budget); 1.37 $/m^3 (100–150%); 2.7 $/m^3 (over 150%)
Time of day rate (agricultural uses)	1	Total average daily demand from 21 to 6 h: 0.03 $/m^3; total average daily demand occurring at a continuous, constant level over a 24-hour period: 0.31 $/m^3

Source: Adapted from Ref. 7.

9.15 CONCLUSIONS

This chapter has reviewed the financial and economic benefits and costs of wastewater recovery for irrigation, concluding that these vary according to local circumstances such that generalizations are difficult and potentially misleading. A key message is that the value of recycled water is determined by the use to which it is put. Full cost recovery is a desirable objective but depends on ability to pay and the importance of other management objectives, including social and environmental criteria.

Given increasing pressure on water resources, especially from the urban sector, wastewater recovery and reuse for high consumptive applications such as irrigation can release pressure on available water supplies, whether fresh water or treated potable water. However, the use of potable-quality water for irrigation probably represents an inefficient use of water (except where available fresh water supplies are very limited, are polluted, and deemed unfit for use on food crops). For this reason, it is important that the benefits of wastewater use are judged in terms of the benefits derived from actual use.

It is a spurious argument to justify the supply of recycled, relatively low-quality water in terms of savings in fresh water or potable water supplies, because if these latter resources are limited they should be put to their best possible uses and not be directed to uses that could manage with lower-quality water. However, in practice, due to imperfections in water markets, water demand, supply, and prices may not encourage wise use of water: expensive, often subsidized water is put to low-value use. Thus, providing alternative, lower-quality water for irrigation through recycling schemes can relieve pressure on existing supplies, at least in the short term.

REFERENCES

1. Mara, D.D., Appropriate wastewater collection, treatment and reuse in developing countries, *Munic. Eng.*, 4, 299, 2001.
2. World Health Organization (WHO), *Health Guidelines for the Use of Wastewater in Agriculture and Aquaculture*. Report of a WHO Scientific Group, Technical Report Series 778, World Health Organization, Geneva, Switzerland, 1989.
3. Ackers, M.L., et al., An outbreak of *Escherichia coli* O157: H7 infections associated with leaf lettuce consumption, *J. Infect. Dis.*, 177 (6), 1588, 1998.
4. Tyrrel, S.F., Knox, J.W., Burton, C.H., and Weatherhead, E.K., Assuring the microbiological quality of water used to irrigate salad crops: an assessment of the options available, unpublished report, Cranfield University, Bedford, UK, 2004.
5. Hide, J.M., Hide, C.F., and Kimani, J., Informal irrigation in the peri-urban zone of Nairobi, Kenya—an assessment of the surface water quality used for irrigation, Report OD/TN 105, HR Wallingford, 2001.
6. Lazarova, V. (ed.), *Role of Water Reuse in Enhancing Integrated Water Resource Management*, Final Report of the EU project CatchWater, EU Commission, Brussels, 2001.

7. Mantovani, P., Asano, T., Chang, A., and Okun, D.A., Management practices for non-potable water reuse, WERF, *Project Report 97-IRM-6*, ISBN: 1-893664-15-5, Alexandria, VA, 2001.
8. Lindow, D., and Newby, J., Customized cost-benefit analysis for recycled water customers, in *Proc. WEF/AWWA Water Reuse Conf.*, Orlando, Florida, 207, 1998.

10

Community and Institutional Engagement in Agricultural Water Reuse Projects

Paul Jeffrey

CONTENTS

10.1 Introduction 285
10.2 Public perceptions of water reuse for agricultural production 289
10.3 Institutional barriers 292
10.4 Models for participative planning 298
10.5 Participative planning processes for water reuse projects 301
References 305

10.1 INTRODUCTION

In exploring opportunities and developing options for water recycling in agricultural contexts, policy makers, planners, and system designers face a number of problems that do not have simple technological or legislative remedies. While the development of technologies that provide opportunities for water recycling has moved on apace over the past decade, their practical application will not depend solely on effective and reliable engineering performance. Successful employment of preferred strategies and technologies will require an understanding of the social and institutional environment in which they are to be applied. For example, the forces that promote involvement in recycling may vary between households and cultures and will certainly be different for domestic, agricultural, commercial, and industrial users.

1-56670-649-1/05/$0.00 + $1.50

In particular, the application of water-recycling systems (i.e., locating and operating them) within, or for the advantage of, communities can be severely disrupted if some understanding of key factors such as perceived need and benefit is not acquired.

Although careful natural resource management is widely accepted as a central contribution to sustainable development, there remain wide gaps in our knowledge about how individuals and communities value natural resources and how these valuations might support or hinder certain types of resource management. An understanding of the underlying incentives for and perceived barriers to different kinds of resource use is essential if we are to develop workable environmental management policies. In particular, we need to be able to identify the richness of individual agendas (informed as they are by social, economic, cultural, and sometimes very parochial contexts) to provide a counterweight to institutional, governmental, and commercial agendas. As an example of this richness, Table 10.1 illustrates the variety of stakeholders (in a simple supply chain) who might be involved in an agricultural reuse project and identifies their temporal perspectives, major motivations for action, the regulatory pressures on them, and concerns they may have about water reuse for agriculture. The diversity of incentives for action depicted in Table 10.1 illustrates why policy, whether applied through the use of technological, financial, or legal tools, is often ineffective because the diversity of possible reactions has been concealed by assumptions about "average" or "optimum" behaviors.

It might be helpful at this early stage to review the types of issue that will be discussed in more detail below. We should remember that communities are composed of individuals, and from both a community and an institutional perspective, the major concerns about water reuse schemes are driven by potential risks. These risks are perceived by different types of actors such as individuals, communities, and institutions (both public and private). With specific regard to reuse for agriculture, the main elements of risk can be categorized as follows:

Infection during irrigation water application/storage
Infection during harvesting/processing/packing
Infection as a result of product handling or consumption
Damage to the quality of the produce
Damage to the productive potential of the land or farming business
Damage to the environment
Damage to the reputation of the community and its produce
Risk that people will not purchase the final product
Financial risk of investment in the system

These are not merely concerns about personal exposure, but are transposed onto family members, friends, colleagues, visitors to the area, and other members of one's community. Avowed concerns may also relate to future generations as well as the current one. In this respect, the growing literature on environmental risk has a significant contribution to make to the field of

Table 10.1 Influences on the Perspectives of Different Actors in an Agricultural Reuse Scheme

Actor	Temporal perspective	Motivations for action	Focus of regulatory activity	Representative concerns
Sewage treatment works	Hourly, daily, and seasonal	Environmental protection – commercial success	Water quality or minimum treatment function	Ability to maintain water quality
Recycled water supplier	Annual	Maximizing water sales	Water quality	Stability of recycled water demand
Farm unit	Seasonal	Maximizing agricultural production (quality + quantity)	Crop quality	Consistency of supply + perspective of retailers and consumers
Retailer	Annual	Maximizing sales	Service levels and quality of produce	Quality of product and perception of consumers
End user of product	Daily	Consumption	N/A	Personal and family health

water recycling. For example, research in this field has demonstrated that a range of factors can influence the level of perceived risk of a project.[1] These descriptors could be used as a template for selecting low-resistance reuse projects and as a guide to managing the public outreach aspects of reuse projects.

Voluntary risks are considered less risky.
Natural risks are considered less risky.
Familiar risks are considered less risky.
Memorable risks or risks associated with signal events are considered more risky.
Risks with dreaded outcomes (e.g., cancer) are considered more risky.
Risks that are well understood are considered less risky.
Risks controlled by self are considered less risky.
Risks perceived as unfair are considered more risky.
Morally corrupt risks (i.e., "evil" phenomena) are considered more risky.
A risk controlled or caused by an institution that is not trusted is considered more risky.

It is worth noting that in the context of water reuse, perceived risk is known to decrease with both trust in institutions[2,3] and the provision of examples with which to describe and demonstrate reuse schemes.[4]

Understanding how individuals, communities, and institutions might react to water-reuse schemes is, however, just a starting point. There are now numerous social and legislative entities that promote wider participation by the public and institutions in water-resource planning and management (e.g., Article 14 of the Water Framework Directive[5] and the Aarhus Directive[6]). Participative planning and stakeholder-engagement processes are intended to both improve the quality of planning (via a social learning process) and promote the democratic principles of informed consent, social justice, and open government. We shall have an opportunity to discuss some of these concepts in more detail later. The objectives of this Chapter are (1) to provide the reader with an understanding of the principal dimensions of individual, community, and institutional attitudes towards reuse projects in the agricultural sector, (2) to discuss how constructive dialogues can be nurtured between different interest groups, and (3) to propose a framework for planning and implementation of appropriate participative planning processes.

It should be noted that empirical evidence upon which one could base prescriptive advice is scarce. Case study material, while available, is of inconsistent detail and rarely comprehensive in its coverage of the relevant problems. The next two sections therefore draw on a range of sources, both published and unpublished, to illustrate the public perception and governance issues surrounding water reuse for agriculture. These are followed by a review of stakeholder-engagement models and techniques, the objective being to critically examine the range of stakeholder-engagement instruments available and to assess their suitability for deployment. Finally, we describe and discuss an integrated stakeholder-engagement process that is suited for the particular case of water reuse for agriculture. We would note that, in order to provide a structure for reporting current knowledge in Sections 10.2 and 10.3, the distinction between the public and institutional stakeholders has been adopted. The usefulness of this distinction diminishes somewhat in the last two sections of the chapter as we move from diagnosis to prescription.

10.2 PUBLIC PERCEPTIONS OF WATER REUSE FOR AGRICULTURAL PRODUCTION

The use of recycled wastewater in any context can quite understandably be a source of concern for communities who have no previous direct experience of similar schemes. Irrespective of what conclusions scientific enquiry leads to, the impressions and attitudes that the public holds can speedily and effectively bring a halt to any reuse scheme. The central dilemma for anybody attempting to understand how individuals respond to change is that people interpret their surroundings in a highly personal manner. Not only is interpretation individualistic, it is also dynamic (i.e., changes over time) and as such is extremely difficult to monitor. A fundamental choice of approach has to be made when attempting to understand how individuals react to, and interact with, their environment. This choice is between the insights that can be gained, on the one hand, from an improved understanding of individual actions or, on the other hand, the process of change itself in which individuals are but one element. In other words, should we concentrate upon readily identifiable aspects of behavior and the attitudes that can be identified in support of decisions relating to them, or do we attempt to disentangle the maze of physical and social interactions in which the subject matter and the decisions relating to it are embedded? There is no unambiguous solution to this dilemma, and different investigators have adopted both approaches in their enquiries into perceptions of water reuse.

The particular issues, that require understanding here are both complex and complicated, having to do with beliefs, attitudes, and trust. Studies of public attitudes to water reuse have been carried out since the late 1950s (originally in the United States, but more recently in Europe, Central America, and Africa). A valuable summary of research[7] was provided during the early years, reporting that individuals who consider their potable supplies to be under threat (in terms of either quality of quantity) or perceive an economic benefit are generally more positive towards the idea of recycling water. Other work has demonstrated that acceptance of water-recycling schemes in general is influenced by the degree of human contact associated with the reuse application. Uses such as garden irrigation and toilet flushing are consistently preferred over uses such as food preparation and cooking.[8] More recent studies[9] have considered other determinants of attitudes to reuse schemes, including the scale of the scheme (e.g., single house/multiple house) and the context of the scheme (e.g., domestic, commercial, or public premises).

The source of recycled water as well as the environment in which it is to be used are likely to influence attitudes towards the system as a whole. Source waters perceived as clean or safe will be more willingly accepted than those that appear dirty or dangerous. However, the "use history" of the water is likely to be a moderating factor in willingness to use. Ask yourself if you would prefer to use your own wastewater to irrigate your garden vegetables or your next-door-neighbor's wastewater.

However, just as individuals vary in their attitudes towards water reuse, so communities and societies also differ. Indeed, the dangers inherent in ignoring cultural (ethnic/historical/religious) norms have been recently demonstrated,[10] and the benefits provided by public education have been pointed out.[11,12]

As part of the project on which this volume is based, several studies were conducted of the perspectives and attitudes of different stakeholders to reuse for agriculture. Surveys were conducted in France, Italy, and the United Kingdom to identify the major barriers to water reuse across a range of reuse project types and cultural contexts. The major findings of these studies were as follows:

Communities across Europe are sensitive to water-reuse issues, although this is more evident in the northern part of the continent than in the south.

Many corporate stakeholders are nervous about supporting reuse projects in the absence of clear and legally binding water-quality guidelines.

Use of a water-recycling system where the source and application are located within their own household is acceptable to the vast majority of the population as long as they have trust in the organization that sets standards for water reuse. Using recycled water from second-party or public sources is less acceptable, although half the population show no concern, irrespective of the water source.

Water recycling is generally more acceptable in nonurban areas than in urban areas. (This disparity is most pronounced for systems where the source and use are not within the respondent's own residence.)

Willingness to use recycled water, particularly from communal sources, is higher among metered households than among nonmetered households, and higher among those households that take water-conservation measures than among those who do not.

The use of recycled waters for irrigation is widely accepted by farmers who believe them to be safer than river waters.

There are strong concerns over the sale of products that have been irrigated with recovered wastewater, especially vegetables. Farmers can overcome resistance through positive evidence from the consumers and the retailers that there will be a market for the products cultivated with the recovered water.

The establishment of standards for the reuse and management of monitoring programmes promotes confidence in reuse schemes.

In concluding our remarks on stakeholder attitudes to water recycling, we would reemphasize the contextual sensitivity of these attitudes. Regional, local, and personal histories together with recent water-related events in the area produce a unique context for each recycling scheme (see Box 1). Under such conditions, a checklist approach to eliciting public attitudes is unlikely to be successful in all circumstances.

Box 1 Stakeholder Attitudes to Reuse for Irrigation in Sardinia

Scheme Description

The scheme for reuse of the effluent from the Is Arenas wastewater treatment works in Sardinia foresees the realization of a tertiary treatment line located downstream from the Is Arenas plant for the reduction of phosphorus and bacterial content in the treated effluent. The effluent will then be discharged into the Simbirizzi Reservoir, which will act as storage basin prior to the reuse of the polished effluents. The wastewater accumulated in the reservoir will then be used to irrigate the irrigation district of southern Sardinia (about 7900 ha in area). In general, the wastewater from the purification plant of Is Arenas can either be destined for direct reuse in irrigation (without storing the treated wastewater in the reservoir) or used indirectly. The system will be capable of providing about 43 Mm^3/year of Is Arenas effluents to the irrigation area.

Focus Groups in Support of Project Design and Implementation

Stakeholder engagement in support of the Is Arenas reuse project was primarily concentrated on the use of focus groups. A number of meetings were organized that began with an explanation of the characteristics of the project. Subsequent discussion was managed with the objective of identifying the perceived strengths and weaknesses of the project. The contributions of the participants were summarized as key words and immediately written on stickers of different colors: green for strong points, red for weak points, and blue for concrete proposals. These stickers were immediately placed on a thematic meta-plan, allowing a record to be kept of the perspectives and relationships between ideas. Spoken contributions were also recorded on tape.

The following issues provided an agenda for the focus group discussions:

1. Farm practices and water use
2. Perceived risks and advantages, especially in the distribution of products
3. Quality standards and control systems

Reuse Project Strengths

The most significant strength of the project was perceived to be the secure availability of an additional resource (as one participant said, "having been through so many years of drought, we farmers look forward to it, as since 1989 we have been through 5 dry years with no water at all"[11]). Another strong point of the project was considered to be the high quality of the initiative's technological aspects as well as its positive impact on the environment in general. The words of one organic farmer are illustrative in this context: "For us the concept itself of water recycling is fundamental: organic agriculture is not only to free oneself of chemical compounds, but also to approach the environment in a more sustainable way. Therefore, to transform what used to be waste into a useful resource is one of our main principles."

Reuse Project Weaknesses

Identified project weaknesses included the following concerns:

The need to change farm practices was perceived negatively by the farmers, as it represents an extra cost. As one regional governance representative said, "I have already heard that when the Is Arenas plant gives us water, we will have to stop using sprinklers and start using drip irrigation."

The impact on horticultural businesses where crops require a large amount of water and develop quickly. The concern here is one of nonacceptance of the products by consumers.

Organic farmers are in a sensitive situation with regard to the project, as it is essential for them to be able to use the recycled water without compromising their ability to obtain either organic farming support funds or quality certification.

Finally, some participants raised concerns about the cost of the resource and the fact that legislation on the use of such waters is still uncertain.

10.3 INSTITUTIONAL BARRIERS

The ways in which institutional characteristics and relationships influence the success or failure of water-management projects is, perhaps, the least understood aspect of sustainable water use. Involving as it does a combination of organizational, political, and economic science, it is not difficult to understand how the study of the institutional issues has lagged behind identification of the technical, social, and environmental determinants of successful reuse schemes. In particular, more work is needed which characterizes the life cycle of reuse schemes in institutional terms: at what stage different institutional actors become involved, what their roles are, to what extent arrangements between actors are formal or informal, how responsibilities are demarcated and adopted, etc.

There exists very little domain-specific knowledge on which to base best practice in this area, and many reported studies only allude to institutional relationships as influences on scheme success or failure. Perhaps the only direct evidence comes from a recent study,[13] which through the use of case study material, have emphasized that the planning and development of the institutional framework that monitors, controls, and delivers treated wastewater (particularly where there are many institutions working in the same or similar areas) is vital for the safe and efficient exploitation of the resource.

One may ask, quite justifiably, why institutional issues are of importance in this context. The answer lies in the distribution of power and influence within our communities. Most of us live in societies where responsibility for different aspects of our environment (in its widest sense) has been distributed between a range of different political, regulatory, and community-based institutions, which use a mixture of legal, financial, and educational instruments to influence, and hopefully modify, behavior. Natural resource-management projects as extensive and multifaceted as water reuse schemes require planning and control across a range of professional and institutional boundaries. Within the context of water-reuse for agriculture, the key institutional responsibilities that we might be interested in will cover subjects such as water quality, treatment plant design and operation, water distribution, cost recovery, agricultural product promotion, and quality control. Responsibility for these aspects of a specific scheme will normally lie with a number of bodies and will doubtless vary by nation, state and maybe even regionally. In addition, there will be social and economic groupings who, while they have no legal responsibilities, nevertheless have an economic or other interest in a reuse scheme. We can thereby list a supplementary set of stakeholders who may seek influence in the design, construction, and operation of a reuse scheme; local residents, environmental protection groups, farmers organizations, wholesalers, retailers, and consumer groups. Finally, we should not overlook the organization (which may be from the private or public sector) that will build and operate the reuse scheme. It may be the primary beneficiary of the scheme, but it is a key institutional actor. Different institutions have different incentives, different objectives, different viewpoints, and different ways of articulating and arguing about the issues. How can the often competing and incompatible aspirations of such a wide variety of factions be reconciled?

There are indeed many institutional factors that can cause reuse schemes to falter before they are even implemented or fail to achieve their ambitions. We may speculate that, as has been noted with many human activities, novelty generates a conservative or even openly negative response from existing institutions.[14] This is not to imply that such a reaction is necessarily unconstructive or harmful. It helps to remember that institutions, like individuals, have both purpose and principle; they react to propositions for reuse schemes for a reason, and we do well to try and understand the stimuli that generate institutional perspectives and attitudes. In broad terms, institutional barriers to the implementation of reuse schemes in the agricultural sector are very similar to those found in schemes that provide recycled water for other purposes. Primarily, they revolve around issues of legality, legitimacy, responsibility, and trust.

Legality is an important consideration for institutional entities. Innovation in any form presents a challenge to existing legislation, particularly where the integrity and strength of petitions is judged against precedent. In countries where there is little or no regulatory guidance for reuse schemes and there has been no previous litigation to base precedent on, institutional actors (both public and private) are understandably wary about taking on new responsibilities. They are, in legal terms at least, being encouraged to sail in uncharted waters.

The extent to which institutions can claim legitimacy to act is partly a function of their legal standing and partly a function of how they are themselves perceived by other institutions and actors. The obvious problem here is that an institution's legitimacy profile will vary across other actors, making it difficult for all parties to reach a consensus about which actors are justified to play which role or take which responsibility.

Perhaps the most open source of disagreement between institutional actors in planning reuse projects is the distribution of responsibilities. To be truthful, "disagreement" is a misnomer in this context; a more appropriate noun might be "deliberation." Legal and regulatory arrangements are typically concerned with rights and responsibilities. Therefore, it often takes significant effort to take on new responsibilities and integrate their implications into existing administrative practices and procedures. Care must also be taken that any new responsibilities do not clash with existing ones or create inconsistencies or contradictions in the institution's activities.

As noted above, trust in those institutions that either manage or set the quality standards for reuse schemes can promote public confidence in the scheme itself. Just which institutions are trusted by the public (or indeed by other institutions) will vary greatly between national and even regional contexts. Figure 10.1 shows the results of a survey carried out in the United Kingdom to ascertain an institutional trust profile with respect to water reuse, while Figure 10.2 shows the results from a similar survey conducted in Sardinia.

Finally, some of the more common institutional issues that have been observed to restrict enthusiasm for water-reuse projects in agricultural contexts can be listed as follows:

Lack of agreement between institutional actors on appropriate regulations, standards, and/or monitoring procedures

Difficulty in identifying a win-win strategy

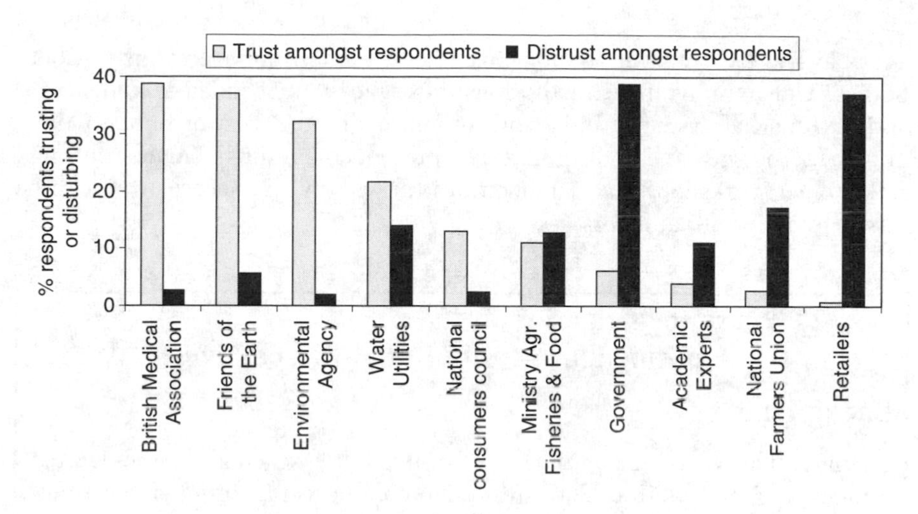

Figure 10.1 Trust in institutions to set quality standards for recycled water: results from a U.K. survey.

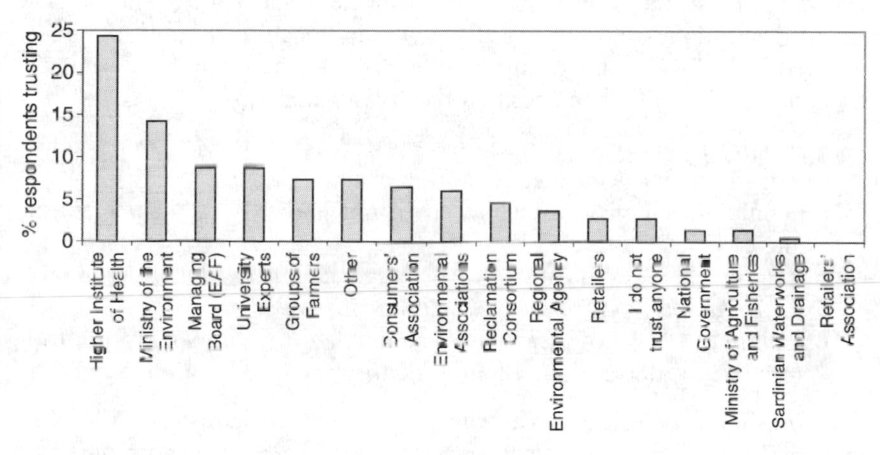

Figure 10.2 Trust in institutions to set quality standards for recycled water: results from a survey in Sardinia.

Late or nonentry of influential institution
Waiting for reconfiguration of incentives to take effect
Inability to envisage a resolution
Sensitivity to negative publicity
High perceived financial risk of the project

These points emphasize the importance of developing a "consortium of the willing" for any type of water-reuse initiative. Our experiences suggest that institutions are perhaps more pack oriented in their behavior than might be thought. Key regulatory and commercial actors like to keep abreast of each other's opinions and intentions. Hence, reuse initiatives can fail to gain momentum if a common understanding of the problem and consensus about possible feasible solutions is not engendered among important institutional bodies. Unlike individuals, institutions are typically embedded in wider legal and/or financial systems, and their commitments/level of exposure to these must be recognized and addressed. Box 2 presents the results of an institutional stakeholder workshop, that demonstrates the variety of concerns that can be raised by institutional actors.

Box 2 Reuse for Irrigation in the United Kingdom—Institutional Perspectives

Background

A major barrier to the initiation of water-reuse schemes in the United Kingdom is the lack of cross-institutional consensus regarding viable applications, standards, and responsibilities. A recent workshop brought together key actors from the water supply, agricultural, and food processing/retailing sectors to discuss the issue of treated wastewater quality for irrigation. The various institutions involved had their own apprehensions about the issues involved. The broad aim of the workshop was to provide a forum for discussion and debate with specific objectives being:

- To identify the barriers to reclaimed water use in the food-production industry, from field to supermarket shelf
- To examine specific issues such as environmental risk, public perception, and economic costs
- To identify where responsibility for setting reclaimed water use standards for irrigation in the United Kingdom should lie
- To explore the range of mechanisms available for setting and monitoring reclaimed water-use criteria and identify preferred options

The workshop was designed as a structured interactive discussion. A briefing document was distributed to participants prior to the workshop, outlining the objectives and ground rules of the session and a list of themes for discussion. Attendance was by invitation, the general aim being to ensure that representatives from as many major stakeholders as possible were present.

Workshop Design

In deciding a format for the workshop, three particular considerations were foremost. First, an informal atmosphere to the session was needed, so as to encourage debate and allow as broad a range as possible of themes and topics to be discussed without undue constraint. Second, an inclusive debate was encouraged, where all participants contributed in equal measure. Finally, the workshop's role as a scoping activity meant that an emphasis on topic exploration where participants sought to understand the issue and others' attitudes to it rather than debate solutions to specific problems was desirable.

Ensuring that the workshop reflected these ambitions involved effective management of both the composition and the proceedings of the workshop. This was achieved first through a careful balancing of the various interest groups represented and, more importantly, of the personalities actually involved. A second key influence on the successful running of the workshop was sensitive supervision of the discussions through the chair. The research team briefed the chairperson (who was selected on the grounds of his non-partisan and impartial position) at some length, emphasizing the need to maintain a balanced discussion while keeping the participants focused on specific issues. The final element in effective workshop management was the provision of a formal but loosely structured agenda. However, following opening statements by the participants, and as the debate evolved along the lines anticipated, a decision was made by the chair and the research team not to impose the formal agenda on the group. It was considered more important to maintain the flow of discussion rather than punctuate the session with a sequence of artificially imposed topic changes. As it turned out, very few topics that were identified in the formal agenda remained unexplored by the end of the session.

Workshop Findings

Public perception and the lack of a reuse standards framework for the United Kingdom were confirmed as perhaps the most significant barriers to recovered water reuse for irrigation.

The problems of both securing public trust and setting/monitoring/enforcing these standards were seen as issues that affect, and should be addressed by, all concerned parties (water industry, farming, retailing, government agencies).

Reuse schemes focused on inland treatment works were considered to be inappropriate.

The supply of irrigation water will allow the farming community to manage the risk associated with climate change rather than increase productivity per se.

Standards should be as rigorous as possible but should not necessarily be derived from existing examples.

Two distinct options for monitoring standards were suggested: one focused on the process (i.e., the water-treatment train) and the other on the product (i.e., the quality of water at the irrigation node, the quality of the crop).

A large measure of commonality was recognized between the two issues of sludge application to land and the use of reclaimed water for irrigation. However, it was generally felt that, at least as far as public perception and standards are concerned, they should be treated independently, though the debates around each should continue to inform each other.

10.4 MODELS FOR PARTICIPATIVE PLANNING

Understanding the features of individual and institutional attitudes towards water reuse provides a diagnosis without a prescription. People may have concerns about water recycling for agriculture, but these need not necessarily prevent the design and operation of reuse schemes. Having identified the issues and concerns, we need to go on to consider what kinds of tools are available to support dialogue between stakeholders.

A logical extension to the use of social-enquiry methods that seek to gauge public attitudes to an issue such as water recycling is the involvement of the public (and other stakeholders) in the planning and management of reuse schemes. Traditional, nonparticipatory processes such as top-down direction and instruction have been shown to not work. History also shows that coercion does not work. The results are clear in the decline in the state of the environment, the increase in social exclusion, and the lack of trust of the public in their governments and industry. On the one hand, public participation benefits both planning and management institutions, and at the same time it benefits the public in general. The general objectives of such engagement processes are to:

Let the community know what is going on in their street neighborhood/village/town
Give the local community an opportunity to get involved in the project—input to planning, employment, etc.
Make sure that the planned development project will be supported locally
Help to identify potential problems early
Identify key personalities, stakeholders, and interest groups who may help or hinder the project
Identify community needs and prioritize them

The case in support of wider participation in natural-resource management has traditionally been worded in terms of the links between economic efficiency, equity, justice, and environmental concerns[15] and more recently in terms of decision quality, shared responsibility, and extended democracy. However, it is worth noting that these are reasons why one might engage in the process rather than being a list of demonstrable, realized benefits. Indeed, although there is a widespread belief that participation is intrinsically good as a process (motivated by a set of normative considerations) and while there is some evidence that participation generates broadly better outcomes[17] and that the additional costs of such processes are not inhibitory,[18] there have been surprisingly few empirical studies that evaluate the benefits of participation in either qualitative or quantitative terms. Those that are reported have emphasized the importance of engaging both stakeholders and public at an early stage[19] and the need to include information sharing and education of the community as integral parts of the process.[20] Perhaps more usefully from a practitioner perspective, other work has highlighted the ability of participation to "alleviate an initial uneasiness" among the public about planners' and politicians' intentions[21] and helped to move from a situation where people are viewed as part of the problem (their behavior/responses to be optimized to restrict undesirable impacts on the technical system) to one where people are viewed as part of the solution.[22]

The expansion in opportunities for participation has also generated new tools to support the process itself,[23] and well over 50 such methods have been identified.[24] These include frameworks for organizing face-to-face dialogue and debate, consultation techniques based on interviews or questionnaires, and, increasingly, the deployment of customized ICT platforms and Internet applications.[25–27] As is the case with evaluating the benefits of participation (see above), few useful evaluations of the various available methods have been conducted, eminent exceptions being those pertaining to Citizens' Juries and Community Advisory Committees,[28] a review[29] and an evaluation of participatory tools from a user's perspective.[30]

The objective of engagement is also a theme that has attracted much comment. However, although there has been a significant increase in the amount of participative planning carried out since the publication of Shelly Arnstein's influential work[31] in 1969, there have been few better propositions

Table 10.2 Levels of Involvement in Participative Planning

Level 1	Manipulation	Assumes a passive audience, which is given information that may be partial or constructed.
Level 2	Education	
Level 3	Information	People are informed as to what has been decided or has already happened. Alternatively, participation is used to gather information from those involved to develop solutions based on their knowledge. However, decisions are made by those initiating the participation process.
Level 4	Consultation	People are given a voice, but the process does not concede any share in decision making, and professionals are under no obligation to take on board members' views.
Level 5	Involvement	People's views have some influence, but institutional power holders still make the decisions.
Level 6	Partnership	People negotiate with institutional power holders over agreed roles, responsibilities, and levels of control. People participate in the joint analysis of situations and the development of plans to act. Such a process involves capacity building–the formation or strengthening of local groups or institutions.
Level 7	Delegated power	Some power is delegated.
Level 8	Citizen control	Full delegation of all decision making and actions. People participate by taking initiatives independently to change systems, such as plans and policies. They may have contacts with external institutions to obtain the resources and technical advice they need but retain control over how those resources are used.

Source: Adapted from Ref. 31.

for classifying levels of participation. Table 10.2 presents a classification based on Arnstein's structure.

Numerous techniques and methods are applicable to participative planning/management processes, some of which are restricted to one-way information flow (e.g., adverts, surveys) and some of which facilitative dialogue and deliberation between stakeholders (e.g., citizens' juries, public meetings). The following list is not exhaustive, but it does provide comment on the strengths and weaknesses of the major techniques currently deployed across Europe.

Community liaison groups (long term): The following section will emphasize the importance of nurturing long-term relationships with communities. Liaison groups are one way in which a continual but low-key communication can be maintained. They are typically no larger than 10–12 persons strong and meet three or four times a year to discuss issues of concern or interest to any of the attendees. Notwithstanding their obvious benefits, liaison groups do have their drawbacks, principal among which is that their community members can be viewed as having become institutionalized.

Opinion polls/surveys: These may be used to find out citizens' views on specific issues. Opinion polls are generally used to obtain immediate reactions. A "deliberative opinion poll" would be used to compare a group of citizens' reactions before and after they have had an opportunity to discuss the issue at hand. These can provide detailed, comprehensive information on the considered views of respondents based on accurate information. However, low response rates can be a problem, and written documents can put some people off commenting. Additionally, the costs of printing, distributing, and analyzing the forms can be significant, and the time scale of the activity will be longer than for some other methods of consultation.

Interactive web sites: These may be based on the Internet or on a local authority-specific intranet, inviting e-mail messages from citizens on particular local issues or planning matters. Alternatively, a discussion forum might be used to elicit a long-term and more detailed picture of citizens' concerns.

Public meetings: These are widely used to facilitate debate on broad options for a specific planning application, strategy, or development plan. They may be initiated by the local authority (or a particular department) or be convened in response to citizen or community concerns. These types of event provide local opportunities for people to comment on matters that affect them directly or indirectly, offers a convenient and transparent way to demonstrate public consultation/build up good relationships, and can be used to inform the public at the same time as getting views. However, those who attend are unlikely to be representative of the local population, attendees' ability to contribute can be limited by a lack of knowledge and possible lack of interest, and there is a danger that the agenda on the day is dominated by local, topical, or personal concerns.

Citizens' juries: A citizens' jury is a group of citizens (chosen to be a fair representation of the local population) brought together to consider a particular issue. Citizens' juries receive evidence from expert witnesses and cross-questioning can occur. The process may last several days, at the end of which the group reaches a decision or prepares a short report setting out the views of the jury, including any differences in opinion. Benefits of this method are that it provides informed feedback, can be used to explore reactions to a range of issues, and promotes a sense of inclusion among jury members. However, the process can be an expensive one to organize and manage, is not suitable for all issues, and usually works best where organizations have already made substantial progress in their consultation.

Citizens' panels: These are ongoing groups who function as a sounding board for a planning team. Panels focus on specific service or policy issues or on wider strategy. The panel is ideally made up from a statistically representative sample of citizens.

Focus groups: One-off focus groups are similar to citizens' juries in that they bring together citizens to discuss a specific issue. Focus groups need not be representative of the general population, however. The major strength of this type of tool is that it provides an opportunity to explore not only what stakeholders' opinions are, but how those opinions are formed and how they might be influenced by specific factors (information, previous experiences, assumptions, etc.). Group settings are also effective means of encouraging debate and generating ideas. The drawbacks include the fact that sessions need to be managed by an experienced facilitator (so can be expensive) and the difficulty in prioritizing issues during the debate.

Open house/exhibition: This is an event at which the public is invited to drop by to speak with staff about a particular planning issue, view displays, and perhaps break into small discussion groups. These types of event gives the public the flexibility of when to attend, can arouse interest by giving the public something to see or do, provides ad-hoc feedback on services and ideas for change, and can, if effectively managed, be a source of suggestions and comments. However, those who attend a specific event may not be particularly representative of users and nonusers, and feedback may be limited to responses to the information presented/on display.

10.5 PARTICIPATIVE PLANNING PROCESSES FOR WATER-REUSE PROJECTS

The objective of this final section is to propose a framework for the planning and implementation of appropriate participative planning processes within the context of water reuse for agriculture. The preceding pages have spelled out in some detail the current state of knowledge regarding public perceptions of, and institutional attitudes towards reuse schemes and reviewed a range of tools available to support participative planning. It now remains to fuse these elements together into a prescriptive program for action. The suggested engagement framework presented here draws on both literature sources and case study experience but is particularly influenced by two recent documents produced by the Water Environment Research Foundation[32] in the United States and the documentation supporting the Queensland Water Recycling Strategy[33] in Australia.

As noted above, there is little empirical evidence available on which to base recommendations. One reuse scheme operations manager recently complained that the only guidelines available for managing stakeholder and public participation were personal experience and the mistakes of others. Consequently, the first piece of advice to be offered here is to go and consult with people who have direct experience of participative planning action. Some countries such as Australia, the United States, and Spain perhaps have more experience than others in the water field, but lessons can indeed be learned from analogous sectors such as sludge use,[34] energy,[35] and natural resource management.[36] It must be remembered, however, that participative planning methods and techniques should always be tailored to the sensitivities, knowledge, and culture of target communities.

A central premise of the program specified below is the need to integrate an ongoing program of stakeholder and public contact on the one hand, with project- or action-specific consultation and involvement on the other. Case study evidence from both the water and agricultural sectors suggests that existing relationships with the public or stakeholders are important precursors of successful projects.[37] A growing emphasis on broader participation in natural-resource governance is creating more positive and continuous relationships between political, regulatory, commercial, and citizen bodies. It is against this background of a more extensive social and governance contract, wider acceptance of the validity of different types of knowledge, and the increasingly heterogeneous nature of contributions to decision making that participation in reuse schemes is occurring.

Assuming that there is some level of ongoing contact between the various types of actor (at a minimum these should include regulators, communities, and the water supply organization), any specific project will have a number of stages that provide a basis for planning and structuring participative planning activities (see Table 10.3).

The issues listed in Table 10.3 are representative and assumptions should not be made about the focus or level of concern that might be expressed by local actors. Indeed, a major element of a participative planning action should be to identify or elicit an agenda of beliefs and opinions from different constituencies. Such opinions might be positive or negative with respect to a reuse scheme, be based on knowledge or ignorance, or be held by an individual or a group. They will certainly be variable across a population

Table 10.3 Stages of a Reuse-for-Irrigation Scheme with Associated Objectives and Representative Issues

Project stage	Objectives	Representative issues
Conception	Explore desirability and feasibility of a reuse scheme	Profile of benefit and loss from the project—i.e., who and how Risks that might be associated with the proposal.
Design	Identify best option (technological, system configuration etc.) and make detailed plans for project implementation	Safety of operation and protection of public health Cost allocation and financing Customer acceptance of products produced with recycled water
Decision and construction	Make application to authorities for formal permission to construct and implement the project; also implement agreements on responsibilities and rights of consortium members (e.g., water suppliers, farmers, wholesalers etc.); build technological components of the scheme	Facility sitting and operational characteristics Environmental and social impacts
Operation	Manage and run the project	Quality control Deviations from planned operations regime Incident management
Decommission	Dismantle physical infrastructure and remediate landscape	

and over time. With respect to structuring a dialogue between the various parties, attention should be paid to the norms of local communication and decision making. Comprehensive guidance on the design of engagement processes is difficult to formulate when local circumstances play such a crucial role in mediating the applicability of different approaches. However, at a minimum, an effective engagement process will:

1. Provide opportunities for all interested parties to make their views known
2. Ensure that decision makers are aware of these views
3. Support communication processes, which promote education, understanding, transparency, and debate

Figure 10.3 illustrates the types of activity that might be used to structure such a process. This particular configuration of actions allows both institutional and individual perspectives to be collected, compared, and debated.

The type of process depicted in Figure 10.3 cannot guarantee that all perspectives will be captured. However, because it involves feedback of elicited opinions to the different groups, it provides a means for ongoing communication within which new concerns can be declared and debated. As such, effective participation is a function of process rather than product.

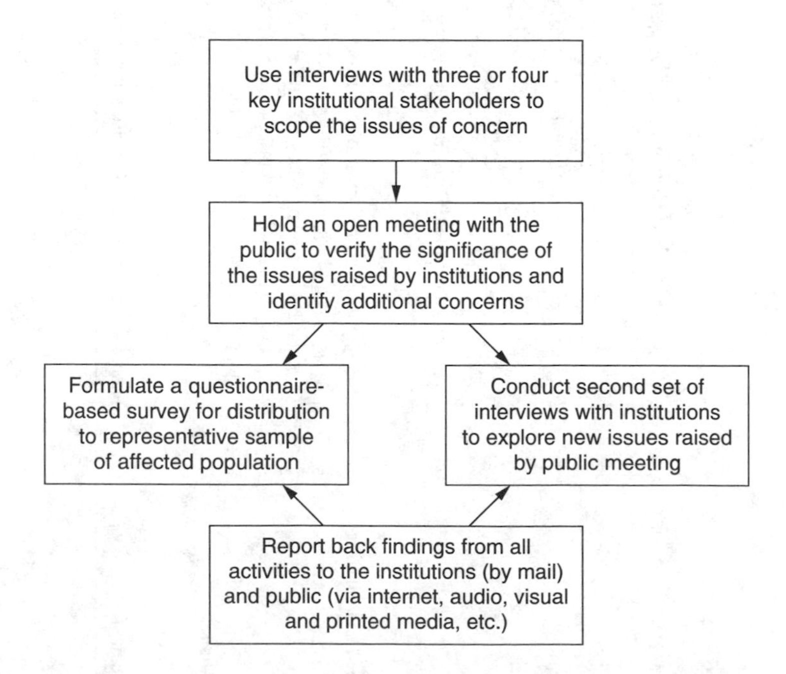

Figure 10.3 Illustrative engagement process incorporating both institutional and individual actors.

We have spoken much about the need to recognize the context of a reuse scheme before designing an engagement strategy. However, some characteristics of successful participation processes are not at the mercy of local conditions. For example, irrespective of context and locality, effective public participation exercises typically entail:

A clear purpose and time scale for the process
Adequate resources
Mechanisms for addressing power issues among stakeholder groups
Sensitivity to cross-cultural issues
Mechanisms for communicating to participants how their input was, or was not, utilized
Turning cooperation (working together for individual benefit) into collaboration (working together for mutual benefit)

In conclusion, participative planning for water reuse in agriculture should provide opportunities for dialogue and debate at all stages, make the whole process transparent (not only with regard to what is being discussed, but also concerning what level of influence participants can have on the process at each stage), and maintain a flexible stance with regard to the tools that are used to support the participative process. In a broader sense, successful participation seems to avoid a sales pitch approach to promoting reuse and rather seeks to understand the conditions under which a particular scheme can be accepted and effectively adopted. In doing this we challenge the assumption that people are simply mute targets, waiting to be influenced by pieces of information or new technologies. An engagement program focused on dialogue and social learning moves us towards a more equitable and consensual agenda for change characterized by mutual understanding.[38] In a world where participative processes are becoming a requirement rather than just a recommendation, new questions as well as new answers are required.

REFERENCES

1. Canter, L.W., Nelson, D.I., and Everett J.W., Public perception of water quality risks—influencing factors and enhancement opportunities, *J. Environ. Syst.*, 22 (2), 163–187, 1994.
2. Kromm, D.E., and White, S.E., Reliance on sources of information for water-saving practices by irrigators in the high plains of the U.S.A., *J. Rural. Stud.*, 7 (4), 411–421, 1991.
3. Hofmann, N., and Mitchell, B., The RESPECT model: evolving decision-making approaches in water management, *Water. Pol.*, 1 (3) 341–355, 1998.
4. Gibson, H.E., and Apostolidis, N., Demonstration, the solution to successful community acceptance of water recycling, *Water Sci. Technol.*, 43 (10), 259–266, 2000.
5. CEC (Commission of the European Communities). Directive 2000/60/EC of the European Parliament and of the Council Establishing a Framework for the Community Action in the Field of Water Policy, Published at Official Journal (OJ L 327) on December 22, 2000.

6. CEC (Commission of the European Communities). Directive 2003/35/EC of the European Parliament and of the Council providing for public participation in respect of the drawing up of certain plans and programmes relating to the environment and amending with regard to public participation and access to justice, Council Directives 85/337/EEC and 96/61/EC. 2003.
7. Bruvold, W.H., and Crook, J., What the public thinks: reclaiming and reusing wastewater, *Water Eng. Man*, April, 65, 1981.
8. Bruvold, W.H., Obtaining public support for reuse water, *J. Am. Water Works Assoc.*, 77 (7), 72, 1985.
9. Jeffrey, P., Influence of technology scale and location on public attitudes to in-house water recycling in England & Wales, *J. CIWEM*, 16 (3), 214–217, 2002.
10. Mancy, K.H., Fattal, B., and Kelada, S., Cultural implications of wastewater reuse in fish farming in the Middle East, *Water Sci. Technol.*, 42 (1–2), 235–239, 2000.
11. Sbeih, M.Y., Recycling of treated water in Palestine: urgency, obstacles and experience to date, *Desalination*, 106 (1–3), 165–178, 1996.
12. Crites, R., Regional water reuse in California, in *Proc. Hawaii Water Reuse Conference*, Kauaii, Hawaii, November 7–8, 2002.
13. Lawrence, P., Adham, S., and Barrott, L., Ensuring water re-use projects succeed—institutional and technical issues for treated wastewater re-use, *Desalination*, 152 (1–3), 291–298, 2003.
14. Schon, D., *Beyond the Stable State*, Norton, New York, 1973, chap. 2.
15. Grimble, R., and Wellard, K., Stakeholder methodologies in natural resource management: a Review of principles, contexts, experiences and opportunities, *Agric. Sys.*, 55 (2), 173–193, 1997.
16. CEC (Commission of the European Communities), Directive 2003/35/EC of the European Parliament and of the Council providing for public participation in respect of the drawing up of certain plans and programmes relating to the environment and amending with regard to public participation and access to justice, Council Directives 85/337/EEC and 96/61/EC, 2003.
17. Isham, J., Deepa, N., and Pritchett, L., Does participation improve performance? *World Bank Econ. Rev.*, 9 (2), 175–200, 1995.
18. Hentschel, J., *Does Participation Cost the Bank More? Emerging Evidence*, Human Resources Development and Operations Policy Working Papers 31, Washington, DC, World Bank, 1994.
19. Grima, A.P., (1983) Shaping water quality decisions: an evaluation of a public consultation program, *Water Int.*, 8 (3), 120–126, 1983.
20. Pena, S., and Cordova, G., Public participation and water supply: the case of two communities on the USA–Mexico border, *Water Int.*, 26 (3), 390–399, 2001.
21. Moorhouse, M., and Elliff, S., Planning process for public participation in regional water resources planning, *J. Am. Water Works Assoc.*, 38 (2), 531, 2002.
22. Wegner-Gwidt, J., Winning support for reclamation projects through pro-active communication programs, *Water Sci. Technol.*, 24 (9) 313–322, 1991.
23. Sanoff, H., *Community Participation Methods in Design and Planning*, John Wiley & Sons, New York, 2000.
24. Carman, K., and Keith, K., *Community Consultation Techniques: Purposes, Processes and Pitfalls*, Queensland Dept. of Primary Industries, Queensland, Australia, 1994.
25. Guimarães-Pereira, Â., Gough, C., and De Marchi, B., Computers, citizens and climate change—the art of communicating technical issues, *Int. J. Environ. Poll.*, 11 (3), 266–289, 1998.

26. Kingston, R., Carver, S., Evans, A., and Turton, I., Web-based public participation geographical information systems: an aid to local environmental decision-making, *Comput. Environ. Urban.*, 24 (2), 109–125, 2000.
27. Al-Kodmany, K., Online tools for public participation, *Gov. Inform. Q.*, 18 (4), 329–341, 2001.
28. Petts, J., Evaluating the effectiveness of deliberative processes: waste management casestudies, *J. Environ. Plan. Manage.*, 44 (2), 207–226, 2001.
29. Hinchcliffe, F., Guijt, I., Pretty, J.N., and Shah, P., *New horizons: The Economic, Social and Environmental Impacts of Participatory Watershed Development*. Gatekeeper Series No. 50, International Institute for Environment and Development (IIED), London, 1995.
30. Halvorsen, K.E., Assessing public participation techniques for comfort, convenience, satisfaction, and deliberation, *Environ. Manage.*, 28 (2), 179–186, 2001.
31. Arnstein, S.R., A ladder of citizen participation, *J. Am. Inst. Plan.*, 35, 216–224, 1969.
32. Water Environment Research Foundation, *Water Reuse: Understanding Public Perception & Participation*, Water Environment Federation, Alexandria, VA, 2003.
33. State of Queensland, *Queensland Water Recycling Strategy*, The State of Queensland, Environmental Protection Agency, Australia, 2001.
34. Tyson, J.M., Perceptions of sewage sludge, *Water. Sci. Technol.*, 46 (4–6), 2002.
35. Gregory, R., Fischhoff, B., Thorne, S., and Butte, B., A multi-channel stakeholder consultation process for transmission deregulation, *Energy Policy*, 31 (12), 1291–1299, 2003.
36. Ravnborg, H.M., and Westermann, O., Understanding interdependencies: stakeholder identification and negotiation for collective natural resource management, *Agr. Syst.*, 73 (1), 41–56, 2002.
37. de Garis,Y., Lutt, N., and Tagg, A., Stakeholder involvement in water resources planning, *J. CIWEM.*, 17 (1), 54–58, 2003.
38. Berkhout, F., Hertin, J., and Jordan, A., Socio-economic futures in climate change impact assessment: using scenarios as learning machines, *Global Environ. Change.*, 12 (2), 83–95, 2002.

11

Institutional Issues of Irrigation with Recycled Water

Eric Rosenblum

CONTENTS

11.1 Introduction 310

11.2 Ownership of water, wastewater, and recycled water 311
- 11.2.1 Water rights 311
- 11.2.2 Water use limits 316
- 11.2.3 Rights to recycled water 318

11.3 Wastewater regulations 319
- 11.3.1 Effluent regulations 320
- 11.3.2 Pretreatment to protect recycled water quality 322

11.4 Planning and implementation issues 324
- 11.4.1 Land use planning 324
- 11.4.2 Environmental regulations 326
- 11.4.3 Construction issues 328
- 11.4.4 Wholesaler/retailer issues 329
- 11.4.5 Customer agreements 332

11.5 Program management 333
- 11.5.1 Integrated planning 333
- 11.5.2 Matrix analysis of institutional issues 333
- 11.5.3 Summary of institutional guidelines 337

References 339

1-56670-649-1/05/$0.00+$1.500

11.1 INTRODUCTION

Recycled water is used to irrigate agricultural crops or landscape when water from rivers, lakes, and aquifers is in short supply. Sometimes the lack of water is due to lack of rain; other times water is in short supply because existing supplies are already allocated to other users or because water must be left in the stream to sustain an ecosystem. From the other end of the cycle, recycled water is used for irrigation when alternative forms of wastewater disposal are prohibited or because the expense of conveying wastewater for treatment is greater than the cost of reuse. In both instances the decision to reuse wastewater is heavily influenced by local, state, and federal water laws, as well as the agencies and boards responsible for managing the environment from which water is drawn and to which wastewater is ultimately discharged.

The term "institution" as used in this chapter refers to any organization of rules or individuals that impacts the reuse of treated effluent. In this broad context, institutional issues can range from the statutes that govern ownership of water to the makeup of agencies that administer environmental regulations. Practices related to financing water and wastewater treatment projects and land use development can also have a direct influence on the decision to implement water reuse.

Managers are well advised to evaluate institutional issues at the early stages of project development to identify gaps in the institutional structure that may impede the development of their water reuse projects. Since water reclamation affects nearly all phases of water supply and use, it is important to recognize and involve each of the groups collectively responsible for the coordinated management of water, wastewater, and related resources. As the World Water Council[1] recently noted, "Integrated water resources management has to be applied through a complete rethinking of water management institutions, putting people at the centre." In this way, sponsors of water reclamation projects can both contribute to and benefit from the creation of institutions capable of comprehensive resource management.

Yet another reason to address institutional and legal issues at the early stage of the planning process is that legal matters can be quite technical, and the body of statutory and case law in the area of water reuse is relatively small. An early review of the basic institutional and legal issues will allow managers to address problems when they can be most effectively handled, and to obtain counsel when necessary to help them weigh alternatives and risks.

This chapter does not pretend to offer a comprehensive review of all the water-related laws and regulations found in each of the many governmental systems employed by the nations of the world. Rather it seeks to highlight a few of the main issues that both support and challenge the development of recycled water. The chapter is outlined as follows:

Ownership of Water
Water-Reuse Regulations
Planning and Implementation Issues
Program Management

11.2 OWNERSHIP OF WATER, WASTEWATER, AND RECYCLED WATER

Establishing the right to use recycled water is one of the first priorities in developing a water-reuse project.[2] Such a right can be asserted or abridged based on a number of legal principles including ownership of surface water and groundwater (or holding rights to their use), responsibility for treating and disposing of wastewater, and even the rights of downstream users to water returned to the environment.

These ownership issues may determine the characteristics of recycled water programs developed in various parts of the world. For instance, in countries where the government owns all the water and allocates it to cities or farmers, large centrally managed water-reuse projects may be built to supplement the overall national supply. On the other hand, where water rights are privately held and the market value of water can stimulate investors, small recycled water projects can be built as a hedge against shortages, especially during dry years. In some cases, downstream landowners who hold rights to return flows can even prevent upstream communities from irrigating with effluent, requiring them instead to return their used water to the stream. Likewise, governments that regulate the discharge of wastewater may subsidize water-recycling projects as alternative means of disposal.

In addition to laws governing the ownership of water and wastewater, the agencies, commissions, and boards that implement these laws can have significant influence on the development of water-reuse projects. As noted earlier, water-reuse is more likely to be promoted by governing boards that have jurisdiction over water and wastewater issues or where these responsibilities are well coordinated between agencies. These issues are explored in more detail in the following sections.

11.2.1 Water Rights

The following description of water rights is based on an explanation contained in a currently unpublished update of the EPA/USAID Guidelines for Water Reuse (1992)

A "water right" is a right to use water and in many cases does not actually involve ownership of the water itself. A water right allows water to be diverted at one or more particular points and a portion of the water to be used for one or more particular purposes. To one degree or another, most of the world's nations claim ownership interest in their water supplies on behalf of the public and have established national or regional boards and commissions to allocate flows to various sectors of the economy or to individual entities. In the United States, the states generally retain ownership of "natural" or public water within their boundaries, and state statutes, regulations, and case law govern the allocation and administration of the rights of private parties and governmental entities to use such water.

Water rights are an especially important issue, since the rights allocated by the state can either promote reuse measures, or they can pose an obstacle to reuse. In water-poor areas, for example, water rights laws might prohibit the use of potable water for nonpotable purposes, but they may also restrict the consumptive use of reclaimed water by requiring its return to the stream. A basic doctrine in water-rights law is that harm cannot be rendered upon others who have a claim to the water.

In the United States, state laws allocate water based on two types of rights—the appropriative doctrine and the riparian doctrine. The appropriative rights system is found in most western states and in areas where water is limited. It is a system by which the right to use water is appropriated, assigned, or delegated to the consumer. Generally, appropriative water rights are defined by comprehensive water codes that govern and control water rights. They are usually acquired by applying to an institution that issues permits or licenses, and the priority of these allocations establishes their relative value.

The basic notion of appropriative rights is "first in time, first in right." In other words, the right derives from beneficial use on a first-come, first-serve basis and not from the property's proximity to the water source. The first party to use the water has the most senior claim and has a continued right to the water that a "late" user cannot diminish in either quantity or quality. This assures that senior users have adequate water under almost any rainfall conditions, whereas later users have some moderate assurance to the water. The last to obtain water rights may be limited to water only during times when it is available (wet season). The right is for a specific quantity of water, but the appropriator may not divert more water than can be used: if the appropriated water is not used, it is lost. The appropriative rights doctrine allows for obtaining water by putting it to beneficial use in accordance with procedures set forth in state statutes and judicial decisions. Nevertheless, in countries where appropriative water rights are allocated through an annual process of arbitration or administrative fiat, the actions of the allocating agency can become intensely politicized as water users compete to establish the priority of their need.

By contrast, water used under a riparian system can be used only on the riparian land and cannot be extended to another property. The riparian water rights system, often used in more water-abundant areas, is based on the proximity to water and is acquired by the purchase of the land. The right of one riparian owner is generally correlative with the rights of the other riparian owners, with each landowner being assured some water when available. However, riparian use can only be for a legal and beneficial purpose, and a riparian user is not entitled to make any use of the water that substantially depletes the stream flow or that significantly degrades the quality of the stream. Riparian rights can present an obstacle to implementation of an integrated water-use plan, since they cannot be traded to allow the market to assist in determining the "highest use" of the water supply. Furthermore, unlike the appropriative doctrine, the right to the unused water can be held indefinitely and without forfeiture, which limits the ability of the water authority to quantify the amount of water that has a hold against it and can lead to water being allocated in excess of that available. This doctrine also does not allow for storage of water (in the United States, California has both appropriative and riparian rights.)

Access to groundwater supplies in the United States is generally governed by various forms of the appropriative water rights system, whereby that water percolating through the ground may be controlled by each of three different appropriative methods: absolute ownership, reasonable use, or specific use. Absolute ownership occurs when the water located directly beneath a property belongs to the property owner to use in any amount, regardless of the effect on the water table of the adjacent land, as long as it is not for a malicious use. The reasonable-use rule limits groundwater withdrawal to the quantity necessary for reasonable and beneficial use in connection with the land located above the water. Water cannot be wasted or exported. The specific-use rule occurs when water use is restricted to one use. During times of excess water supply, storage alternatives may be considered as part of the reuse project so that water may be used at a later date. The ownership of or rights to use this stored reclaimed water will need to be determined when considering this alternative.

Many nations subscribe to a regulatory system of water allocation similar to appropriative rights in that water is recognized as a public good and access to water is ultimately controlled by the state. However, not all countries recognize the relative priority of these allocations ("first in time, first in right"), and they may be freely adjusted on a periodic basis. In some countries (e.g., Germany, France, Israel), a national board grants permits to various groups of stakeholders, including cities and agricultural associations, conveying the right to use specific amounts of water for varying periods, and may change allocations as conditions warrant, e.g., during periods of drought.[3] An alternative system may be used when the government invests in the development of a water supply and then contracts with customers to use the water produced by the specific project for a set period of time.

National or regional agencies that value recycled water as a means of relieving the stress between competing local water users can facilitate water reuse through policies that link allocations to water reclamation efforts.

Case Study: France

Since 1992, France's Ministry of the Environment has been responsible for coordinating national water policy with various technical ministries. National policy in turn has been made responsive to the needs of numerous stakeholders through the National Water Committee, made up of elected officials and representatives from various socioeconomic sectors including an association of 1800 irrigation customers representing a third of French irrigable acreage. In creating an institutional framework with broad responsibility for French water and wastewater issues, the French system appears to be well positioned to implement the policy articulated in the Declaration of Johannesburg (2003) to "introduce measures to improve the efficiency of water infrastructure to reduce losses and increase recycling of water".[4] It should also be noted that France's integrated management of its water resources does not appear to be incompatible with private administration of water facilities as is commonly practiced there.

When water rights are allocated by a central government agency, the political process can work to promote a number of progressive water-development programs to ensure that all sectors have enough water. In many countries with centralized water management there is broad popular support for ecologically protective measures like water conservation, water recycling, and the preservation of stream flows for the environment. These political pressures can serve to counteract or replace market forces that may otherwise spur the development of new supplies whose long-term social, economic, or environmental cost may not be considered.

Case Study: Israel

In Israel, where water is scarce, it is allocated to various sectors of the national economy by the national government based on recommendations from the Israel Water Commission. Central planning has supported water reuse such that by 2000 Israel reused as much as 70% of all wastewater, which comprised up to 20% of Israel's water supply.[5] Recycled water is manufactured largely in the urban areas and transported to agricultural areas. For example, the Dan Region reclamation plant serves a population of 1.7 million (including the city of Tel Aviv) and treats about 120 Mm^3/year of wastewater to secondary standards, after which it is polished by percolation into recharge basins and pumped to agricultural areas on the southern coastal plain and northern Negev through a 120 km pipeline (the "Third Line").

In all, there are about 200 seasonal reservoirs throughout Israel that provide a total recycled water storage capacity of about 120 Mm^3. In 1998 recycled water made up about 22% of all water used for agriculture (276 Mm^3) and is projected to reach 44% (496 Mm^3) by 2010. By comparison, in 2002 the government also authorized construction of seawater-desalting plants sufficient to generate an additional 400 Mm^3 of fresh water to augment existing supplies. Notwithstanding this central approach, the nation has established as a goal "gradual changeover to management of the water supply system based on the economics of supply and demand combined with central supervision, within the framework of the Water Supply System Reform Law."[6] It may be that without the central decision making guiding the construction of desalination facilities, in the future market forces will favor the use of reclaimed water for a variety of purposes.

Even when adequate supplies exist to satisfy agricultural and urban needs, the impact of diverting water from its natural channels may create long-term environmental challenges. This situation is illustrated by recent attempts in the United States to protect the ecosystems of the San Francisco Bay and the delta of the Sacramento and San Joaquin rivers. Collectively, this great riparian system provides water to a population of more than 20 million people and sustains the world's seventh largest economy. However, the health and biodiversity of the watershed has declined significantly over the past century and may have passed the point where it can be restored.[7]

Another important water-rights issue faced by water professionals in the international arena is the equitable allocation of water across national boundaries. Watersheds rarely respect political borders, which results in the need for extraordinary levels of cooperation and a willingness to compromise for the common good. For instance, the watershed located in one country may produce the majority of flow in a river used by another. Water managers in each of the two countries have a unique ability to identify creative solutions that provide disputing parties with "win-win" solutions.[8] This is especially true in the Middle East—between Turkey, Syria, and Iraq, for example, over the waters of the Tigris and Euphrates rivers, and between Jordan and Israel over the waters of the Jordan River.

Case Study: Water in the Middle East

According to Resources for the Future (an environmental policy think tank located in Washington, DC):

> The competition for water in the Middle East is so intense that lasting peace in the region is unlikely in the absence of an agreement over shared water use.... Water has already been the source of armed conflict in the region between Syria and Israel, once in the 1950s and again in the 1960s. Several times over the past thirty years, disputes among Turkey, Syria, and Iraq over the development and use of the Euphrates River have nearly ended in armed conflict. Disputes arose in the 1960s when Turkey, where 90 percent of the water originates, and Syria started to plan large-scale withdrawals for irrigation. The conflicts heated up in 1974 when Iraq threatened to bomb the dam at Tabqa, Syria, and massed troops along the border because of the reduced flows they were receiving in the Euphrates.

To illustrate the role of water as a security consideration, RRF cites the Ataturk Dam (completed in 1990), which gives Turkey a potent weapon to be used against downstream countries, potentially reducing flows to Syria by as much as 40% and to Iraq by 80%. But the dam could also be operated to benefit all countries within the basin by reducing the variability of the river's natural flows. Similarly, the rivalry between Jordan and Israel over the Jordan River is another opportunity for trans-border cooperation.[9]

An impressive example of international collaboration in water resources issues is the Jordanian-Israeli peace treaty "annex" regarding water-related matters. As described in a recent text book on ethics in water industry[7] this document is a clear and concise statement of the obligations of two neighboring countries with regard to efficient use of an important shared resource, while respecting the rights and needs of one another. The annex covers a number of major provisions, including allocation of water supplies, storage, water quality and protection, and groundwater resources. It also sets up a Joint Water Committee to administer the agreement and provide technical and policy level guidance.

The approach to this agreement is direct and comprehensive, addressing distribution of surface waters from both the Yarmouk and Jordan Rivers, as well as the withdrawal of groundwater from the basin each nation knows by a variant pronunciation of the same name (*wadi Araba* in Arabic, in Hebrew *emek Ha'Arava*). Most important, both Jordan and Israel have agreed to coordinate improvements to these water systems to help ensure their continued productivity for both. To implement this treaty, the parties created a Joint Water Committee with three members from each country. Clearly they have their work cut out for them. However, given their careful beginning, they may yet succeed in securing an equitable distribution of vital water supplies that helps, rather than hinders, attempts to bring the parties together in a regional peace.

11.2.2 Water Use Limits

Water-use regulations further limit how an entity with water rights can use water or distribute it to various parties. Over the past decade, it has become increasingly common for federal, state, and even local entities to set standards for how water may be used, including the extent to which it must be conserved or reused, as a condition of supplying water. During times of drought these standards serve to promote reuse by requiring water users to reduce their water use compared to some prior time. Nor are such standards enforced only during periods of extreme shortage. In California, for instance, certain uses of potable water (e.g., irrigation, power plant cooling) are prohibited at any time whenever nonpotable sources are available, environmentally appropriate, and economically sound (California Water Code, Section 13550[10]).

Three main types of water use regulations are water supply reductions, water efficiency goals, and water use restrictions. To different degrees, each can provide a stimulus to the development of reuse projects.

Water supply reductions are often imposed during periods of drought. Where water shortages are common, cutbacks may be imposed by statute or they may be written into water-allocation agreements between the various parties. During such times appropriated water rights may be invoked so that the senior rightsholders receive their full allocations, or have their allocations reduced less than those with more junior rights. Whatever the cause, water shortages often provide a powerful incentive to implement water-reuse projects, especially where less costly methods (e.g., water conservation) have already been implemented.

Water purveyors—that is, water rights holders who in turn resell water to individual water users in a community—may respond to their reduced supply by charging more for water or by implementing tiered rates that increase with water use above a baseline level. They may also establish priority categories among customers to make sure that water is available for firefighting, hospitals, and other critical purposes. It is worth noting that an important economic benefit of water reuse can be calculated by multiplying the probably frequency of water shortages by the increased cost of water to the community during periods of short supply.

Water efficiency goals can be either mandatory or voluntary. Where voluntary goals (or targets) are promulgated, public support for conservation and reuse are usually stimulated by advertising campaigns that underscore the need to protect limited supplies. On a local level, if a water user fails to meet mandatory goals, the water purveyor can impose higher fees or surcharges or even terminate service. On a state level, meeting goals may be a prerequisite for receiving grants or loans that can be used to build recycled water projects.

Water use restrictions may either prohibit the use of potable water for certain purposes or require the use of recycled water in place of potable water. Ordinances of the second type generally allow the prohibited "unreasonable" uses of potable water to occur when recycled water is unavailable, is unsuitable for the specific use, is uneconomical, or when its use would have a negative impact on the environment. Another important consideration in evaluation of water-use restrictions is what type of penalties or consequences they contain. On a local level, failure to comply with use restrictions may be grounds for termination of service; however, other regulations designed to protect water customers from termination may mitigate or even neutralize that penalty. On a statewide level, water-use restrictions allow local jurisdictions a legal foundation for regulating local use. They may also be effective in promoting water recycling when such rules also require state agencies to evaluate alternative supplies for all state-funded projects. A policy requiring all federally funded projects in the United States to evaluate the use of recycled water during the planning process has been discussed in recent years, but no such rule has yet been adopted.

Case Study: Japan

Because of the country's density and limited water resources, Japan is a leader in urban water reuse,[11] using about one third of all recycled water or 40 Mm^3/year (30 mgd) for urban purposes, especially flushing toilets. In Tokyo, the use of reclaimed water is mandated in all new buildings larger in floor area than 30,000 m^2 (300,000 sq. ft.). Initially the country's reuse program required multifamily, commercial, and school buildings to be equipped with an on-site reclamation plant returning treated effluent for use in toilet flushing and other incidental nonpotable purposes. It was later determined that municipal treatment works were more cost-effective than individual reclamation facilities providing effluent to buildings through a dual distribution system. A wide variety of buildings (especially schools and office buildings) have been retrofitted for reclaimed water use. Other examples of large urban systems can be found in Chiba Prefecture Kobe City and Fukuoka City, where recycled water has been used to augment streams and irrigate parks, agricultural areas, and greenbelts outside the city limits of these urban centers. It should be noted that Japan's reclaimed water-quality requirements for unrestricted use are more stringent than U.S. regulations for coliforms.

Finally, local water use restrictions can also serve to encourage reuse when the use of recycled water is generally accepted and readily available at a cost below other supplies (including privately owned wells). In such cases it may not be necessary to test the enforceability of the statutes, since the potential consequences of noncompliance may be sufficient to persuade most customers to use recycled water for appropriate purposes. Otherwise, penalties should be specified at a level adequate to deter violation. These may include disconnection of service and a fee for reconnection with fines and jail time for major infractions (e.g., Mesa, Arizona, and Brevard County, Florida).

11.2.3 Rights to Recycled Water

In arid parts of the world, recycled water may constitute a more reliable supply than either surface or groundwater, and where there is competition for developed supplies it may be the only adequate water available. In these circumstances, the question of who owns recycled water may be fundamental to the success of any project.[2] The downstream water user's right to reclaimed water depends on the state's water-allocation system:[2]

> Some states issue permits to the owners of reclaimed water or to appropriators of it when discharged into a natural water course... treating such discharges into a reclaimed water course as if it has been abandoned and thus available for appropriation. Other states issue appropriation permits containing a provision that clarifies that the permit does not, in itself, give the permittee a right against a party discharging water upstream who may cease to discharge the water to the water course in the future.

In other words, the law can either promote or constrain reuse projects depending on how its system of water rights regards the use and return of recycled water. In general, the owner of a wastewater treatment plant that produces effluent is generally considered to have first rights to its use and is not usually bound to continue its discharge. However, when a discharger's right to reuse is constrained, such restrictions are usually based on one or more related principles of reduced discharge, changed point-of-use, or changed hierarchy of use. These principles are illustrated in recent U.S. court cases[2] that are applicable comparable circumstances elsewhere.

In one case, downstream ranchers opposed several cities in the Phoenix area when they contracted to sell recycled water to a group of electric utilities (Arizona Public Service vs. Long, 773 p. (2d 988, 1989).[2] The ranchers claimed that the cities had no right to sell unconsumed effluent because public surface waters must be returned to the riverbed; the cities countered that reclaimed water had become their property when they expended funds to create it through treatment. The Supreme Court of Arizona validated the contract, holding that the cities were not obligated to continue to discharge effluent to satisfy the needs of downstream appropriators and that reclaimed water was not subject to regulation under either Arizona's Surface Water Code or Groundwater Code. The Court then urged the state legislature to enact statutes in the area.

In related cases, U.S. courts have upheld the rights of a city to move its point of discharge to a different location notwithstanding the expectation of downstream users (Thayer vs. City of Rawlins, Wyoming, 594 p. 2d 951, 1979),[2] but they have also required cities to leave enough flow in receiving streams to maintain existing ecosystems.

These cases serve to point out the potential vulnerability of recycled water projects implemented in areas where there may be little precedent for its use. Wherever downstream water users may claim to suffer from withdrawal of supplies that might otherwise be returned, it is in the interest of the project sponsor to address this issue early in the process and to look for ways to keep all parties whole through water exchanges or other benefits wherever possible.

On the other hand, some courts have held that reusing water is preferable to disposal, as when a California city was required to show cause why they discharged 6000 m^3/d of treated effluent to the Pacific Ocean rather reusing it in the local community. This ruling was founded on a state prohibition against the "unreasonable" waste of potable water used when recycled water is available.[10] The case also illustrates a trend towards regarding water of any quality suitable for some type of reuse, such that its discharge may be limited for the sake of preserving a scarce public resource.[12]

11.3 WASTEWATER REGULATIONS

Where national or provincial governments exercise jurisdiction over public waterways, they usually oversee the public and private discharge of wastewater to maintain minimum water quality standards. In the United States the primary authority for the regulation of wastewater is the Federal Water Pollution Control Act, commonly referred to as the Clean Water Act (CWA).[13] The CWA assigns to the federal government the goal of making all U.S.

surface waters "fishable and swimmable" while at the same time allowing states the right to control pollution to the extent that state regulations are at least as stringent as federal rules.[2]

Primary jurisdiction under the CWA is with the U.S. Environmental Protection Agency (EPA), but in most states the CWA is administered and enforced by the state water pollution control agencies. Similarly, European nations have largely adopted national policies and regulations limiting wastewater discharge into public waterways, while allowing individual treatment works to be managed by regional or local authorities, including private companies.

In Germany, for example, water and wastewater regulations are implemented by the Federal Ministry for the Environment,[14] which is responsible for the Federal Water Act and the Wastewater Charges Act, for resolving basic questions of water-resources management and transboundary cooperation and for provisions to the European Union. It was stated[15] that "the choice of institutional arrangements in a country will depend on the history of its public administration and on its culture and traditions. For example, Latin American nations tend to be more decentralized than say China that has had a long tradition of centralized administration."

11.3.1 Effluent Regulations

Although disinfection of pathogens remains an important goal of wastewater treatment throughout the world, the focus of effluent regulations in developed countries appears to have shifted in recent years to the removal of microcontaminants such as trihalomethanes and various metals like copper and nickel that even at very low concentrations have deleterious effects on aquatic organisms. Other emerging concerns include pharmaceutically active compounds ranging from antibiotics to analgesics and a wide range of chemicals that can interfere with the function of the endocrine system.

When regulations establish limits for trace elements and microcontaminants, local agencies may be required to discharge effluent that is already suitable for reuse without further treatment, reducing the investment needed to meet recycled water standards. This can provide a powerful incentive to reuse treated effluent even in areas where rainfall is plentiful. However, even if the quality standards are comparable, the level of reliability required by effluent regulations may be less rigorous than paying customers expect, so supplementary treatment systems may be needed to ensure continuous production. These issues should be thoroughly explored by those planning water-reuse projects prior to project design and implementation. Another important issue is the distribution of costs between those responsible for wastewater treatment and the water-reuse program, especially when the agencies are governed by different boards of directors. While the cost of meeting effluent regulations may be the responsibility of a given agency, they may require compensation for the water supplied to the reuse program as a means of defraying their sunk costs. Such commercial relationships should be explored early in the development of the program.

Although less common than regulation of water quality, the quantity of treatment plant effluent discharged to a receiving body may also be limited. In some instances the limit is due to the capacity of the discharge facilities, while in other cases discharge may be regulated by policy reflecting the limitation of the receiving water to assimilate effluent. Such regulations may be continuous or seasonal and may or may not correspond to period where recycled water is in demand.

Case Study: Silicon Valley (San Jose), California

State regulators in California required the San Jose/Santa Clara Water Pollution Control Plant (serving the Silicon Valley area of northern California) to recycle treated effluent as an alternative to limiting discharge into the south end of San Francisco Bay during the summer dry weather period (May through October). In this instance the limitation was due not to contaminants or pipeline capacity but because the point of discharge was into a saltwater marsh which was made brackish by the discharge of relatively fresh treated effluent. The salt marsh in question is home to rare wildlife (the California clapper rail and salt-marsh harvest mouse), and the conversion of their habitat from salt to brackish marsh was deemed to violate federal and state laws protecting endangered species.

In response to this restriction, public officials in the community evaluated several options, including water conservation, water reuse, the purchase of mitigating wetlands to offset damage, and construction of a marine outfall pipe to discharge effluent into deep water away from the marsh. The community eventually chose to implement the first three alternatives concurrently, including a nonpotable water reuse program that distributes 8 Mm^3 annually, or up to 40,000 m^3/d during the summer irrigation season. In choosing to expend $140 million (US$) on the construction of 100 km of recycled water pipe, local decision makers balanced project cost against the "no-project" alternative of limiting effluent discharge and the economic impact of the resultant moratorium on new construction, which was estimated as high as US$500 million.[16]

From an institutional perspective, the role of the regulatory body responsible for enforcing effluent discharge regulations is to provide clear and consistent direction with respect to wastewater treatment standards and any other requirements that might facilitate the use of recycled water. The reuse of treated effluent for irrigation presents many formidable challenges to agencies that collect and treat wastewater. As the operations manager of one large (600,000 m^3/d) wastewater treatment plant explained, "It's a big step to go from treating water for discharge to reusing it on schoolyards. We have a whole new set of customers".[17]

By requiring dischargers to meet clearly defined quality standards, regulators help create a positive working relationship between organizations responsible for producing and distributing recycled water. Regulatory agencies can also assist in reassigning liability for the ultimate disposal of treated effluent in the environment as necessary when agencies transfer ownership of the water after distribution. In any event, it will be necessary to define which regulations apply and to determine which of the various regulatory bodies have jurisdiction over the effluent at various stages of treatment, distribution, and reuse. For instance, effluent standards may be regulated by an environmental agency responsible for the quality of water in streams, while reuse standards may be under the jurisdiction of the regional or local health department. It may be wise to obtain the participation of potable water suppliers during these discussions as well if they are not already involved in the development of the reuse program.

11.3.2 Pretreatment to Protect Recycled Water Quality

The quality of recycled water is determined by both the constituents in the wastewater and the type of treatment provided. Although many pollutants (e.g., suspended solids, biochemical oxygen demand, pathogens, and nutrients) can be removed by conventional processes, other contaminants are more resistant to biological treatment, filtration, and disinfection. Detectable concentrations of certain industrial solvents may be present in treatment plant effluent, as are many pharmaceutical compounds and most soluble inorganic salts. Potable reuse projects require additional contaminant removal steps to further purify the water, either mechanical (e.g., reverse osmosis) or natural (e.g., aquifer recharge).

When recycled water is used for nonpotable purposes, it is usually sufficient to control contaminants by limiting their introduction to the treatment plant at the source. Originally instituted to protect treatment plants from upset, all treatment plants serving communities with industrial dischargers are now required to adopt local source control regulations. To the extent that these ordinances reduce the concentration of refractory chemicals in treatment plant effluent, they also provide an effective method of maintaining the quality of recycled water. Where a contaminant is not regulated, or where its discharge limit is above the tolerance of local plants (e.g., boron), in most jurisdictions source control ordinances can be easily revised to accommodate a new limit under broad authority to regulate wastewater quality.

Agencies that provide recycled water for irrigation are usually concerned about its salinity (see also Chapter 5, § 5.5.1). Salinity or total dissolved solids (TDS), especially sodium and chloride, can have a deleterious effect on certain sensitive plants and can otherwise impair the usefulness of recycled water. Salinity problems often result when recycled water is manufactured from wastewater with a significant industrial component, which can contribute acid and base rinses and cooling water blowdown. Other sources that increase the salinity of recycled water include residential and commercial water softeners that discharge high concentrations of salt and seawater that may infiltrate the sewer system in coastal areas.

A detailed analysis of salinity, SAR, and the quality appropriate for irrigation with recycled water is provided in Chapter 2 and Chapter 5. However, from an institutional perspective it should be noted that agencies are challenged to maintain water-quality standards when the agency responsible for distributing recycled water is separated from the agency responsible for treating effluent. These responsibilities may be further complicated by an agency's limited ability to regulate discharges into the sanitary sewer system. Unless they utilize special advanced treatment methods, most wastewater-treatment plants do not reduce the salinity of wastewater prior to reuse and actually increase it through the addition of chemicals like chlorine.

In response, many communities implement source control measures to restrict the discharge of chemical salts into the sanitary sewer system either by requiring their placement in a special brine line or by charging a fee for their treatment and removal.[18] Overcoming the objection of trade groups that manufacture and sell them, some jurisdictions prohibit the use of selfregenerating water softeners as a last resort to attain appropriate recycled water quality.[19]

While source control can be an effective means of ensuring recycled water quality for irrigation, it requires a well-established organization of laws and agencies that are empowered to develop and enforce discharge standards on residential, commercial, and industrial dischargers. Lacking this advanced institutional infrastructure, many communities may find it easier (though more expensive) to investigate various removal options like reverse osmosis or electrodialysis. If that is beyond their means, they can try to isolate the most saline streams from the treatment works and provide a separate, less saline process train for reuse, or they can collaborate with area growers to select plants more tolerant of the average quality of recycled water produced.

Case Study: Los Angeles, California

In a recent study, the Metropolitan Water District of Southern California estimated that selfregenerating water softeners contribute between 5–10% of the total dissolved solids (TDS) found in recycled water in the southern California area. These devices pass water through a bed of ionexchange resins, which picks up the calcium and magnesium ions and exchanges them with sodium ions, after which the spent resin is automatically flushed with a high-salt brine that is ultimately discharged to the sewer system. According to the report, "The discharge of salt from the regeneration of water softeners into the wastewater collection system has a negative impact on recycled water and wastewater plant effluent. Higher salinity increases the treatment costs and reduces the potential for reuse of wastewater for nonpotable irrigation and industrial purposes[20]." Similar studies undertaken in the area by the Sanitation Districts of Los Angeles County (LACSD) indicated that self-regenerating water softeners discharge effluent with an average chloride concentration of 10,300 mg/L and that these softeners, installed in 11% of the residential units within the study area, contribute about 70% of the chloride load. They further determined that eliminating softeners would reduce the overall chloride concentration from 168 mg/L to 113 mg/L.[21]

The issue of regulating water softeners is complicated in California by the fact that state law has exempted such devices installed in private residences from regulation by local authorities. In order to improve recycled water quality, local agencies successfully worked with the state legislature to amend that law such that local codes may now restrict the use of water softeners when "limiting the availability, or prohibiting the installation, of the appliances is a necessary means of achieving compliance with the water reclamation requirements or the master reclamation permit issued by a California regional water quality control board."[22]

11.4 PLANNING AND IMPLEMENTATION ISSUES

11.4.1 Land Use Planning

Unlike water and wastewater laws that are often promulgated by national governments and regional agencies, land use planning regulations are usually developed and enforced by local jurisdictions. Planning regulations and the local agencies that implement them provide a powerful stimulus to water recycling when they permit new residential, commercial, and industrial projects that increase the overall demand for water. Nonpotable recycled water used for irrigation can supplement potable supplies, allowing the construction of projects that might otherwise strain existing resources. In urban areas, where the cost of extending utilities is relatively high, construction of satellite treatment and reuse facilities can reduce the size of collection systems and help control development budgets. And in places where effluent limits are stringent, wastewater-treatment plants are at capacity, or receiving streams cannot assimilate additional discharges, nonpotable irrigation may be the only effluent-disposal alternative that allows a development to move forward through the approval process.

Conversely, planning agencies that fail to consider adequately their community's water supply and wastewater treatment limitations retard reuse programs by effectively subsidizing new developments by deferring to their successors the responsibility of providing those resources in a sustainable manner. In either case it is in the interest of the agency responsible for water recycling to work closely with land use planners so as to integrate reuse goals with overall community development requirements.

Case Study: Sydney, NSW, Australia

The role of land use planning in promoting nonpotable irrigation is illustrated by the case of Rouse Hill, a phased development of up to 70,000 homes near Sydney, Australia. The New South Wales state government had identified the Rouse Hill area northwest of Sydney as a major corridor for expansion of that city, and in the late 1980s targeted a large portion of land to house Sydney's growing population.[23] However, due to general shortage of water and the limited capacity of the local receiving streams to assimilate effluent, the Department of Infrastructure, Planning and Natural Resources (DIPNR) and the local shire council together required the development to use its treated wastewater for irrigation of front and back yard lawns, as well as parks and playgrounds and for flushing toilets in Rouse Hill homes. Under the Rouse Hill Development Project, a group of land owners managed the staged provision of water-related services and initially funded infrastructure through borrowings, and the Sydney Water Corporation (SWC) took over ownership of the assets after the successful commissioning of the works.[24] Referring to the general shortage of capital for construction of reuse facilities to serve existing areas, a recent report form the government of western Australia cited this eastern example and noted that "There is a need for further analysis of the costs and benefits of a variety of reuse options, particularly those that offset scheme water demand, and particularly those that reduce the cost of wastewater service provision in addition to reducing the cost of water supply and energy. The major gains will come from options that either offset demand from scheme supplies, or can provide credits within a tradeable water entitlement arrangement."[25]

Where approved developments will be required to include nonpotable irrigation, the managers responsible for implementing reuse will need to coordinate closely with the land use planning agency in order to ensure that the recycled water distribution facilities are designed to meet all relevant local, state, and federal regulations. In this case it would be helpful to include appropriate specifications in the materials planners provide to developers and to the greatest extent possible incorporate all reuse guidelines in the prevailing planning requirements. Failure to do so will result in the creation of a secondary process in which the reuse agency may struggle to review and approve plans in time with the primary planning schedule. In other instances, the use of nonpotable water occurs in a different location, and the potable water saved through reuse is reallocated to the new development. In that case coordination with both the planning agency and the water authority will be essential to ensure that the reuse project provides the anticipated volume of water and that allocations of water from other sources are executed according to plan.

The local planning process can also pose a challenge to reuse projects by subjecting them to the scrutiny of a public that may have many misconceptions about recycled water. Federal and state environmental assessment regulations (which are often included in the local planning process) require public notice of published plans and advertised hearings to solicit opinion from all parties potentially affected by the proposed project. It is not unusual at such hearings to hear opposition to the use of recycled water for reasons ranging from health effects to growth inducement to environmental justice. These concerns often mask underlying worries about growth or political issues that may be hard to deal with directly. However, unless the specific concerns are thoroughly addressed in the planning process, it is unlikely that the project will proceed to the point that the underlying issues can emerge to be dealt with. Furthermore, a failure to address project conformance with general plan guidelines and local requirements for the preparation of specific plans will render a reuse project vulnerable to challenge in the courts or to appeal before the regulatory bodies even after the project is approved.

11.4.2 Environmental Regulations

Although there are as yet no universally acknowledged international environmental standards, there is considerable agreement among the world's nations about the need for environmental protection. Almost every country has some set of laws, policies, or guidelines which they use to establish national environmental goals and to evaluate the potential impact of planned projects. Within each country, many states and provinces also have rules that mandate environmental assessment and mitigation planning for projects prior to construction, all of which can work to further the development of water reuse.

On one level, regulatory support for nonpotable irrigation and other reuse practices stems from the fact that, along with conservation, water recycling is the among the most environmentally friendly water-supply alternatives. Taken together, laws that protect aquatic, biological, scenic, and cultural resources can result in a virtual moratorium on the construction of new dams or other large-scale water diversions that flood the habitat of protected species, inundate areas of historical significance, or require the displacement of thousands of residents. When such projects can be proposed for evaluation, a comparable supply of recycled water may prove more cost-effective when the economic analysis includes both direct and indirect benefits. Not only do water-recycling projects place less stress on the aquatic environment that conventional water supply schemes such as reservoirs and groundwater mining, they often use less energy, produce less air pollution, and require less maintenance as well. A growing stream of research has developed over the past decade outlining appropriate methods for comprehensive evaluation of environmental benefits.[26] Familiarity with regional and national environmental regulations and the techniques for determining the least-cost alternative can help managers identify appropriate partners for funding reuse projects.

Environmental agencies also support reuse programs directly through grant or loan programs. In view of the multiple environmental benefits that stem from reuse, it is not surprising that many agencies regularly appropriate

Case Study: Northern California CalFed Program

The degradation of northern California's riparian environment was caused by a number of factors but was chiefly due to the diversion of surface water for irrigation. Habitat loss has been accelerated during the last 50 years by urbanization, which has replaced wetlands with hardscape and diverted additional water for municipal and industrial use. From 1994 to 2000 a consortium of federal and state agencies (CalFed) held a series of stakeholder negotiations with the goal of developing an acceptable approach to restoring the biodiversity of the Sacramento-San Joaquin River Delta by balancing the needs of farmers, cities, and the environment. The resultant recommendations included increasing water use efficiency through water conservation and water recycling.* In all, improvements valued at more than US$3 billion have been proposed over a 7-year period, to be funded by the U.S. government, the State of California, and local interests. The CalFed example illustrates the difficulty of satisfying urban, agricultural, and environmental water needs even in a highly affluent society.

funds for research and development of environmentally sustainable programs like water recycling. In a similar vein, international development agencies (e.g., IMF, World Bank) also take into consideration the environmental benefits associated with water reuse when investing their funds.[7] According to the WorldWatch Institute, though followed sporadically, "the [World] Bank's policy that governs adjustment lending stipulates that the environmental impact of these loans should be fully considered as they are prepared, with a view toward promoting possible synergies and avoiding environmentally harmful results."[27] This view is borne out by the World Bank's withdrawal of funding from the massive Sardar Sarovar Project in the Indian State of Gujarat that would have yielded nearly 30,000 million m^3 of water per year by inundating the Narmada River Valley.[28]

On the other side of the equation, environmental assessment regulations also require the careful assessment of any negative impacts of recycled water projects. Examples of potential environmental impacts include the following:

- Visual impact of tanks and resrvoirs
- Disturbance of underground cultural resources and hazardous materials by underground pipelines
- Degradation of groundwater when nonpotable water is applied for irrigation over an unconfined aquifer
- Growth inducement when availability of nonpotable water allows water-limited communities to expand

*"Principles for agreement on Bay-Delta Standards between the state of California and the Federal Government" (December 15 1994) cited in California Deparment of Water Resource website, accessed August 19, 2004 http://calwater.ca.gov/Archives/General Archive/SanFranciscoBayDeltaAgreement.

Environmental justice when undesirable environmental facilities (e.g., wastewater treatment plants, landfills) are routinely sited in economically depressed neighborhoods with proportionally large minority populations

In each case, mitigation for potential impacts must be identified. The manager of a recycled-water project must be familiar with not only the federal and state regulations guiding the environmental assessment process, but also their interpretation by the local jurisdiction.

11.4.3 Construction Issues

Just as certain institutions influence the planning and design of recycled water projects, their construction and implementation are governed by other rules and organizations. In addition to the state and local health departments that require minimum distances between potable and nonpotable pipelines, there are national plumbing codes, local and regional rules with respect to acceptable building materials and construction practices and even contract and labor laws, the review of which is beyond the scope of this chapter. It must suffice to note that prior to and during design, the project manager should become familiar with state and local construction regulations and obtain all necessary permits from local agencies, utilities, and other parties so as not to delay project construction.

In addition to these general rules, many states have rules specifically pertaining to construction of recycled-water systems. These regulations frequently cover setbacks between recycled-water and potable pipelines and include details for pipeline crossings (e.g., nonpotable below potable). Where it is not practical to maintain minimum distances, some states allow nonpotable pipelines to be encased in suitable materials (e.g., PVC). The reader is referred elsewhere in this text for a detailed discussion of health-related requirements. However, it is worth noting that in states where national or provincial legislatures have adopted resolutions formally endorsing the use of recycled water or establishing targets for reuse, the agency promulgating construction regulations may be willing to evaluate alternative designs that provide the same level of protection of public health and safety to meet these goals.

This was recently brought out in a study comparing regulations in two U.S. states, California and Florida.[29] The study contrasted the stricter construction requirements in California with the more cooperative attitude of Florida officials, who were encouraged by state statute to promote reuse. As shown in Figure 11.1, the effect of this difference was to accelerate the pace of development of reuse projects in Florida compared to California, which risks failure in achieving its official goal of 40 m^3/s annual average by 2010.[30]

The issue of funding is also addressed in Chapter 9 of in this guidebook, but it should be mentioned that many programs establish their own

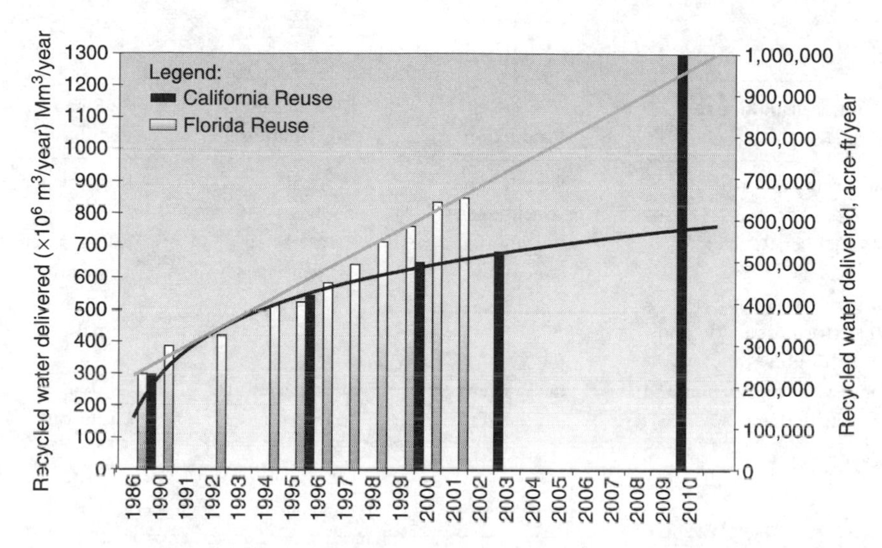

Figure 11.1 Comparison of historical and projected future rates of implementation of water reuse in California and Florida.

construction-related rules that projects must meet to qualify for funding. These could include some or all of the following:

"Value engineering" of the project by professionals not involved in the original design to ensure that plans are "cost-effective" and meet professional standards of practice.

Institution of a revenue program identifying sources of funds to pay for the initial construction (this is especially true when grant funds are provided for construction on a reimbursement basis).

Evidence that customers will use a specific quantity of recycled water once it is supplied.

Once on-site facilities have been constructed, state and local regulations often require that crossconnection tests be performed to ensure complete separation between potable and nonpotable systems. Depending on the quality of the water provided and the type of use, agencies may also restrict the times of use and require periodic inspection and reporting on system operation, even after the on-site system has been installed and approved.

11.4.4 Wholesaler/Retailer Issues

One of the first steps in implementing a water-reuse program is the identification of roles and responsibilities for the manufacture, wholesale and retail distribution of recycled water. Many different types of institutional structures can be utilized for implementation of a water-reuse project and responsibility for recycled water production, wholesale and retail distribution can be assigned to different groups depending on their historic roles and technical and managerial expertise (Table 11.1).

Table 11.1 Some Common Institutional Patterns

Type of institutional arrangement	Production	Wholesale distribution	Retail distribution
Separate authorities	Wastewater treatment agency	Wholesale water agency	Retail water company
Wholesaler/Retailer system	Wastewater treatment agency	Wastewater treatment agency	Retail water company
Joint Powers Authority, JPA (for production and distribution only)	JPA	JPA	Retail water company
Integrated production and distribution	Water/Wastewater authority	Water/Wastewater authority	Water/Wastewater authority

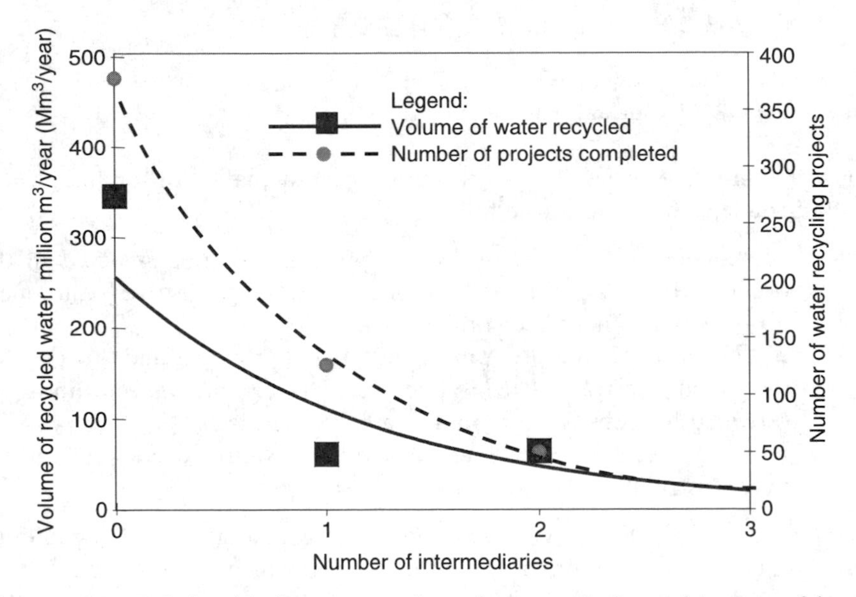

Figure 11.2 Relationship between number of intermediaries and success of imple-

In general, the simpler the structure the better. The various departments and agencies within government can come into conflict over the proposed reuse system unless steps are taken early in the planning stages to find out who will be involved and to what level. Close internal coordination between departments and branches of local government will be required to ensure a successful reuse program. Obtaining the support of other departments will help to minimize delays caused by interdepartmental conflicts.

The challenge of implementing reuse projects with multiple partners was also addressed in the aforementioned study,[29] which observed an inverse relationship between success of water-reclamation projects and the number of intermediate agencies. As shown in Figure 11.2, both the volume and the number of implemented projects declined as the number of participating agencies increased from one to four. Since only a limited number of agencies in California integrate water and wastewater functions, the remaining special-purpose (and single-function) agencies must form partnerships to implement water-reuse projects with the result that the rate of implementation of water-recycling projects in California continues to decline.

Case Study: Irvine Ranch Water District, California

A good example of an integrated authority is the Irvine Ranch Water District in California, an independent, self-financing public entity responsible for all phases of recycled water production and distribution. Under its original enabling legislation, it was strictly a water-supply entity, but in 1965, state law was amended to assign it sanitation responsibilities within its service area. Thus, the district is in a good position to deal directly, as one entity, with conventional potable water and nonpotable water services. This contrasts markedly with other more complex institutional arrangements where the manufacturer of recycled water sells reclaimed water to several purveyors, who then redistribute it to a number of users. Where separate water and wastewater agencies must work together, the boards of directors of the institutions can help staff to overcome contractual hurdles by adopting resolutions indicating their intent to cooperate.

In evaluating alternative institutional arrangements, responsible managers should make a determination of the best municipal organizations or departments to operate a reuse program. For example, even if the municipal wastewater treatment service is permitted by law to distribute reclaimed water, it might make more sense to organize a reuse system under the water-supply agency or under a regional authority (assuming that such an authority can be established under the law). Among the criteria for assigning roles are *financing power*, the ability of an agency to assume bonded indebtedness, and *contracting power*, the agency's ability to contract for goods and services and execute agreements.

A regional authority can often operate more effectively across municipal boundaries and can obtain distinct economies-of-scale in operation and financing.[31] One of the best ways to gain the support of the other agencies is to make sure that they are involved from the beginning of the project and are kept informed as the project progresses. There is, on occasion, an overlap of jurisdiction of some agencies. For example, it is possible for one agency to control the water in the upper reaches of a stream and a separate agency to control the water in the lower reaches. Unless these agencies can work together, there may be little hope of a successful project that impacts both. Any potential conflicts between these agencies should be identified as soon as possible. Clarification as to which direction the overall agency should follow will need to be determined. By doing this in the planning stages of the reuse project, delays in implementation may be avoided. This is especially important when it becomes necessary to enact new legislation to establish a new public entity.

11.4.5 Customer Agreements

The last link in the chain of institutional arrangements required to implement water-recycling projects is the relationship between the water supplier and the water customer. There are two dimensions to this arrangement:

1. The legal requirements established by state and local jurisdictions defining the general responsibilities of the two parties to protect the public
2. The specific items of agreement between the parties, including commercial arrangements and operational responsibilities

Legal requirements for nonpotable water use are usually stipulated in state laws, agency guidelines, and local ordinances and are designed to protect public health. A public agency responsible for distributing recycled water can emphasize the importance of these rules by adopting an ordinance requiring customers to observe them as a condition of use. This is especially important in countries where recycled water is still statutorily considered wastewater effluent, and the recycled water supplier effectively delegates its authority to discharge to its customers whose irrigated acreage is legally considered a "land outfall." Although not included in the customer agreement, recycled water suppliers financing their project with customer revenues may also wish to implement a local ordinance that defines when property owners must connect to the reuse system immediately (e.g., parks, golf courses) and which properties must connect as the system becomes available.

Customer agreements should clearly assign responsibilities for enacting the protective measures required to avoid cross-connection. These measures will usually require at a minimum color-coding pipe to distinguish nonpotable from potable water and may also include the use of backflow preventers and periodic inspection of facilities. The agreement should state which party is responsible for inspection, under what conditions and with what frequency inspection may be required, and the consequences if users refuse to perform or allow inspection (i.e., disconnection of service). A customer agreement might also specify design of the irrigation system (e.g., a permanent below ground system), construction details like pipe materials and quick disconnect fittings for hand watering, and the use of timers for irrigation.

With respect to the commercial arrangements, the customer agreement should specify all rates and charges, fees, rebates, terms of service, and other special conditions of use. Any fees charged for reclaimed water connection and the rates associated with service should be addressed in the customer agreement, either directly or by reference to an appropriate rate ordinance. The agreement may also include details on financing onsite construction to separate the customer's potable and nonpotable systems. It is not uncommon for local agencies to fund all or part of the cost of retrofitting a customer's existing system in order to defray the overall cost of recycled water use.

In addition to the elements presented above, it is often helpful to establish various other terms of service that are particular to the water-reuse program and its customers. For example, the customer agreement may specify a certain level of reliability that may or may not be comparable to that of the potable system. The supplier of recycled water may also wish to retain the right in the ordinance to impose water use scheduling as a means of managing shortages or controlling peak system demands.

11.5 PROGRAM MANAGEMENT

11.5.1 Integrated Planning

In the publication they unveiled in March 2003 at the 3rd World Water Forum in Kyoto, Japan, the World Water Council observed that progress towards the goal of sustainable water use rests on the ability of our global institutions to cooperate. "The water crisis has been called a crisis of management," they wrote in World Water Actions: Making Water Flow for All.[1] "In most countries reforms to improve management in the water sector are under way, often beginning with adjustments in the legal, institutional and regulatory frameworks. The most visible change is towards greater coordination of water concerns across sectors."

Nowhere is the need for interagency coordination more apparent than in the development of water-reuse projects like nonpotable irrigation, and nowhere are the benefits of cooperation more evident. A successful recycling water project is produced by the combined efforts of water-resource managers, wastewater engineers, planners, and regulators. In this way water reuse challenges us to address not only the technical aspects, but also the economic and political dimensions of resource allocation as well. Only by integrating these elements—technical, economic and political—will we attain the societal changes necessary to live in balance with the natural environment.

11.5.2 Matrix Analysis of Institutional Issues

Despite the significant benefits provided by water-recycling, many cities encounter resistance to water-reclamation projects, even for nonpotable use.[32] For example, within the past 5 years at least four major potable reuse projects were abandoned or deferred due to public opposition.[33,34] More recently, even proposals to reuse water for nonpotable irrigation have been rejected or modified in response to public concerns.[37] Among the reasons cited by decision makers for rejecting reuse are health concerns, burdensome regulations, a preference for other "cleaner" water sources, and the high cost of reuse.

As illustrated in Figure 11.3, there appears to be a hierarchy of factors that must be satisfied in order to support the decision to implement a water-reuse project.[38] *Technical feasibility* concerns the ability of treatment facilities

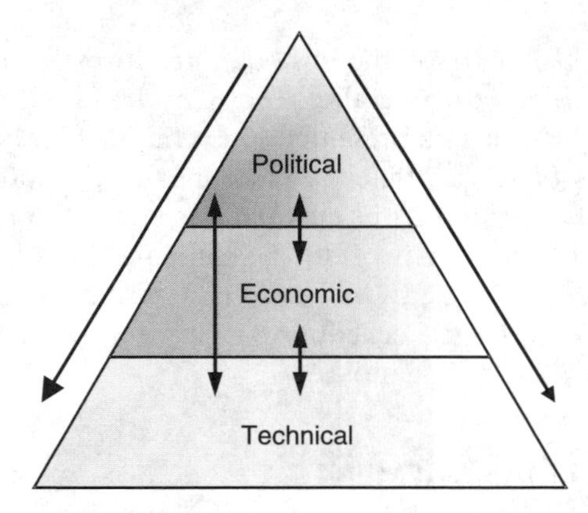

Figure 11.3 Hierarchy of decision-making factors in water-reuse projects (from Ref. 39).

to purify wastewater, as well as all other aspects of the physical setting in which the project occurs, e.g., the health of the ecosystem and the adequacy of water sources to meet projected demand. *Economic incentive* (or disincentive) includes market and non-market costs such as the value of the recycled water and the cost of mitigating negative impacts on the environment.

Political motivation includes values and rationales, both overt and hidden, that incline decision makers or the general public to either support or fight against water reuse. Examples include proposing water recycling as a means of enhancing supply reliability, or opposing reuse as a means of limiting community growth, or any element of water reuse that might be perceived as a salient campaign issue by an elected official. While this category of factors is in some respects the most difficult to identify and analyze, it is also the most critical in determining whether or not a particular project is adopted.

The direction of influence of these factors is primarily from the top down. In other words, when political forces are aligned to support reuse, economic resources are identified and technical solutions purchased or invented. By contrast, a project that is technically and economically feasible but lacks political support will probably not be implemented. However, influence can also occur from the bottom up, as when regulations mandate technical mitigation of environmental impacts or when uncertainty about the safety of recycled water incites public fear. Economic factors are also converted into political issues, for instance, when lower taxes are a political theme and water supplies are translated into economic terms. Broadly considered, the politicization of economic factors may often favor reuse, for instance, when water reclamation is preferred over dam construction with a large environmental mitigation cost.

A diagrammatic approach has been proposed (Figure 11.4) to help analyze various institutional factors as they influence the decisions of agencies

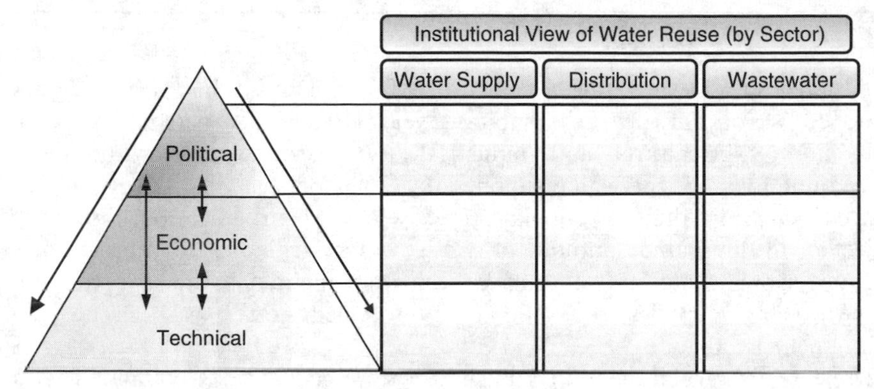

Figure 11.4 Matrix for anlayzing institutional factors influencing water reuse (from Ref. 39).

representing different sectors of the urban water cycle.[39] By evaluating technical, economic, and political issues related to reuse from the perspective of each of the various sectors (water supply, water distribution, and wastewater treatment), one may better understand and respond to the values that determine a community's support. Factors that on balance support water reuse are indicated by a plus ("+") sign in the appropriate cell of the matrix, while factors that lead an agency to oppose reuse are designated by a minus ("−") sign. A completed column of plus signs indicates that an institution is aligned in favor of reuse and may sponsor a reclamation project. Success is unlikely, however, without the cooperation of other agencies in the urban hydrologic cycle, as indicated by a row of plus signs.

Application of this matrix analysis to the evolution of water-recycling decisions in the Silicon Valley area of northern California (see Section 11.2.1) illustrates use of political influence in resolving economic and technical challenges to clear the way for project implementation. In that case, the matrix analysis indicates, a strong mandate from a regulatory agency was able to set an economic value on the protection of a salt marsh degrade by effluent discharge. This economic incentive was in turn sufficient to motivate the discharger to meet with area water supply and distribution agencies and resolve their technical and economic objections to reuse, which had hindered its application for the previous 20 years.

There are many examples in the international water community of this type of cooperation between the various institutions responsible for supplying water and treating wastewater in urban areas. At a minimum, cooperation requires extensive communication between agencies at the earliest stages of project development. In the water industry, this approach has been carried out under the rubric of "integrated resource planning," in which alternatives like water conservation and water reuse are given due consideration along side more conventional water-supply projects.

Case Study: Tunisia[40]

Situated in the arid north coast of Africa, Tunisia has a per capita water use of only 370 m^3/inhabitant year (1000 L per person per day), including agricultural and industrial use. Groundwater withdrawals account for 76% of water supplies; by 2020 deficit withdrawals, which are increasing due to population growth and urbanization, are projected to limit the amount of fresh water available for agriculture. Interregional water transfers have been implemented to make better use of existing supplies, but ultimately recycled water was needed to augment supplies.

Most of the urban areas in Tunisia are sewered (78% as compared with 40% in rural areas), and nearly 60% of sewage collected is treated at 62 treatment plants (160 Mm^3/year of out 240 Mm^3/year of wastewater discharged). Nearly half of the treatment (62 Mm^3/year) is produced by five Tunis treatment plants (oxidation ditches, stabilization ponds, and activated sludge facilities). Notably, Tunisia has developed and implemented sanitation master plans for several towns, integrating treatment and reuse needs as new plants are designed and built. A wastewater-reuse policy launched in the early 1980s favors agricultural and landscape irrigation, and in the Tunis area more than 5000 ha are irrigated by reclaimed water. A new water-reuse project currently planned for the City of Tunis West is projected to provide over 100,000 m^3/d (40 Mm^3/year) of reclaimed water by the year 2016, irrigating about 6000 ha. Another project in the Medjerda catchment area provides the 11 largest towns with sewerage networks, treatment plants, and reclaimed water-irrigation schemes in order to protect the Sidi Salem Dam from wastewater contamination.

Reuse is slowly increasing in newer areas, and by 2020 as much as 30,000 ha may be irrigated with nearly 300 Mm^3/year of reclaimed water. At that point, recycled water use will equal 20% of available groundwater resources, eliminating the excessive groundwater mining that currently causes saltwater intrusion in coastal aquifers. Eight existing golf courses are irrigated with secondary-treated effluent. Other pilot projects have been implemented to study the use of reclaimed water for groundwater recharge, forest irrigation, and wetlands development. Tunisian water standards follow FAO (1985) and WHO (1989) guidelines for restricted irrigation (<1 helminth eggs/L). The Water Law prohibits use of raw wastewater in agriculture and irrigation with reclaimed water of any vegetable to be eaten raw, and the use of secondary effluent is limited to growing nonvegetable crops including fruit trees (citrus, grapes, olives), fodder (alfalfa, sorghum, berseem), sugarbeet, and cereals. There is a significant additional market for recycled water to sustain peri-urban agriculture, but more precise water-quality standards may be needed since these areas are mostly devoted to vegetable production.

One immediate benefit of integrated planning is that it allows agencies to perform a "full cost accounting" of alternative projects in which all the benefits and costs are adequately recognized. Planners can most effectively compare alternative projects when the full range of benefits and costs are identified and expressed in monetary terms. However, the benefits associated with water-reuse projects include watershed protection, reduced risk of water shortages, enhanced economic development, and other factors not readily quantified for consideration by traditional cost-benefit techniques. As a result, recycled water projects are often undervalued when compared to other projects and significant opportunities for beneficial reuse are lost.[41]

To solve this problem, various techniques have been developed to assign a monetary value to a range of benefits associated with water recycling, including effluent reduction, pollution reduction, water supply, infrastructure and energy savings, flood control, ecosystem improvements, and public support and other related effects. Where benefits were not directly quantified by market values or replacement cost, indirect methods such as contingent valuation were used to establish a monetary equivalent.[42]

11.5.3 Summary of Institutional Guidelines

The following guidelines can assist managers in addressing legal and institutional issues during the planning and implementation of a reuse system.

11.5.3.1 Identify the Legal and Institutional Drivers for Reuse

Understanding the laws governing water supply and wastewater treatment in the project area will allow the project manager to identify which parties are responsible for each aspect of water supply, distribution, and wastewater treatment and disposal, as well as any special constraints that support water reuse. This background will allow the project to address the most pressing local issues and encourage participation in the program.

11.5.3.2 Establish Ownership of Recycled Water

Water professionals in responsible positions in upstream watersheds should seriously consider their responsibilities to downstream inhabitants in terms of the quality and quantity of water leaving their part of the basin. This is equally true for diversion of recycled water as for other water supplies. Prior to developing a nonpotable irrigation program, ensure that the effluent targeted for reuse is available for diversion from its current point of discharge without impacting downstream users. If other users can claim rights to the effluent, include them as stakeholders in the planning process and identify appropriate compensation. If the discharge is into a saltwater estuary or bay, determine whether or not the receiving stream requires a minimum flow to maintain existing habitat.

11.5.3.3 Identify All Relevant Institutions and Contact Participating Stakeholders

Identify the institutions affected by a water-recycling program, and consider in detail the alternative institutional structure for operation of the water-reuse system. Work with all parties to evaluate the advantages and disadvantages of different working relationships, and identify as early as possible any legislative changes that might be required to create the necessary institutions and the level of government at which the legislation must be enacted.

Throughout development of the reuse project, contact should be maintained with the federal, state, and local agencies involved, as well as customer groups, public interest organizations, and others motivated to participate in the process. Promote their understanding of the project and its goals, and keep them informed of pending milestones, including permit reviews or the enactment of new legislation. Continued contact and an open flow of information can keep the process from becoming an obstacle.

11.5.3.4 Select Appropriate Reuse Technology

Alternative technologies should be fairly explored, and water professionals in developed countries should participate in programs that allow them to extend their expertise to the developing world. For example, in developing a nonpotable water project in a developing country, a reasonable consideration of the relative merits of gravity filtration, upflow mixed media filtration, and membrane filtration should take into account not only water quality and treatment reliability, but also long-term maintenance (e.g., availability of parts and supplies), energy costs, and the ability to establish training programs necessary to ensure a qualified work fore.

11.5.3.5 Develop a Realistic Schedule: Assess Cash Flow Needs

An accurate assessment of cash flow needs is required to anticipate funding requirements, formulate contract provisions, and devise cost-recovery techniques. Create a revenue plan with short-term and long-term business goals for the program. Cost analysis of water resources projects should include mitigation costs to reduce impacts on the environment and on human communities to acceptable levels, and managers should attempt to ensure that the beneficiaries of these improvements participate in the overall financing of the project. Funding by international aid agencies in developing countries should be equally cognizant of the human, environmental, and economic impacts of the projects on all affected parties.

11.5.3.6 Prepare Contracts

Formal contracts are usually required to establish usage of the reuse system and to govern its operation. Provisions relating to the quality and quantity of the reclaimed water are essential and may include a range in which each can fluctuate, and the remedies, should the quantity or quality go outside that range. Responsibility for any storage facilities and/or supplemental sources of water should be defined. There must be an explicit statement as to how the reuser will pay for the recycled water, to what extent, and for what reasons he is responsible and liable for costs. Both parties must be protected explicitly in case either party defaults, either by bankruptcy or by the inability to comply with the commitments of the agreement. Finally, the ownership and maintenance of the facilities must be stated, particularly for the transmission and distribution facilities of the reclaimed water. The point at which the water-conveyance facilities become the property and responsibility of the user must be explicitly stated. In the case where the user is a private enterprise, that statement should be reasonably straightforward. However, in the case where the user is another municipal entity, it is especially important that each party knows its responsibility in the operations and maintenance of the facilities.

11.5.3.7 Ensure Follow-Through

Long-term monitoring responsibility must be specified, especially if a monitoring program is required as a condition to use reclaimed water for irrigation purposes. For example, if the crops grown are not to be utilized for human consumption, it may be appropriate to assign the responsibility for compliance with such regulations to the user. Specific compliance with environmental regulations must be assigned to each party, along with a schedule of performance and consequences for failing to abide by the terms of the agreement.

11.5.3.8 Develop and Maintain a Public Education Program

Except in the most limited cases, implementation of a water-reuse program will require the support of the public and acceptance by prospective customers. Early development of an educational outreach program providing accurate information about the nature of recycled water and the benefits and costs of water reuse will facilitate public review required by federal and state environmental regulations and help gain approval through the local planning process.

REFERENCES

1. Guerquin, F., et al., *World Water Actions: Making Water Flow for All.* World Water Council/Japan Water Resources Association/UNESCO, Marseille, France, 28, 2003.
2. Cologne, G., and MacLaggan, P., Legal Aspects of Water Reclamation, in *Wastewater Reclamation and Reuse*, Takashi A., ed. American Water Works Association, Denver, CO, 1995.
3. Howe, C., Sharing water fairly, reprinted in *Our Planet 8.3* (October 1996), accessed October 15, 2003, http:||www.ourplanet.com/imgversn/83/howe.html
4. Coulomb, R., French Water Stakeholders are mobilizing, Société Hydrotechnique de France-Suez (2003), attachment in Guerquin, F., et al., *World Water Actions: Making Water Flow for All*, World Water Council/Japan Water Resources Association/UNESCO. Marseille, France, 2003.

5. Lazarova, V., et al., Role of water reuse for enhancing integrated water management in Europe and Mediterranean countries, *Water Sci. Technol.*, 43, 10, 23, 2001.
6. Israel Water Commission, Transitional Master Plan for Water Sector Development In the Period 2002–2010—Executive Summary, 2003.
7. Shiekh, B., Rosenblum, E., and Pawson, M., "Ethical Dilemmas in Planning and Design of International Water Projects," in *Navigating Rough Waters: Ethical Issues in the Water Industry*, Davis, C.K., ed., AWWA, Denver, CO, 2001.
8. Gleick, P., Water conflict chronology, Pacific Institute for Studies in Development, Environment, and Society, http://www.worldwater.org/conflict.htm
9. Frederick, K.D., Water as a source of international conflict, Resources (Spring 1996) Resources for the Future, www.rff.org
10. California Water Code §§174, 275, 13550 and California Code of Regulations Title 23, §§856, 1977.
11. Ogoshi, M., Suzuki, Y., and Asano, T., Water reuse in Japan, in *Proc. 3rd International Symposium on Wastewater Reclamation, Recycling and Reuse*, 1st World Congress of the International Water Association (IWA), Paris, France, 2000.
12. California State Water Resources Control Board Order 84-7, The Matter of the Sierra Club, San Diego, 1984.
13. Clean Water Act, Federal Water Pollution Control Act, Pubic Law 92–500, 33 U.S.C. 1251, 1992.
14. Federal Ministry for the Environment, Nature Conservation and Nuclear Safety, International Insitutional Stuctures to Support Water Resource Planning and Water and Wastewater Quality—Germany, p.2 attachment in *World Water Actions: Making Water Flow for All*, Guerquin, F., et al., World Water Council/ Japan Water Resources Association, UNESCO, Marseille, France, 2003.
15. Alaerts, G.J., Strategy options for sewage management to protect the marine environment, p.19 UNEP/ Global Programme of Action, Coordination Office P.O. Box 16227, 2500 BE The Hague, The Netherlands, 2002, accessed October 15, 2003, http://www.gpa.unep.org
16. Rosenblum, E., Selection and Implementation of nonpotable water recycling in Silicon Valley (San Jose area) California, *Wat. Sci. Technol.*, 40, 4–5, 51, 1999.
17. Rosenblum, E., From discharge to recharge: the changing role of operations, *Operations Forum, November*, 1996.
18. Shiekh, B., and Rosenblum, E., Economic impacts of salt from industrial and residential sources, in *Proc. AWWA/WEF Water Sources Conference: Reuse, Resources Conservation*, Las Vegas, NV, January 27–30, 2002.
19. California Health and Safety Code §116786.
20. Bookman-Edmonston Engineering, Water softener issues, Technical Appendix 6, Salinity Management Study Final Report, Metropolitan Water District of Southern California and U.S. Bureau of Reclamation, June 1999.
21. Sanitation Districts of Los Angeles County, Santa Clarita Valley Joint Sewerage System Chloride Source Report, pp. 6–1, October 2002.
22. California Health and Safety Code, Chapter 172, Section 116786, AB334 chaptered August 2003.
23. Shelter New South Wales (NSW), Land supply and housing affordability in Sydney—a background paper, Sydney Australia, September, 2003, accessed October 16, 2003 at http://www.shelternsw.infoxchange.net.au/Publications/sb03land.pdf.
24. Independent Pricing and Regulatory Tribunal of New South Wales, Prices of Water Supply, Sewerage and Drainage Services for Sydney Water Corporation, 1996.

25. Water and Rivers Commission, State Water Conservation Strategy (Western Australia), 2002, 45.
26. Sheikh, B., Rosenblum, E., Kasower, S., and Hartling, E., Accounting for the benefits of water reuse, in *Proc. AWWA/WEF 1998, Water Reuse Conference*, Orlando, FL, February, 1998.
27. French, H., Coping with ecological globalization, in *State of the World 2000*, WorldWatch Institute, Norton, New York, 2000, 197.
28. Roy, A., The greater common good, in *The Cost of Living*, Modern Library Paperbacks, New York, 1999.
29. Sheikh, B., York, D., Hartling, E., and Rosenblum, E., Impact of institutional requirements on implementation of water recycling/reclamation projects, in *Proc., Water Sources Conference*, San Antonio, TX, January 2004.
30. California Water Code §13577, Added by Statutes 1991, c. 187 (A.B. 673).
31. Okun, D.A., Principles for water quality management, *J. Environ. Eng.*, ASCE 103, EE6, 1039, 1977.
32. Hartley, T., Water reuse: understanding public perception and participation, WERF Research Report 00-PUM-1, Water Environment Reuse Foundation, Alexandria, VA, 2003, 1.
33. VanWagoner, W., Snow, T., Butler, B., Bailey, B., and Richardson, T., Use of recycled water to augment potable supplies: an economic perspective, A White Paper by the Potable Reuse Committee, WateReuse Association, T. Richardson, Chair, Available from WateReuse Association (Alexandria, VA) or online at http://www.watereuse.org/Pages/information.html.
34. Wegner-Gwidt, J., Public support and education for water reuse, in *Wastewater Reclamation and Reuse*, Asano, T., ed., Technomic Publishing Co., Lancaster, PA, 1998, 1432.
35. CH2M Hill Tampa Water Resource Recovery Project Pilot Studies Tampa, FL CH2M Hill, 1993 cited in *Issues in Potable Reuse: The Viability of Augmenting Drinking Water Supplies with Reclaimed Water*, National Research Council, National Academies Press, Washington, DC, 1998.
36. Pinellas County Board of County Commissioners, Principles and Position Statements on Water Resources, November 19, 2002, available at http://www.pinellascounty.org.
37. Recycled Water Task Force, Meeting Minutes, February 26, 2003, California Department of Water Resources (DWR), accessed March 8, 2004, http://www.owue.water.ca.gov/recycle/docs/Feb2603Minutes_Draft.pdf.
38. Kasower, S., personal communication, 1999.
39. Rosenblum, E., The water reclamation matrix: a framework for sustainable urban water use, in *Proc. Nuevas tecnologias en la applicacion de la reutilizacion de aguas residuales y trataminetos de biosolidos*, Marbella, Spain, May 24–25, 2004.
40. Bahri, A., The experience and challenges of reuse of wastewater and sludge in Tunisia, *Water Week 2000*, World Bank, Washington, D.C., April 3–4, 15, 2000.
41. Sheikh, B. et al., Accounting for the benefits of water reuse, in *Proc. AWWA/WEF Water Reuse Conf.*, Orlando, FL, 1998.
42. Jordan, J., Incorporating externalities in conservation programs, *JAWWA*, June, 49, 1995.

12

Case Studies of Irrigation with Recycled Water

CONTENTS

12.1 El Mezquital, Mexico: the largest irrigation district using wastewater 345
12.1.1 General description 345
12.1.2 Wastewater quality 347
12.1.3 Effects of wastewater reuse on agriculture and health 348
12.1.4 Mexican legislation for agricultural irrigation 348
12.1.5 Helminthiasis 349
12.1.6 Removal of helminth eggs 351
12.1.7 Filtration step 352
12.1.8 Bacteria removal 353
12.1.9 Sludge treatment and disposal 355
12.1.10 Costs of recycled water 355
12.1.11 Unplanned aquifer recharge by irrigation with wastewater 356

References 361

12.2 Water reuse for golf course irrigation in Costa Brava, Spain 363
12.2.1 History 363
12.2.2 Tips for adequate recycled water management in golf course irrigation 364
12.2.3 Electrical conductivity (EC) 364
12.2.4 Nutrients 365
12.2.5 Maturation pond design and management 368
12.2.6 Conclusions 372

References 373

12.3 Monterey County water recycling projects: a case study in irrigation water supply for food crop irrigation 374
12.3.1 History and motivation 374
12.3.2 Project overview 374
12.3.3 Public perception 377

1-56670-649-1/05/$0.00 + $1.50

12.3.4 Current project status and operation 378
12.3.5 Conclusions, lessons learned and recommendations 378

References 379

This chapter illustrates the role of irrigation with recycled water for different purposes and under different economic and geographical conditions:

1. The choice of appropriate treatment to improve the health safety of agricultural irrigation with wastewater in the Mezquital Valley, Mexico
2. The main concerns of golf irrigation with recycled water in Costa Brava, Spain
3. The challenges of irrigation of food crops with recycled water in Monterey, California

In developing countries such as Mexico, the major issue is to improve the health safety of wastewater reuse for agricultural irrigation. The existing practice of using untreated wastewater leads to significant health problems, and for this reason significant research efforts have been made to identify the most cost-competitive treatments to disinfect urban wastewater without removal of fertilizing elements such as carbon and nitrogen. In this case, the major technical challenge is the removal of helminth eggs, which are present in very high concentrations (10–90 helminth eggs/L) in the municipal wastewater in Mexico. Promising results have been obtained with advanced chemically enhanced primary treatment followed by sand filtration and combined disinfection using UV irradiation and chlorination.

The case study of Mexico City provides very interesting information on the efficiency of soil-aquifer treatment (SAT). Irrigation with untreated wastewater for almost a century did not greatly affect the quality of groundwater used for the potable water supply in the region of the Mezquital Valley.

The second case study illustrates the main issues associated with the implementation of golf course irrigation with recycled water in Europe, in particular in northeastern Spain (the tourist coastal region of Costa Brava). In this case, secondary wastewater treatment is mandatory for all medium and large wastewater-treatment works, and the main challenge is to keep and optimize the fertilizing capacity of recycling water, in particular nitrogen content. It was demonstrated that long-term storage in natural ponds leads to significant loss of nitrogen. Other important disadvantages of storage of recycled water in ponds are algae growth, odors, and increased maintenance costs.

The third case study demonstrates the feasibility of irrigation of food crops, including vegetables eaten raw, with recycled water in Monterey, California. Despite the high level of treatment and extensive water-quality monitoring, this project was difficult to initiate and implement because of public apprehension. The lessons learned demonstrate the effectiveness of good management practices and communication strategy among stakeholders.

12.1

El Mezquital, Mexico: The Largest Irrigation District Using Wastewater

Blanca Jimenez

With their economic capacity, developed countries treat a high percentage of their wastewater to a secondary level. The effluent produced has a high quality, and its reuse is promoted to make treatment profitable. In contrast, in many low-income countries water shortage is the reason for reusing wastewater. Untreated wastewater is often used for irrigation in agriculture due to high water demand, facility to convey wastewaters from cities to the agricultural fields, and, most of all, farmers' interest in its fertilizing properties. To take advantage of this situation in Mexico, water reuse is considered to reconcile the interests of farmers with appropriate practices for health protection. According to the recent modification of water-reuse legislation, appropriate treatment schemes have been investigated for implementation in the Mezquital Valley, one of the largest projects of agricultural irrigation with wastewater in the world.

12.1.1 General Description

In Mexico 102 m^3/s of sewage are used to irrigate more than 250,000 ha.[1] Currently the country has the biggest continuous irrigation area of the world to apply this practice, known as "El Mezquital." Since 1896 this area has received the wastewater from Mexico City (Figure 12.1.1). Of the total volume, 80% is sewage and 20% rainwater. However, rainwater is only available from May to October during certain hours of the day and at very high flow (up to 300 m^3/s).

In 1920 the benefits of irrigating agricultural crops with wastewater were well known, and a complex hydraulic system was implemented to regulate water distribution according to crops' needs. Nowadays this system is very complex and consists of nine dams (three with clean water and six with wastewater), three rivers (Tula, Actopan, and Salado), and 858 km of channels, which convey 60 m^3/s of sewage produced by 19 million inhabitants (Figure 12.1.2).

The Mezquital Valley is part of the Tula Valley, which is located 100 km north of Mexico City at an average altitude of 1900 m. The irrigation area

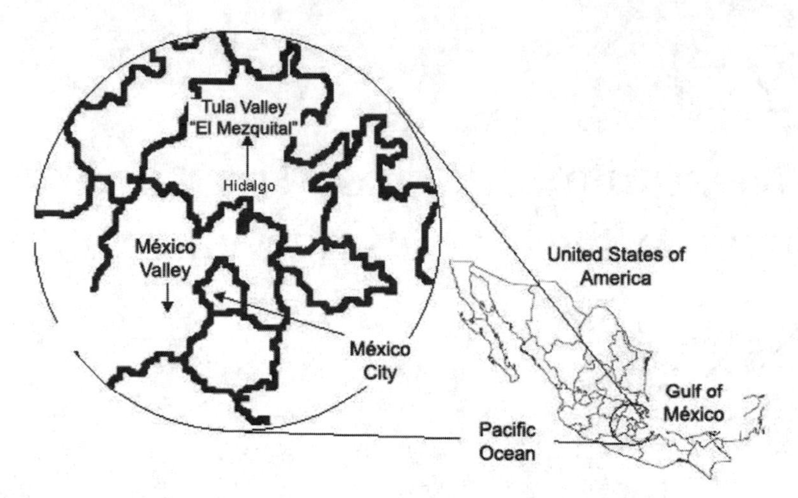

Figure 12.1.1 Location of Mexico City and the El Mezquital Valley in Tula. (From Ref. 22, with permission.)

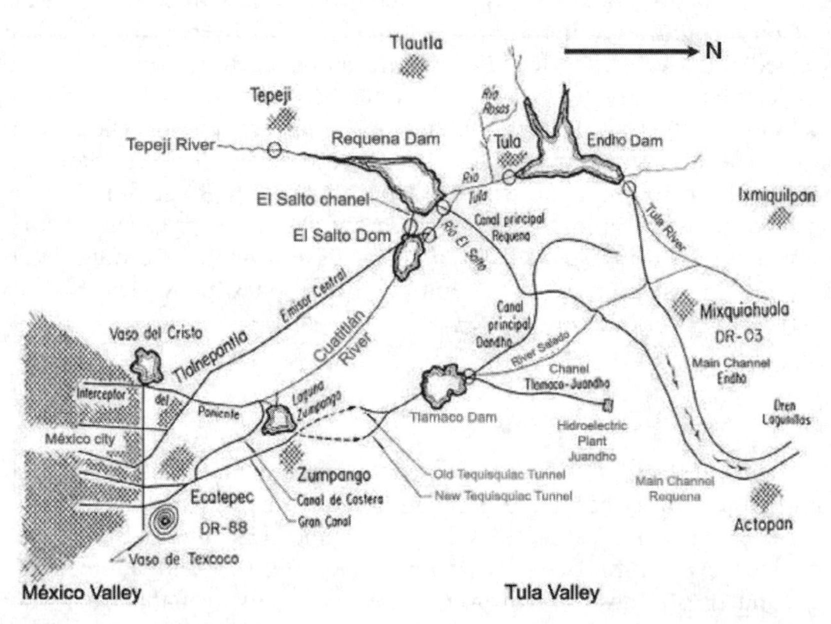

Figure 12.1.2 Main hydrological system used to distribute wastewater within the Tula Valley.

"El Mezquital" of 90,000 ha is divided into three irrigation districts: 03 (Tula), 100 (Alfajayucan), and 25 (Ixmiquilpan). The climate is sub-arid, with rainfall only during 4 months of the year. Annual average precipitation is 550 mm, whereas the evapotranspiration is 1700 mm. There are three types of soil in the region: eutric Vertisols, rendzic and melanic Leptosols, and calcic and haplic Phaeozems.[3] At the present time the natural vegetation limits itself to

Table 12.1.1. Quality of Municipal Wastewater from Mexico City

	Dry season			Rainy season		
Parameter	Mean	Minimum	Maximum	Mean	Minimum	Maximum
COD, mg/L	578	245	1492	475	168	1581
BOD, mg/L	240	20	330	180	40	420
TSS, mg/L	295	60	1500	264	52	3383
N-NH_4, mg/L	23	16	43	17	6	57
NTK, mg/L	26	18	47	17	2	61
Ptot, mg/L	10	1	19	8.3	0.2	27
Helminth eggs, eggs/L	14	6	23	27	7	93
Fecal coliforms, MPN/100 mL	4.9×10^8	1.2×10^8	5.2×10^9	7.4×10^8	7.1×10^7	2.4×10^9

Table 12.1.2 Metal Content in Mexico City Wastewater in Comparison with USEPA Guidelines for Agricultural Reuse

		USEPA Water Reuse Guidelines, 1992	
Parameter (mg/L)	Mexico City wastewater	Long-term use	Short-term use
Aluminum	9.2	5.0	20.0
Arsenic	0.001	0.1	2.0
Boron	0.84	0.75	2.0
Cadmium	<0.0035	0.01	0.05
Chromium	0.037	0.1	1.0
Copper	0.06	0.2	5.0
Fluorides	0.55	1.0	15.0
Iron	3.75	5.0	20.0
Lead	0.05	5.0	10.0
Manganese	0.14	0.2	10.0
Nickel	0.05	0.2	2.0
Selenium	<0.002	0.02	0.02
Zinc	0.25	2.0	10.0

Source: Ref. 4.

mountainous areas, whereas the valleys are dedicated to agriculture, mainly corn and alfalfa.

12.1.2 Wastewater Quality

Table 12.1.1 shows the quality of Mexico City wastewater. During the rainy season sewage is also polluted, and parameters such as suspended solids and helminth eggs have even significantly higher values. However, heavy metal content is lower than the USEPA criteria[4] for agricultural water reuse in both seasons (Table 12.1.2).

Table 12.1.3 Increase in Productivity by Use of Wastewater for Agricultural Irrigation in the Mezquital Valley

	Productivity (ton/ha)		
Crop	Wastewater	Fresh water	Increase (%)
Corn	5.0	2.0	150
Barley	4.0	2.0	100
Tomato	35.0	18.0	94
Forage oats	22.0	12.0	83
Alfalfa	120.0	70.0	71
Chile	12.0	7.0	70
Wheat	3.0	1.8	67

Source: Adapted from Ref. 5.

Figure 12.1.3 Mezquital areas with the same type of soil with and without irrigation with wastewater.

12.1.3 Effects of Wastewater Reuse on Agriculture and Health

With the use of untreated wastewater, agricultural productivity has been increased up to 150% for some crops (Table 12.1.3 and Figure 12.1.3).[5] Nevertheless, serious health problems have been observed: gastrointestinal diseases caused by helminths occur 16 times more often in children 5–14 years of age than in the equivalent zones using freshwater (Table 12.1.4).[6]

Similar health problems associated with poor microbiological quality of reclaimed untreated wastewater have been reported in other developing countries (Table 12.1.5).[7] The mean reason is the low wastewater-treatment level: the percentage of treated wastewater varies from 0 to 30% in some countries in Latin America, Africa, and Asia.[8]

12.1.4 Mexican Legislation for Agricultural Irrigation

Due to the particular situation in Mexico, the local standard that controls water quality for irrigation was modified in 1996 to consider a value of

Table 12.1.4 Comparison of Morbidity in the Mezquital Area and a Similar Zone that Uses Clean Water for Irrigation

		Rate of morbidity		
Microorganism	Affected population by age	Zone irrigated with wastewater (A)	Zone irrigated with clean water (B)	A/B ratio
Ascaris lumbricoides	0–4	15.3	2.7	5.7
	5–14	16.1	1.0	16.0
	>15	5.3	0.5	11.0
Giardia lamblia	0–4	13.6	13.5	1.0
	5–14	9.6	9.2	1.0
	>15	2.3	2.5	1.0
Entamoeba histolytica	0 4	7.0	7.3	1.0
	5–14	16.4	12.0	1.3
	>15	16.0	13.8	1.2

Source: Adapted from Ref. 6.

Table 12.1.5 Microbial Content in Wastewater from Different Countries

Parameter	Concentration	Country
Helminth eggs, eggs/L	6–98	Mexico
	1–8	US
	166–202	Brazil
	up to 60	Ukraine
	up to 9	France
	up to 840	Morocco
Salmonella spp., MPN/100 mL	10^6–10^9	Mexico
	10^3–10^6	US
Protozoan cysts, cysts/L	978–1814	Mexico
	28.4	US

Source: Adapted from Ref. 7.

<1 helminth egg/L and <1000 fecal coliforms (FC)/100 mL for all types of crops[9] and <5 eggs/L and 1000 FC/100 mL for crops that are consumed cooked. The value of 5 eggs/L was established on the basis of the feasible value that conventional biological or physicochemical processes could achieve without filtration.[10,11] Organic matter and suspended solids are not limited in this regulation, since the former is mandatory in soils, whereas suspended solids must be removed to meet the required helminth egg concentration.

12.1.5 Helminthiasis

Helminthiases are common diseases in the developing world. It is estimated that an average of 27% of the population is infected (650 million people), with reported maximum values up to 90% in poor areas.[12,13] In contrast,

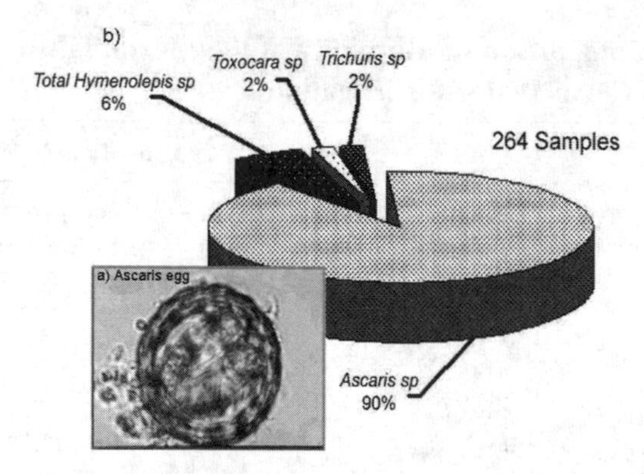

Figure 12.1.4 (a) Typical *Ascaris* sp. eggs and (b) helminth eggs species distribution in Mexico City wastewater. (From Ref. 2, with permission.)

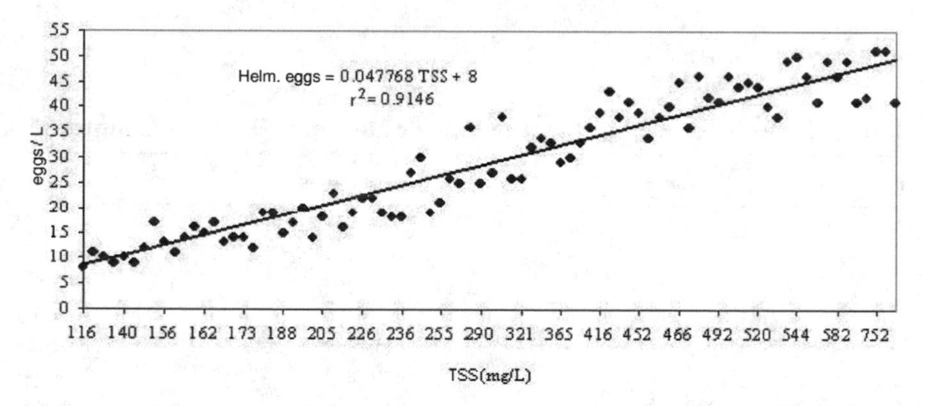

Figure 12.1.5 Correlation between helminth eggs and suspended solids for Mexico City wastewater. (From Ref. 2, with permission.)

in developed countries helminthiases reach a maximum of 1.5%.[14] These diseases are endemic in Africa, Central America, South America, and the Far East, where poverty and unsatisfactory sanitary conditions are common. They are transmitted through helminth egg ingestion from vegetables irrigated with polluted water. Helminth eggs are resistant to chlorine, ultraviolet (UV) light, and ozone. Infective doses are very low (1–10 eggs/L) compared to those for bacteria. *Ascaris* (Figure 12.1.4a) is the most common helminth found in wastewater and sludge (Figure 12.1.4b), and it is also the most resistant to wastewater treatment and medications.

The physical properties of helminth eggs (20–80 μm, specific density 1.036–1.238) greatly influence their removal from wastewater.[15] As a part of suspended solids (Figure 12.1.5), helminth eggs are removed by means of treatment processes such as settlers, lagoons, coagulation-flocculation, and filtration.

Table 12.1.6 Expected Microbial Removal from Sewage by Different Treatment Systems

Process	Log removal			
	Bacteria	Helminth eggs	Viruses	Cysts
Primary sedimentation[a]	0–1	0–2	0–1	0–1
Advanced primary treatment or chemically assisted primary sedimentation[b]	1–2	1–3	0–1	0–1
Activated sludges	0–2	0–2	0–1	0–1
Trickling filter	0–2	0–2	0–1	0–1
Aerated lagoons[c]	1–2	1–3	1–2	0–1
Oxidation ditch	1–2	0–2	1–2	0–1
Disinfection[d]	2–6[g]	0–1	0–4	0–3
Stabilization lagoons[e]	1–6[g]	1–3[g]	1–4	1–4
Dams[f]	1–6[g]	1–3[g]	1–4	1–4

[a]Filtration not considered by USEPA.
[b]Research required to confirm efficiency.
[c]Includes primary sedimentation.
[d]Chlorination and ozonation.
[e]Efficiency depends on number of lagoons and environmental conditions.
[f]Depends on hydraulic retention time.
[g]Performances depend on retention time, which varies with demand.
Source: Adapted from Ref. 4.

12.1.6 Removal of Helminth Eggs

Literature is quite scarce concerning helminth egg removal. In 1992 the USEPA listed some processes that could remove them from wastewater (Table 12.1.6). However, it was underlined that the removal efficiency of physicochemical processes needs to be demonstrated. For this reason, a 5-year study was performed in Mexico City.[2,5,16,17] The physicochemical process investigated is known as advanced primary treatment (APT) or chemically enhanced primary sedimentation (CEPT). The chemicals used for coagulation are mainly aluminum sulfate or ferric chloride, as well as low doses of organic flocculants with high charge and molecular weight.

Advanced primary treatment has been successfully used to treat municipal wastewater in Norway, Sweden, France, Spain, and the United States.[18–24] This process is very advantageous when influent has great variability in quantity and quality, as well as when a low degree of BOD removal is required (residual concentrations >30 mg/L). APT successfully removes heavy metals (about 70%) and phosphorus to produce an effluent suitable to be discharged into the ocean and sensitive areas. In general, APT removes 70% of suspended solids, 50% of nutrients, and 60% of BOD. Different APT technologies are commercially available.

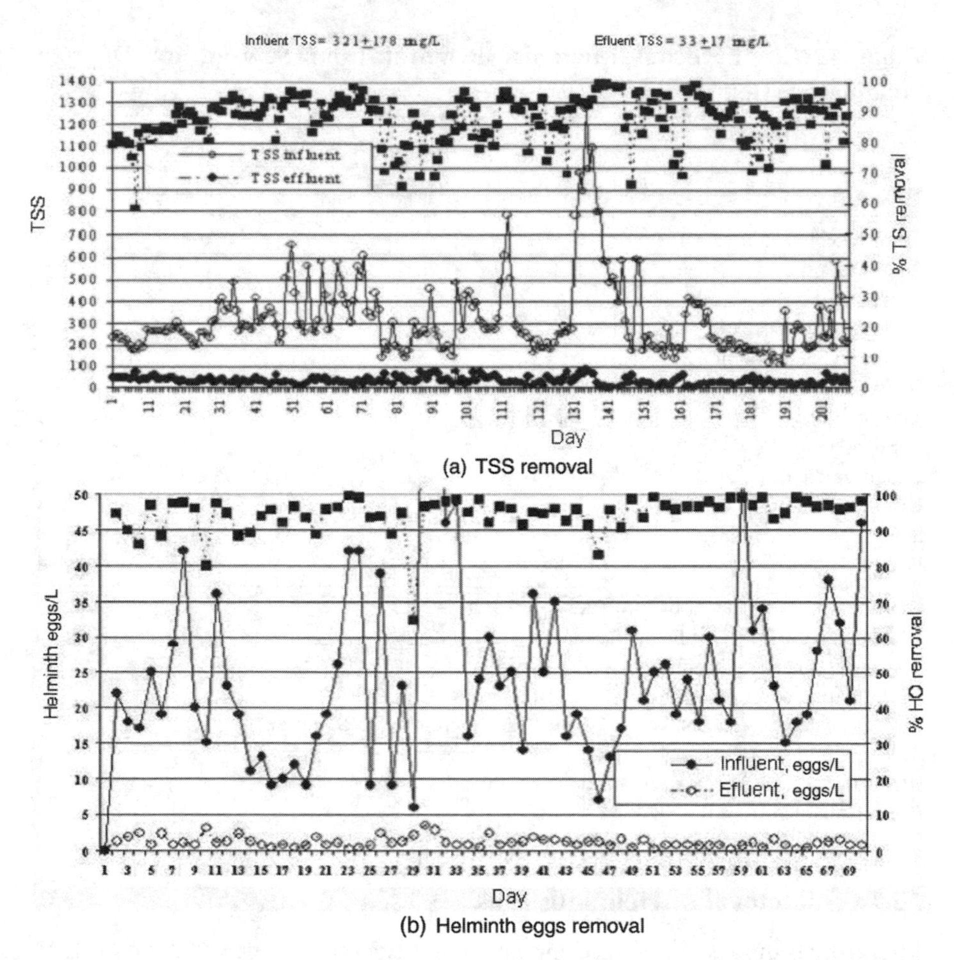

Figure 12.1.6 Efficiency of advanced primary treatment system (APT) with chemically enhanced high-rate sedimentation for (a) TSS and (b) helminth egg removal. (From Ref. 2, with permission.)

In search of the most appropriate treatment scheme for water reuse in Mexico, a 2-year full-scale study with several kinds of industrial prototypes was performed in Mexico City.[5,10,17,25] Despite the high variability in influent suspended solids (TSS 150–5000 mg/L), an effluent with residual TSS of 28 ± 16 mg/L and 2 ± 3 helminth eggs/L was obtained (Figure 12.1.6). The APT processes are very compact, allowing high hydraulic loads from 432 to 4320 m^3/m^2 d.[2,10]

12.1.7 Filtration Step

As mentioned previously, helminth egg content in APT effluents varied from 0 to 5 eggs/L, whereas in an activated sludge process it varied from 3 to 10 eggs/L. Consequently, to reach <1 egg/L it was necessary to add a filtration step (Figure 12.1.7). This step was designed to remove parasites and not necessarily

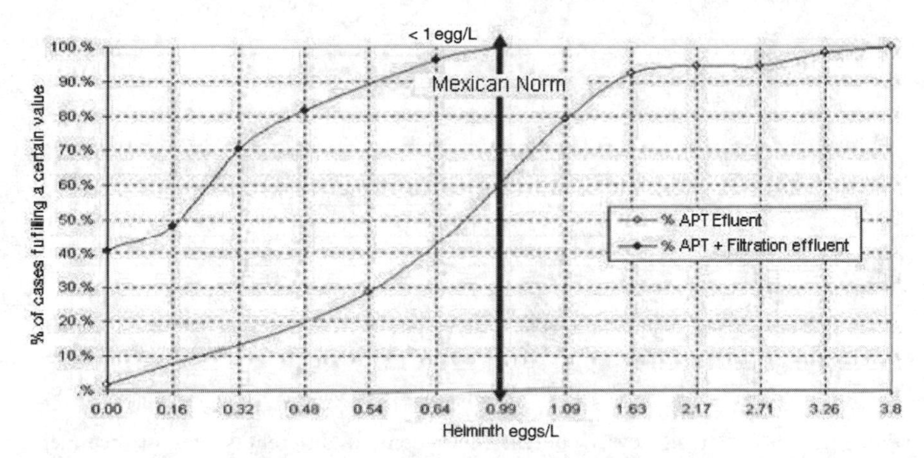

Figure 12.1.7 Percentage of cases with a specific helminth egg content in the APT and APT + filtration effluents. (From Ref. 2, with permission.)

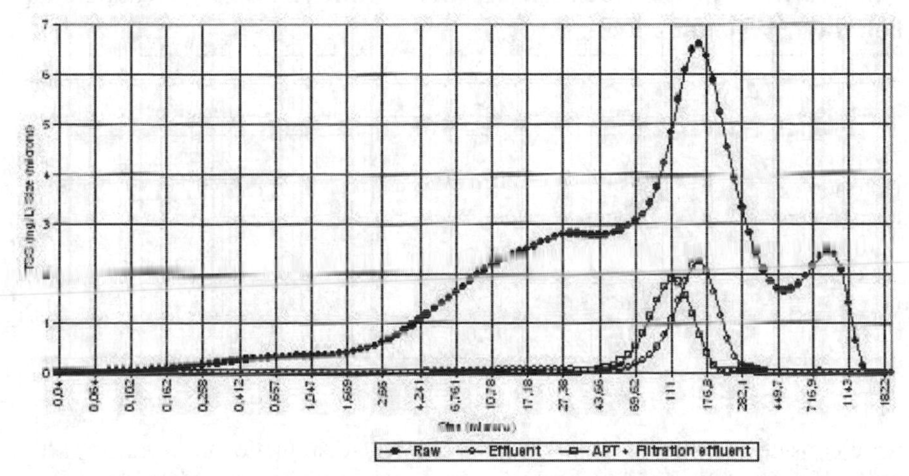

Figure 12.1.8 Particle size distribution in sewage, APT effluent, and APT + filtration effluent.

to fully eliminate suspended solids. Thus, filters with > 1 m depth packed with sand to (1.2–1.6 mm) were installed.[11] Helminth eggs were removed in 100% of the samples to less than 1 egg/L, while TSS efficiency was 50%. Removed particles ranged in size between mainly 20 and 80 μm (Figure 12.1.8).

12.1.8 Bacteria Removal

Filtered APT effluent still contained fecal coliforms in concentrations higher than 1000 MNP/100 mL, and thus a disinfection step was added. Two options

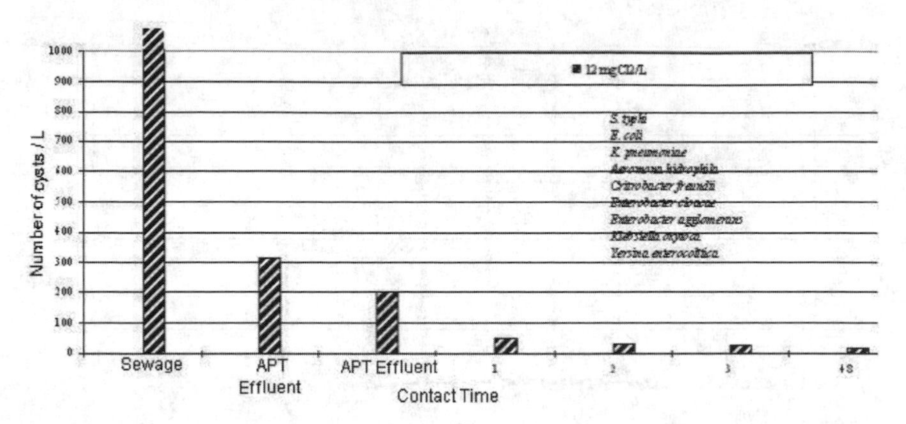

Figure 12.1.9 Parasite cysts content in effluent at different stages of treatment. (Modified from Ref. 25, with permission.)

Table 12.1.7 Effluent Water Quality After Advanced Primary Treatment and Filtration (APT + filtration)

Parameter	Influent	APT	Filtration	Chlorination	Efficiency (%)
TSS, mg/L	350–380	27–40	12–21	6–12	98
Helminth eggs, eggs/L	25–30	0.8–3.0	0.35–0.28	0.35–0.28	98.7
Fecal coliforms, MPN/100 mL	10^8–10^9	10^7–10^8	10^7–10^8	10^2	7 log removal
NTK, mgN/L	22–25	15–20	13–15	13–15	44
Ptot, mgP/L	10–7	5–3	1.0–2.7		61
COD, mg/L	420–505	150–200	150–193	160–186	63

were considered: chlorine and UV plus chlorine disinfection. For chlorination alone, a dose of 10 mg Cl_2/L and a 3-hour contact time were needed to remove 6 log of fecal coliforms and reach the threshold value of 1000 FC/100 mL. Under these conditions bacterial regrowth was avoided during the 22-hour transport through open channels before reaching the irrigation area. The high contact time needed for disinfection was explained by the presence of ammonia in the effluent with the associated production of chloramines.

For combined disinfection, high-intensity, medium-pressure UV lamps were used, with an average UV dose of 20 mJ/cm^2. Chlorine was added at a low dose of 2–4 mgCl_2/L to the irradiated effluent. The combination of filtration with the appropriate chlorine doses allowed good inactivation of various parasites (Figure 12.1.9).

Table 12.1.7 illustrates final effluent quality. Nutrient concentrations (N and P) matched the local crop needs, and residual organic matter was appropriate for the regional soil types.

Table 12.1.8 Microbiological Quality of Mexico City Sludge Compared to That in the United States

Parameter	Mexico	United States
Fecal coliforms, MPN/g TS	10^{10}	10^{8}
Salmonella, MPN/g TS	10^{8}	10^{2}
Helminth eggs (eggs/gTS)	50–120	<1

Source: Refs. 26–29.

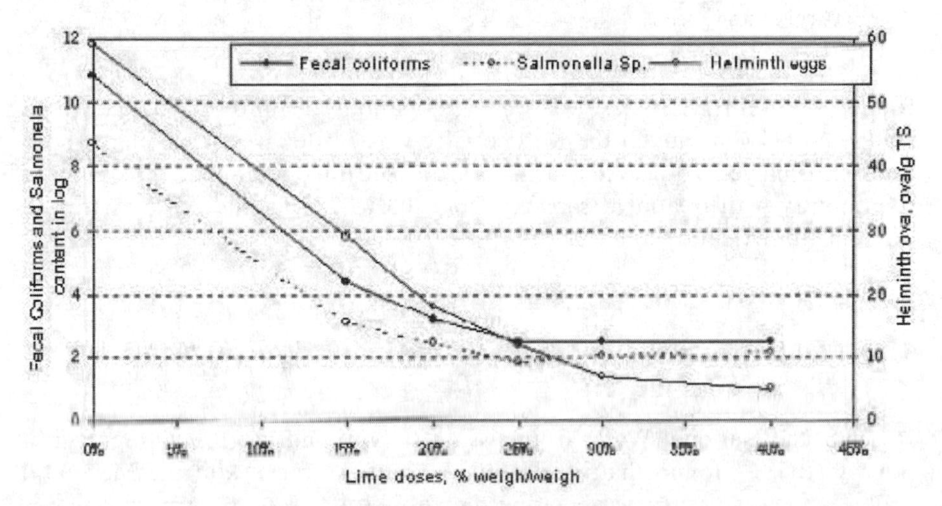

Figure 12.1.10 Fecal coliforms, *Salmonella*, and helminth egg inactivation in wastewater sludge with lime stabilization. (From Ref. 2, with permission.)

12.1.9 Sludge Treatment and Disposal

Sludge treatment is problematic because of the high content of pathogens (Table 12.1.8). Actually the concentration is higher than that found in other countries.[26–29] Lime stabilization treatment was selected because of its high capacity to inactivate microorganisms and its low cost compared to thermophilic anaerobic digestion.[28] Composting was also discarded, because in Mexico City there is not enough area to operate it. Lime stabilization had the advantage of producing biosolids that can be reused to remediate a 5000 ha area of very saline soil near the city.[30] With a 20–30% dry weight lime dose in the treated sludge, an average of 9 log of fecal coliforms, 7 log of *Salmonella*, and 95% of the helminth eggs were inactivated (Figure 12.1.10).

12.1.10 Costs of Recycled Water

The total cost of the APT process with and without filtration compared with activated sludge is presented in Table 12.1.9. It can be observed that APT

Table 12.1.9 Mexico City Wastewater-Treatment Costs Depending on Effluent Quality

Process	Helminth eggs (eggs/L)	TSS removal (%)	COD removal (%)	Cost (US$/m^3)
Advanced primary treatment (APT)	5–10	25	10	0.03
APT with filtration	1–5	75	65	0.05
Activated sludge	1–8	85	85	0.15

Source: Adapted from Ref. 2.

followed by filtration has a lower cost—about a third of the activated sludge cost. Even though the Mexico City wastewater treatment plants have not been built yet, 5 other APT facilities are already in operation in three cities of the country with treatment capacity from 0.3 to 2 m^3/s and reuse of treated effluents for irrigation and other purposes.

12.1.11 Unplanned Aquifer Recharge by Irrigation with Wastewater

In 1998, the National Water Commission (CNA) and the British Geological Survey (BGS)[31] found that, due to the irrigation practices in the Mezquital Valley, raw wastewater was entering the aquifer at a rate of 25 m^3/s, which is 13 times the value of freshwater replenishment.[32] Infiltration is produced by the high irrigation rates (1.5–2.2 m/ha year) applied to wash out the salts from the regional soils. Additional infiltration occurs from the 858 km of unlined wastewater-transportation channels (Figure 12.1.11).[33,34] As a consequence, the water table has risen and several springs have appeared with flow rates of 400–600 L/s. This water is the only supply for 500,000 inhabitants in the Valley of Tula. Table 12.1.10 illustrates the results of monitoring[32] of 153 parameters (8 microbiological, 23 physicochemical, 1 toxicity test, 18 metals, 8 nonmetals, 7 inorganic compounds, and 72 organic ones) performed by private and university laboratories in parallel during the two seasons of the year (dry and rainy). The results show that during wastewater transportation and infiltration, many reactions take place and water is decontaminated. Some pollutants are photolyzed, desorbed to the air, adsorbed in the soil, biodegraded, precipitated, or absorbed by transportation, soils, and plants. Thus, the Mezquital Valley is in fact behaving as a SAT system with good efficiency, as shown in Table 12.1.10. However, it is also observed that during transport through the soil, water salinity is increased (calcium, magnesium, bicarbonates, sulfates, nitrates, nitrites, hardness, and alkalinity content). Because it is not known when this unconventional treatment will reach its saturation level, it is imperative to investigate the

Figure 12.1.11 Unlined channel used to transport wastewater for irrigation in the Mezquital Valley.

fate of all the polluting agents as well as the mechanisms that take part into their removal.

In order to evaluate drinking water quality, 34 wells that supply 83% of the population and represent 57% of the regional water withdrawal were monitored between 1997 and 1998.[32] Six of these wells, which showed high COD and TOC concentrations, were selected to measure 246 semi volatile trace organic compounds. Table 12.1.11 shows some of the results and compares them with the Mexican drinking water standard. From the 34 sources, 21 exceeded the total dissolved solids limit, 13 nitrates and fecal coliforms, 11 sodium and total hardness, 6 sulfates, 4 barium, and 1 cadmium, copper, nitrites, and zinc. Although pesticides commonly used in the region were looked for (atrazine, carbofurane, and 2,4-dichlorofenoxiacetic acid), they were not found in detectable levels. No evidence of acute toxicity with *Photobacterium phosphoreum* (Microtox) was revealed. Of the 246 trace organic compounds measured, none was found in concentrations above the detection limit. Nevertheless, some peaks of nonidentified compounds were detected that seem to be related to humic and fulvic substances.

Surprisingly, the situation in the Mezquital Valley is not unique. Similar cases have been reported in Asia, Latin America, in the Middle East, and northern Africa.[35] Agricultural reuse strategies must take into account such natural purification mechanisms.

Table 12.1.10 Comparison Between Quality of Municipal Wastewater from Mexico City and Aquifer from the Mezquital Valley in Tula

Parameters (unit in mg/L unless specified)	Wastewater	Site 1 Concentration	% Removal	Site 2 Concentration	% Removal	Site 3 Concentration	% Removal
E. histolytica, No./L	0–1.5	ND	100	ND	100	ND	100
Fecal coliforms, MPN/100 mL	10^4–10^{10}	1–4	99.9	0–29	99.9	0–330	99.9
Helminth eggs, HO/L	12–24.5	0	100	0	100	0	100
Salmonella (3 varieties), CFU/mL	0–positive	ND	100	ND	100	ND	100
Shigella, CFU/mL	0–positive	ND	100	ND	100	ND	100
Conductivity, μmhos/cm	1437–1689	1481–1730	−3.4	1535–1801	−11.3	1513–2090	−25.7
pH	7.2–7.4	6.8–8.3		7–8.5		7–8.1	
Redox potential, mV	−16	−78 to −23	−215	−78 to −23	−173	−69–34	−222
TSS, mg/L	83–153	ND–12	97	ND–12	98	ND–12	97
Turbidity, NTU	100–249	0.1–2	99	0.03–2.5	99	0.3–5	99
Bicarbonates, mg $CaCO_3$/L	485	418–942	−21	447–850	−12	430–925	−18
Chlorides, mg/L	155–248	131–180	26	160–216	11	142–317	−31
Fluorides, mg/L	0.7–4	0.3–1	74.0	0.8–1.3	53	0.04–0.8	86
Sulfides, mg/L	3–3.5	ND < 3.4	65	ND < 3.4	70	ND < 3.4	50
Total hardness, mg $CaCO_3$/L	210–220	265–376	−50	438–484	−109	481–530	−128
BOD, soluble, mg/L	166–167	2.4–5	98	1–4.5	98	0.4–5	98
COD soluble, mg/L	274–276	4– < 10	97	4– < 10	97	5.8–23	96
TOC, mg/L	35–188	5.2–30	84	5–73	75	4.7–19	90
Aluminum, mg/L	1.3–5.5	0.03–0.1	98	ND–0.14	96	0.03–0.1	98
Arsenic, mg/L	ND–0.008	ND–0.005	71	ND–0.01	56	ND < 0.005	82
Boron, mg/L	1.–1.2	0.4–0.7	49	0.8–0.7	41	0.08–0.5	82
Calcium, mg/L	41–445	57–90	−82	41–83	−71	69–132	−156
Chromium, total, mg/L	ND–0.04	ND–0.01	90	ND–0.01	91	2	90
Copper, mg/L	0.05–0.07	ND–0.07	77	ND– < 0.02	67	ND– < 0	82

Cyanides, mg/L	0.005–0.01	< 0.005– < 0.018	13	ND– < 0.018	33	ND– < 0.02	17
Iron, mg/L	1–1.2	< 0.003–0.073	96	< 0.003–0.94	86	< 0.02–0.34	92
Lead, mg/L	0.09–0.1	ND–0.04	78	ND–0.08	78	ND–0.038	84
Magnesium, mg/L	24–29	23–47	−13	28–33	−140	26–75	−76
Manganese, mg/L	0.03–0.2	ND– < 0.01	95	ND–0.06	88	ND– < 0.01	95
Mercury, mg/L	ND–0.001	ND–0.002	36	ND–0.005	−13	ND–0.001	64
Sodium, mg/L	198–206	80–317	13	75–264	17	97–384	−7
Ammonia nitrogen, mg N/L	24–32	ND–4.5	97	ND–0.2	100	ND–0.2	100
Nitrates, mg N/L	ND–1	1.5–77	−2785	1.4–56	−2007	1.5–50	−2107
Nitrites, mg N/L	ND–0.001	ND–0.02	−741	ND–0.023	−521	ND–0.036	−1091
Total nitrogen, mg N/L	37–38	ND–6	96	< 0.1–7	96	< 0.1–4.4	96
Phosphorus, mg/L	2.7–3	ND–0.2	95	ND– < 0.5	93	ND < 0.05	93
Chloroform, μg/L	0.2–0.8	ND	100	ND	100	ND	100
p-Cresol, μg/L	46.5	ND	100	ND	100	ND	100
Ethylbenzene, μg/L	1.2	ND– < 5	100	ND < 5	100	ND– < 5	100
Tetrachloroethylene, μg/L	2	ND	100	ND	100	ND	100
m-Xylene, μg/L	9.2	ND	100	ND	100	ND	100
O-Xylene, μg/L	3.8–4	ND– < 5	100	ND < 5	100	ND < 5	100

ND: not detected.

Source: Adapted from Ref. 32.

Table 12.1.11 Quality of Groundwater Supplied for Potable Use in the Mezquital Valley in Tula

Parameter	Drinking water standard	Well 1	Well 2	Well 3	Well 4	Well 6	Well 7	Well 8
Population supplied (inhabitants)		4000	72413	4000	25975	34003	5402	5959
Volume, L/s		100	330	151	50	33	22.5	31
Helminth eggs, egg/L	NA	0	0	0	0	0	0	0
Enteric viruses	NA	ND	ND	ND	ND	ND	ND	ND
Fecal coliforms, MPN/100 mL	0	0	544	0	0	36	6	0
COD, mg/L	NA	17	158	0	5	48	0	0
TOC, mg/L	NA	0	80	0	0	0	0	0
Aluminum, mg/L	0.2	< 0.0005	ND	0.005	0.002	< 0.0005	< 0.0005	< 0.0005
Barium, mg/L	0.7	ND	ND	0.1	ND	ND	1.8	ND
Cadmium, mg/L	0.005	< 0.00001	0.0003	0.00007	< 0.003	< 0.00045	< 0.0003	< 0.001
Copper, mg/L	2	0.008	0.004	0.006	0.004	0.002	0.002	0.002
Chromium, mg/L	0.05	0.002	0.002	0.0002	0.005	0.006	0.001	0.007
Total hardness, mg CO_3/L	500	571	465	562	250	382	340	580
Iron, mg/L	0.3	0.02	0.05	0.054	0.02	0.02	< 0.02	0.002
Fluorides, mg N/L	1.5	0.3	0.8	0.04	0.5	0.53	0.4	1.4
Nitrates, mg N/L	10	29	13	14	5.6	14	9	6.1
Nitrites, mg N/L	0.05	0.0024	0.005	0.003	0.036	0.002	0.009	< 0.000001
Lead, mg/L	0.025	0.0003	0.0006	< 0.0002	< 0.0002	0.003	< 0.0002	0.0002
Sodium, mg/L	200	363	210	224	95	200	56	185
Total dissolved solids, mg/L	1000	1609	1142	1278	604	1070	768	1186
Sulfates, mg/L	400	245	167	128	74	131	60	270
MBAS, mg/L	0.5	0.144	ND	0.112	< 0.001	< 0.001	< 0.001	< 0.001
Zinc, mg/L	5	< 0.00127	ND	2.7	0.007	0.0013	115	0.001

ND: not detected.

Source: Adapted from Ref. 32.

REFERENCES

1. CNA (Comisión Nacional del Agua). Feasibility study for the sanitation of the Valley of Mexico, National Water Commission, Mexico City, Mexico. Final Report (in English), 1995.
2. Jimenez, B., and Chavez, A., Low cost technology for a reliable use of Mexico City's wastewater for agricultural irrigation, *Technol.*, 9, 1–2, 95, 2002.
3. Siebe, C., Nutrient inputs to soils and to their uptake by alfalfa through long-term irrigation with untreated sewage effluent in Mexico, *Soil Use Manage.*, 14, 119, 1998.
4. U.S. Environmental Protection Agency, USEPA, *Guidelines for Water Reuse*, Report EPA/625/R-92/004, Cincinnati, OH, 1992.
5. Jimenez, B., and Chavez, A., Removal of helminth eggs in an advanced primary treatment with sludge blanket, *Environ. Technol.*, 19, 1061, 1998.
6. Cifuentes, E., Blumenthal, U., Ruiz-Palacios, G., and Bennett, S., Health Impact evaluation of wastewater use in Mexico, *Public Health Rev.*, 19, 243, 1991/1992.
7. Jimenez, B., Health risks in aquifer recharge with recycle, in *State of the Art Report on Health Risks in Aquifer Recharge Using Reclaimed Wastewater*, Aertgeerts, R. and Angelakis, A., eds., World Health Organization, Geneva, 2003, Chapter 3.
8. WHO/UNICEF, Global water supply and sanitation assessment report, Joint Monitoring Program for Water Supply and Sanitation, Geneva, 2000.
9. World Health Organization (WHO), *Health Guidelines for the Use of Wastewater in Agriculture and Aquaculture.*, Report of a WHO Scientific Group, Technical Report Series 778, World Health Organization, Geneva, 1989.
10. Jimenez-Cisneros, B., and Chavez-Mejía, A., Treatment of Mexico City wastewater for irrigation purposes, *Environ. Technol.*, 18, 721, 1997.
11. Landa, A., Capella, A., and Jimenez, B., Particle size distribution in an effluent from an advanced primary treatment and its removal during filtration, *Wat. Sci. Techn.*, 36, 4, 159, 1997.
12. Bratton, R., and Nesse, R., Ascariasis: an infection to watch for in immigrants, *Postgrad. Med.*, 93, 171, 1993.
13. Wani, N., and Chrungoo, R., Biliary ascarisis: surginal aspects, *World J. Surg.*, 16, 976, 1992.
14. World Health Organization, Amoebiasis—an expert consultation, *Weekly Epidemiol. Record*, 14, Geneva, 1997.
15. Ayres, R., Enumeration of parasitic helminths in raw and treated wastewater. A contribution to the international drinking to water supply and sanitation decade 1981–1990, Leeds University, Department of Civil Engineering, 1989.
16. Jimenez-Cisneros, B., Wastewater reuse to increase soil productivity, *Wat. Sci. Techn.*, 32, 12, 173, 1995.
17. Jimenez, B., Chavez, A., and Hernandez, C., Alternative treatment for wastewater destined for agricultural use, *Wat. Sci. Techn.*, 40, 4–5, 355, 1999.
18. Gambrill, M., Mara, D., Oragui, J., and Silva, S., Wastewater treatment for effluent reuse: lime induced removal of excreted pathogens, *Wat. Sci. Techn.*, 21, 79, 1989.
19. Karlsson, I., Chemical phosphorous removal in combination with biological treatment, in *Proc. Int. Conf. on Management Strategies for Phosphorus in the Environment*, Lisbon. Lesterd, J. and Kirk, P., eds., Selper Ltd, London, 261, 1985.
20. Shao, Y., Jenkins, D., Wada, F., and Crosse, J., Advanced primary treatment: an alternative to biological secondary treatment: The City of Los Angeles, CA, the

Hyperion treatment plant experience, in *Proc. 66th WEFTEC Annual Conference & Exposition*, Anaheim, CA, October 3–7, Water Environment Federation, 181, 1995.

21. Shao, Y., Cross, J., and Soroushian, F., Evaluation of chemical addition, *Wat. Environ. Techn.*, 3, 66, 1991.
22. Harleman, D., Morrissey, S., and Murcott, S., The case for using chemically enhanced primary treatment in a new cleanup for Boston Harbor, *J. Boston Society*, Section/ASCE 6, 69, 1991.
23. Harleman, R., Chemically enhanced primary treatment of municipal wastewater. Flocculants, coagulants and precipitants for drinking water and wastewater treatment, in *Proc. Intertech Conf.*, October 29–30, Am. Society of Eng. and Tech. Education, 1, 1992.
24. Ødegaard, H., Gisvold, B., Helness, H., Sjovold, F., and Zuliang, L., High rate biological/chemical treatment based on the moving bed biofilm process combined with coagulation, in *Chemical Water and Wastewater Treatment IV*, Hahn H., Hoffmann E., and Odegaard H., eds, 9th Gothenburg Symposium, Springer-Verlag, Berlin, 2000, 251.
25. Jimenez, B., Chavez, A., Maya, C., and Jardines, L., Removal of microorganisms in different stages of wastewater treatment, *Wat. Sci. Techn.*, 43, 10, 155, 2001.
26. Lue-Hing, C., Zenz, D., and Kuchenrither, R., *Municipal Sewage Sludge Management: Processing, Utilization and Disposal*, Technomic Publishing Company, Lancaster, PA, 1992.
27. Barrios, J.A., Rodriguez, A., Gonzalez, A., Jimenez, B., and Maya, C., Quality of sludge generated in wastewater treatment plants in Mexico: meeting the proposed regulation, in *Proc. IWA Spec. Conf. on Sludge Management: Regulation, Treatment, Utilization and Disposal*, Acapulco, Mexico, 54, 2001.
28. Méndez, J.M., Jimenez, B., and Barrios, J., Improved alkaline stabilization of municipal wastewater sludge, *Wat. Sci. Techn.*, 46, 10, 139, 2002.
29. Jimenez, B., Barrios, J., and Andreoli, C., Biosolids management in developing countries: experiences in Mexico and Brazil, *Water 21*, 59, 56, 2002.
30. Garciapiña, T., Jimenez, B., Barrios, J., and Garibay, A., Application of limed biosolids to improve saline-sodic soils from northern Mexico, in *1–6 Seminar in 5th European Biosolids and Organic Residuals Conference*, Aqua Enviro Consultancy Services, Wakefield, UK, 2000, paper 25.
31. National Water Commission (CNA) and the British Geological Survey (BGS), *Impact of Wastewater Reuse on Groundwater in the Mezquital Valley*, Hidalgo State, Mexico, Comisión Nacional del Agua, final report (in Spanish), 1998.
32. Jimenez, B., and Chavez, A., Quality assessment of an aquifer recharged with wastewater for its use as drinking source "El Mezquital Valley" case, *Wat. Sci. Techn.*, 2, 50, 269, 2004.
33. Downs, T., Cifuentes, M., Ruth, E., and Suffet, I., Effectiveness natural treatment in a wastewater irrigation District of the Mexico City region: a synoptic field survey, *Wat. Environ. Res.*, 72, 1, 4, 2000.
34. Jimenez, B., et al., Feasibility of using the Mezquital Valley aquifer to supply Mexico City, Engineering Institute Report prepared for the National Water Commission, Project 8384 (in Spanish), Mexico City, Mexico, 1999.
35. Foster, S., Groundwater recharge with urban wastewater reconciling resource recovery and pollution concerns in developing nations, Experts meeting on health risks in aquifer recharge by recycled water sponsored by WHO, Budapest, Hungary, 2001.

12.2

Water Reuse for Golf Course Irrigation in Costa Brava, Spain

Lluís Sala and Xavier Millet

12.2.1 History

In 1989, the region of Costa Brava, which is located in northeast Spain, was in the midst of a serious drought. Local water resources were already overdrafted because of the increasing demand due to the development of tourist activities. The inland water transfer from the Ter River to Central Costa Brava was still 4 years ahead. Adding to this already difficult situation, the newly constructed Mas Nou Golf Course scheduled planting of the turfgrass for the month of September 1989, just when the availability of local water resources was the lowest.

One year before, foreseeing such a scenario, the mayor of the municipality of Castell-Platja d'Aro approved the construction of the golf course only under the condition that recycled water be used for irrigation. A workshop on wastewater reclamation and reuse organized in this coastal village in 1985 left him—and the technical staff at the local water agency, the Consorci de la Costa Brava (CCB)—with the idea that using recycled water for this purpose is totally feasible.

Water reuse in the Castell-Platja d'Aro wastewater treatment plant (WWTP) allowed the Mas Nou Golf Course, today called Golf Course d'Aro, to become a reality (Figure 12.2.1). At the same time, it served as a demonstration project for the subsequent supply of recycled water to other golf courses in the area. Detailed monitoring, including physical, chemical, and microbiological parameters, was conducted from September 1989 until the end of 1992, and was aimed at producing useful information for the appropriate sanitary, agronomic, and aesthetic management of golf courses irrigated with recycled water.

In early 2004, four golf courses in Costa Brava were using recycled water as the sole source of water for irrigation, while two other courses and two pitch and putt facilities are scheduled to retrofit for irrigation with recycled water in the next few years.

Figure 12.2.1 Views of the Golf Course Mas Nou (now Golf Course d'Aro) irrigated with recycled water since 1989.

12.2.2 Tips for Adequate Recycled Water Management in Golf Course Irrigation

One of the defining characteristics of recycled water is its variable nature, so a program for the adequate monitoring of its quality is a must. Variations on electrical conductivity (EC), an indirect measurement of salinity, require active management of both recycled water at the outlet of the WWTP and in the distribution system to minimize any potential negative impacts on turfgrass. In addition, variations in nitrogen and phosphorus concentrations would have a notable influence on agronomic management practices, as demonstrated by the data gathered from the four golf courses in the Costa Brava irrigated with recycled water. The information presented herein has mostly been drawn from previous publications.[1,2]

12.2.3 Electrical Conductivity

This parameter is not affected by conventional wastewater treatment, so EC values in collected wastewater are very similar to those in the final recycled water. EC in wastewater is mostly related to the EC values of the drinking water in the area connected to sewers. Except in the case of heavy rains and single sewage system, EC in wastewater is slightly higher than EC in drinking water, since the municipal usage of water adds some salts, most notably sodium and bicarbonate. EC in wastewater is also affected by uncontrolled or illegal discharge into sewers or by seawater intrusion in coastal communities. When the source of salts is drinking water, improvements in its quality will not only benefit the consumers, but will allow treated wastewater to be recycled. In contrast, when the source of salts is undesired discharge or intrusion events, they should be repaired and solved in order to protect this valuable alternative water resource. Figure 12.2.2 shows the reduction in the contribution of sodium by recycled water in the Golf Course Les Serres de Pals (Figure 12.2.3) after the source of drinking water to the area was changed for another with a lower salinity. This significant reduction is due not only to the actual decrease in sodium in drinking water, but also to the lower domestic use of sodium for hardness removal in the area.

Another element related to EC is potassium (K), which is one of the main plant nutrients together with nitrogen (N) and phosphorus (P). Potassium is not affected by conventional wastewater treatment or by reclamation treatment, unless the latter includes reverse osmosis. Thus, potassium contribution will ultimately depend on the concentration of potassium in drinking water and on what is added during the municipal use of water.

12.2.4 Nutrients

Nutrients (N and P) are removed to different extents by wastewater-treatment facilities, so their concentrations may vary markedly between different treatment plants or even in the same treatment plant, depending on organic loads. As a rule, in tourist areas such as Costa Brava, organic loads are several times greater in summer than in winter, which also results in higher concentrations of nutrients, especially nitrogen, in the treated wastewater.

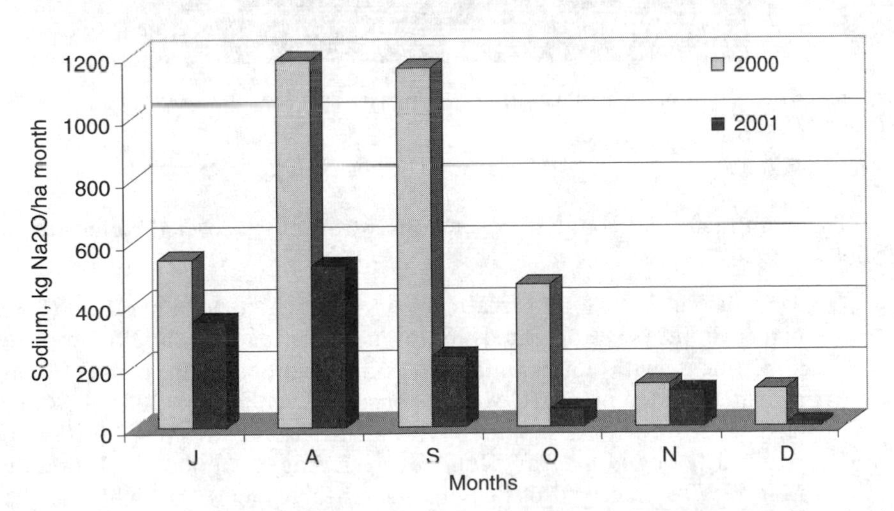

Figure 12.2.2 Reduction of sodium contributions to the Golf Course Les Serres de Pals by recycled water from the Pals WWTP after the sources of drinking water were replaced by others with lower salinity.

Figure 12.2.3 View of the Golf Course Les Serres de Pals, irrigated with recycled water since 2000.

The main factors affecting nitrogen concentration in recycled water are as follows:

1. Type of the WWTP and treatment processes: Extended aeration (EA) plants, if properly operated and/or not overloaded, should produce an effluent with lower nitrogen concentrations than conventional activated sludge plants (CAS). For instance, whereas in 2001 the Golf Course Costa Brava, supplied by Castell-Platja d'Aro WWTP (CAS), received 183 kg N/ha year, Golf Course L'Àngel, supplied by Lloret de Mar WWTP, received 60 kg N/ha year—three times less. This can be attributed entirely to differences in nitrogen concentrations in the recycled water, with an average of 39 mg N/L in the Castell-Platja

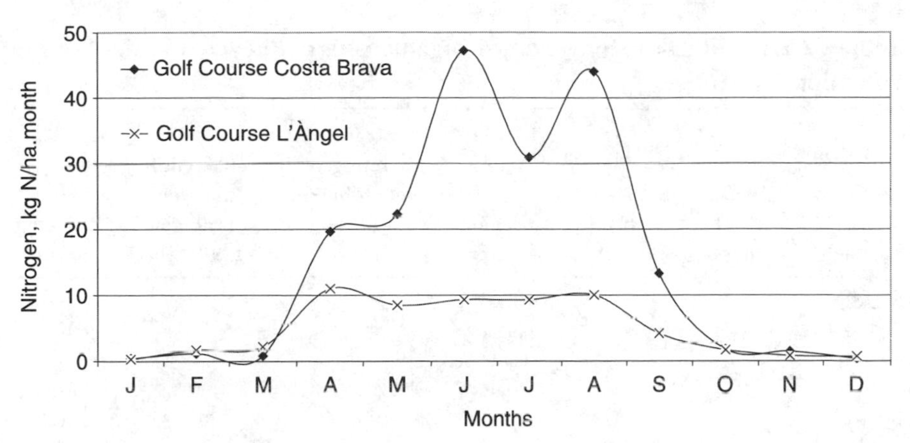

Figure 12.2.4 Differences in nitrogen contribution of irrigation with recycled water due to the type of wastewater-treatment plant.

d'Aro WWTP and of 13 mg N/L in the Lloret de Mar WWTP during the irrigation season, from April to September 2001 (Figure 12.2.4).

2. Storage of recycled water: If recycled water is stored in ponds before its use, it is likely to lose some nitrogen because of the processes fostered by the detention of water in the pond system. The greater the detention time, the greater the reduction of nitrogen. This reduction is also affected by the nitrogen species and is observed to be greater when nitrate is the main nitrogen compound.[3] When using recycled water for irrigation, appropriate design of storage facilities will play an important role in fertilization practices and will also save a considerable amount of money over the years by reducing undesired nutrient losses.[4] Table 12.2.1 summarizes the effects of storage on nitrogen species during irrigation with recycled water, and Figure 12.2.5 shows the decrease in nitrogen contribution at the two golf courses in the Costa Brava that have storage ponds. At the Golf Course d'Aro, ammonia is the main nitrogen species, whereas in the case of the Golf Course Les Serres de Pals this role is played by nitrate. The latter is more readily lost even at low storage detention times.

The factors affecting phosphorus concentration in recycled water are as follows:

1. Type of WWTP: Phosphorus concentration in effluents from different treatment plants will vary, depending on whether they have any kind of phosphorus-removal system, either chemical or biological. Both conventional activated sludge and extended aeration plants have limited phosphorus removal.
2. Storage of recycled water: When recycled water is stored in ponds, longer detention times usually result in lower concentrations of

Table 12.2.1 Effect of Storage on Nitrogen Losses in Recycled Water Prior to Irrigation

Golf course	Average HRT (days)	Minimum HRT (days)	Source of feeding water	Measured nitrogen contribution (kg N/ ha.year)	Theoretical nitrogen contribution[a] (kg N/ha.year)	Reduction factor
Serres de Pals	7	3	Nitrified/ partially denitrified effluent	33.2	100.8	3.0
d'Aro[b] 1st pond	34	13	Conventional activated sludge effluent	162.6	308.1	1.9
d'Aro 2nd pond	42	17	Water from 1st pond	97.8	162.7	1.7

[a]Contribution that would happen if water was not stored in ponds prior to irrigation. Theoretical contribution for Golf d'Aro 2nd pond was calculated as if the nitrogen concentrations in irrigation water were those of 1st pond (no effect of 2nd pond).
[b]Ponds in Golf d'Aro are serially connected.

phosphorus in the irrigation water, since the growth of algae raises pH and fosters phosphorus precipitation. However, the rate of removal is lower than for nitrogen. Thus, phosphorus is better conserved in water, as shown by the lower reduction factors presented in Table 12.2.2, compared to those in Table 12.2.1 for nitrogen.

It is important to state that, either in wastewater treatment plants or in storage ponds, nitrogen is removed more easily and more rapidly than phosphorus, and that the relationship between these two elements will vary depending on the degree of treatment. As a general rule of the thumb, the more intense the treatment, the lower the N/P ratio (Table 12.2.3; Figure 12.2.6). Variations in this ratio produce nutritional situations that need to be managed specifically in order to achieve a proper equilibrium of nutrients.

12.2.5 Maturation Pond Design and Management

Another important issue related to the use of recycled water for golf course irrigation relates to whether storage ponds must be constructed. The common belief is that a good storage pond system is a must, and the larger the better in order to achieve a good degree of autonomy in case of failure or poor

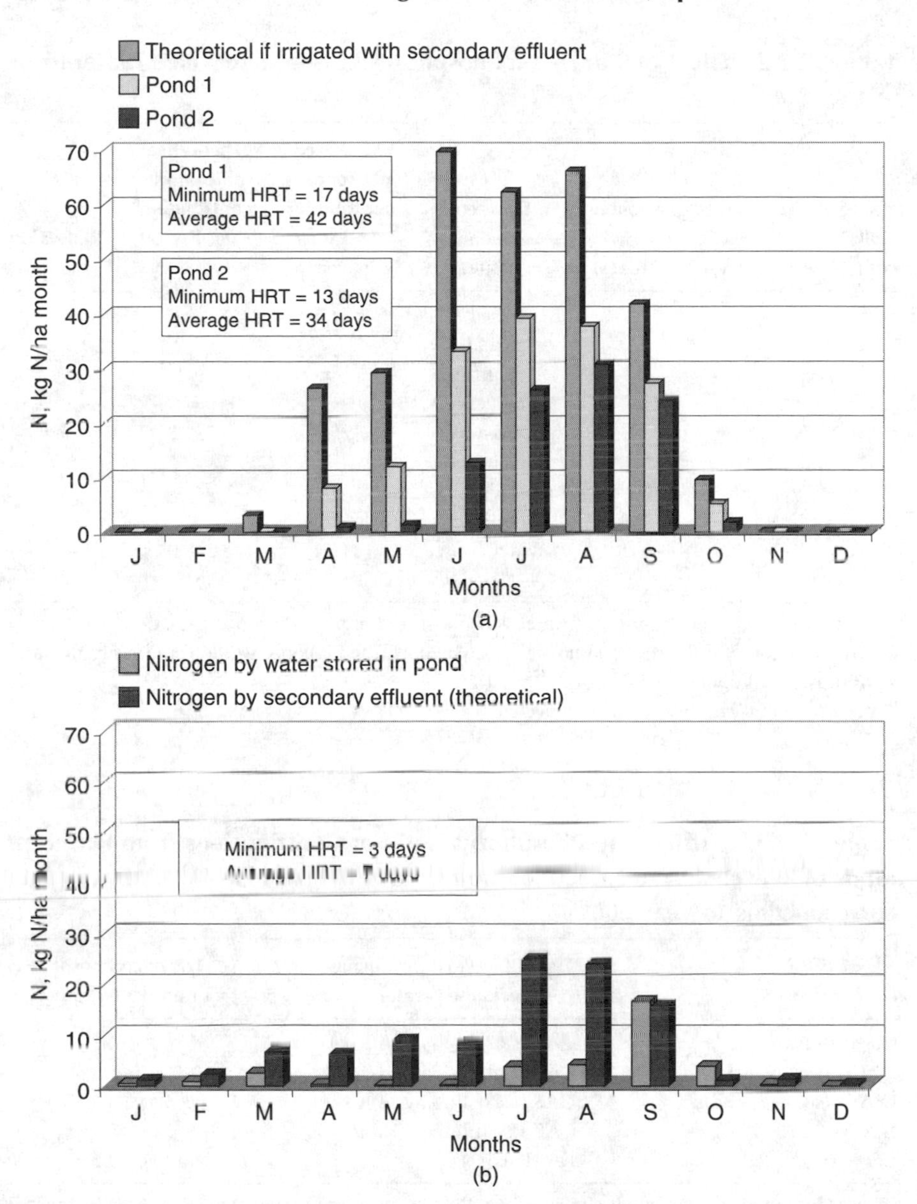

Figure 12.2.5 Decrease in nitrogen contributions at (a) the Golf Course d'Aro and (b) the Golf Course Les Serres de Pals due to the storage of recycled water in ponds with large detention times.

performance of the WWTP. A large pond system would also render a high retention time, so it would have the capacity to remove pollutants still contained in recycled water. However, this is a rather simplistic approach and should be considered only under special circumstances, such as small reclamation facilities producing only a tiny portion of the water needed in

Table 12.2.2 Effect of Storage on Phosphorus Losses in Recycled Water Prior to Irrigation

Golf course	Average HRT (days)	Minimum HRT (days)	Source of feeding water	Measured phosphorus contribution (kg P_2O_5/ ha.year)	Theoretical phosphorus contribution[a] (kg P_2O_5/ ha.year)	Reduction factor
Serres de Pals	7	3	Nitrified/ partially denitrified effluent	67.6	110.8	1.6
d'Aro[b] 1st pond	34	13	Conventional activated sludge effluent	125.6	139.2	1.1
d'Aro 2nd pond	42	17	Water from 1st pond	119.2	128.0	1.1

[a]Contribution that would happen if water was not stored in ponds prior to irrigation. Theoretical contribution for Golf d'Aro 2nd pond was calculated as if the P concentrations in irrigation water were those of 1st pond (no effect of 2nd pond).
[b]Ponds in Golf d'Aro are serially connected.

Table 12.2.3 Nitrogen-to-Phosphorus Ratio of Contributions from Different Kinds of Recycled Water Used for Golf Course Irrigation in Costa Brava (data corresponding to year 2001)

Golf course	Type of WWTP producing reclaimed water	Nitrogen:Phosphorus ratio(kg N/kg P_2O_5)
Costa Brava	CAS (Castell-Platja d'Aro)	2.1
D'Aro (first pond)	CAS (Castell-Platja d'Aro)	1.3
D'Aro (second pond)	CAS (Castell-Platja d'Aro)	0.8
L'Àngel	CAS-PN (Lloret de Mar)	1.4
Les Serres de Pals	EA (Pals)	0.5

CAS, Conventional activated sludge; CAS-PN, conventional activated sludge with partial nitrification; EA, extended aeration.

the peak season or in the case of insufficient quality of treated wastewater. If a given golf course is supplied from a WWTP with sufficient production of recycled water and there is no history of periodic severe disturbances (such as those caused by toxic discharges in the sewage system) or poor-quality events, there is no reason not to use simpler storage tanks, closed and placed in the surroundings of the golf course.

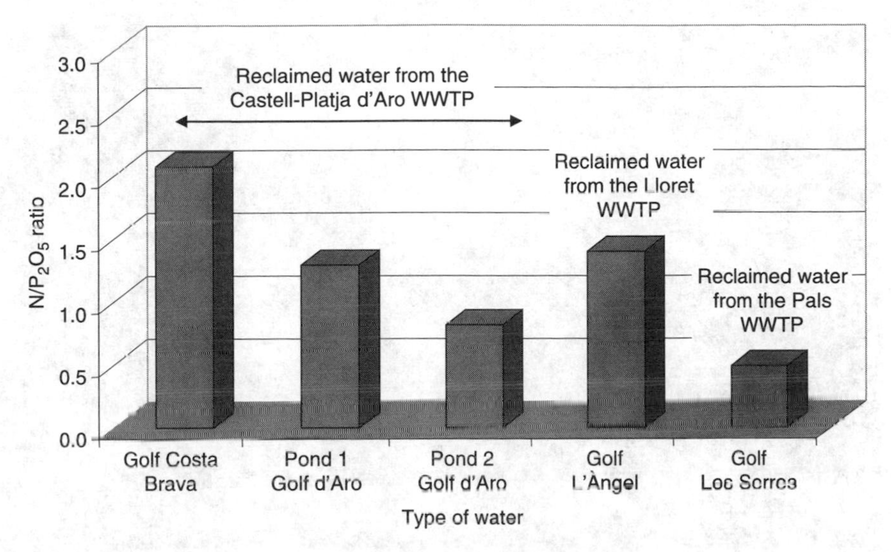

Figure 12.2.6 Variations in the nitrogen-to-phosphorus ratio in recycled water depending on detention time in storage ponds and the degree of wastewater treatment.

When good-quality recycled water is stored in ponds, maintenance costs can be significantly increased. The main disadvantages of storage in ponds are as follows:

Recycled water has low suspended solids concentration and turbidity, and microalgae growing in the ponds produce a loss of quality for these two parameters. The addition of chemicals to kill algae and prevent their growth is not recommended because: (1) they may be transferred to the soil through irrigation, (2) dead algae at the bottom of the pond are likely to contribute to the establishment of anoxic conditions (and odors), and (3) they are associated with extra maintenance costs.

High coliform regrowth potential, either by natural recovery or by wild or introduced bird droppings. If water is kept in a closed storage tank, microbiological quality is also preserved.

As mentioned previously, long detention times reduce nitrogen and phosphorus concentrations in irrigation water, reducing the amount of money that could be saved by the fertilization capacity of irrigation with recycled water.

Increased monitoring costs, since nutrient concentrations must be specifically analyzed in each pond providing water for irrigation. Since detention times are related to water demand, nitrogen and phosphorus concentrations vary greatly during the year and the analysis of concentrations in the recycled water is useless, since they may differ greatly from the reality.

If water for irrigation is taken from the bottom of ponds that have a depth of >2 m, the algae growing in the upper layers of the pond might shade

Figure 12.2.7 View of the Golf Course Costa Brava, irrigated with recycled water since 1998.

bottom layers, so that algae there would not be able to sustain oxygen production, potentially leading to anoxic conditions and odors during irrigation with this stored recycled water.

Increased maintenance costs in order to keep the ponds in good condition, including mosquito abatement, regular cleaning, vegetation control in the borders, etc.

Increased construction costs compared to smaller storage tank due to increased earthworks.

Summarizing this issue, when treated wastewater does not require further treatment and/or when flows are adequate to cope with daily irrigation demands, there are simply too many disadvantages to counteract the aesthetic and strategic benefits of having large pond systems to store recycled water before irrigation.

The Golf Course Costa Brava in Santa Cristina d'Aro (Figure 12.2.7), underwent retrofitting with recycled water in 1998. A conscious decision to not construct open-surface, long-term storage pond was made. The results, 5 years later, are simpler and cheaper maintenance and greater fertilizer savings. In parallel, ponds have been built for aesthetic enhancement, and they are supplied with groundwater just to compensate for evaporation losses in summer.

12.2.6 Conclusions

The Spanish region of Costa Brava has been pioneering the use of recycled water for golf course irrigation since 1989. The supply of this kind of water for the irrigation of four golf courses has provided an opportunity to compare the quality of water supplies, and its effect on agronomic practices.

Recycled water is naturally variable in quality, and adequate monitoring is required for parameters with agronomic significance such as electrical conductivity and concentrations of nitrogen, phosphorus, and potassium. EC and potassium are mostly affected by drinking water salinity and the presence of uncontrolled discharge to the sewer lines, whereas nitrogen and phosphorus are affected by the type of WWTP and its operation, as well as by the detention time in storage facilities. Since nitrogen is more readily removed from water than phosphorus, the ratio between these two elements tends to lean towards the latter as the intensity of treatment or the detention time increase. This should be taken into account for golf courses where water to be used for irrigation is stored in more than one pond, especially if the storage ponds are serially connected.

Ponds for the storage of recycled water for golf courses are a tricky issue. Though the common belief is that the larger these ponds the better, in a quest for greater autonomy and improved treatment, a more detailed analysis reveals several disadvantages that surpass these benefits. However, there is no general rule, and decisions should be made on the case-by-case basis specifically for the given water-reuse project.

REFERENCES

1. Sala, L., and Millet, X., Basic aspects of reuse of reclaimed wastewater for golf course irrigation, Consorci de la Costa Brava, ed., Barcelona, Spain, 126 pp. (in Spanish), 1997.
2. Sala, L., Sala, J., Millet, X., and Mujeriego, R., Variability of reclaimed water quality and implications for agronomic management practices, in *Proc. of the IWA Regional Symposium on Water Recycling in the Mediterranean Region*, Iraklio, Crete, Greece, September 26–29, 2002.
3. Mujeriego, R., Sala L., and Turet, J., Nutrient losses in two landscape ponds used for golf course irrigation, in *Proceedings of the Second IAWQ International Conference on Waste Stabilisation Ponds and the Reuse of Pond Effluents*, Berkeley, CA, November 30–December 3, 1993.
4. Sala, L., Turet, J., and Mujeriego, R., The use of reclaimed water for fertirrigation: effects of pond storage, in *Proc. IV Euroregion Workshop on Biotechnology*, Toulouse, France, November 8–9, 1994.

12.3

Monterey County Water Recycling Projects: A Case Study in Irrigation Water Supply for Food Crop Irrigation

Bahman Sheikh

12.3.1 History and Motivation

The Monterey Regional Water Pollution Control Agency was formed in the early 1970s as a joint-powers agreement among eight public entities (Salinas, Pacific Grove, Monterey, Castroville, Moss Landing, Del Rey Oaks, Seaside, Marina, portions of unincorporated Monterey County and Fort Ord converted from military to civilian use) to provide wastewater treatment, water reclamation, and effluent disposal for the entire northern Monterey region situated along the Pacific Ocean coastline (Figure 12.3.1). The USEPA planning and construction grants that resulted in the regional wastewater-management scheme included a strong provision for reclamation and reuse of the wastewater effluent for agricultural irrigation. This was motivated by the relatively rapid rate of the advance of seawater intrusion into the two confined aquifers supplying freshwater for domestic and agricultural needs in northern Monterey County. An 11-year pilot project was conducted to determine and demonstrate the safety of using disinfected tertiary recycled water for irrigation of such raw-eaten vegetable crops as celery, lettuce, broccoli, cauliflower, and artichokes (Figure 12.3.2). The demonstration project was successfully concluded in 1987, with a final report and publications and numerous presentations at international conferences.[1]

12.3.2 Project Overview

The Monterey County Water Recycling Projects comprise a partnership between the Monterey County Water Resources Agency (MCWRA) and the Monterey Regional Water Pollution Control Agency (MRWPCA). The partnership was formed in 1992, resulting in a $75 million project, including tertiary treatment facilities, a 72 km pressurized distribution system, and 22 supplemental wells (Figure 12.3.3). The purpose of the projects is to supply irrigation water to about 5000 ha of farmland in the northern part of Salinas Valley. The project began full-scale operation in 1998 and currently

Figure 12.3.1 Monterey County Water Recycling Projects' area (in the foreground in the surf zone along the Pacific Ocean coastline).

Figure 12.3.2 Artichoke plant irrigated with recycled water: the City of Castroville prides itself as "The Artichoke Capital of the World"; it is estimated that 70% of all artichokes produced in the United States are irrigated with recycled water.

Figure 12.3.3 View of the irrigated area in Monterey County: recycled water is blended with well water to ensure adequate supply during peak summer months when demand cannot be met by recycled water supply alone.

Figure 12.3.4 View of strawberries, a highly salt-sensitive crop, grown extensively with recycled water in Monterey County, producing abundant crops year after year.

provides about 16 Mm^3/year of recycled water, with a peak production rate of almost 120,000 m^3/d. The project is designed for an ultimate capacity of 25 Mm^3/year, with future provisions for storage of a portion of the winter flows for summer use. Crops grown currently include strawberries (Figure 12.3.4), lettuce (Figure 12.3.5), broccoli, fennel, celery, cauliflower, and artichokes.

Figure 12.3.5 Iceberg lettuce irrigated with recycled water in Monterey County.

12.3.3 Public Perception

In the 1970s, extensive research[2] was conducted throughout California to determine the level of public acceptance of various uses of recycled water. It was found an inverse correlation between acceptance and the level of intimacy of use of reclaimed water. For example, use for drinking was least acceptable (44%) and irrigation of golf courses most acceptable (98%). Irrigation of vegetables was acceptable to 88% of the respondents. Although there is no more recent survey on public acceptance of irrigation with recycled water in California, in particular for irrigation of vegetables, some new trends and concerns must be taken into account.

Initially, the majority of the farming public was skeptical, with a few vocal and active opponents. Therefore, a pilot project, known as the Monterey Wastewater Reclamation Study for Agriculture (MWRSA), was planned and implemented, aimed at demonstrating to the farmers that recycled water meeting California's strict Title 22 regulations would be safe and acceptable for use in the irrigation of food crops and for the long-term productivity of their soils. The potential impact of use of recycled water on the sale of crops to the public was a more complicated concern to address. A market analysis, focusing on major wholesale buyers in large metropolitan areas in the United States (New York, Chicago, Los Angeles, San Francisco), discovered that the buying market was not affected by the type of irrigation water used as long as the irrigation water met regulatory requirements and as long as no labeling of the produce was required. Over the 6 years since the project became fully operational, there have not been any negative impacts on the sale of crops to the wholesale or retail markets. Neither has there been a need for labeling the produce as having been irrigated with recycled water.

The agencies producing and distributing reclaimed water in Monterey County have a detailed and well-rehearsed emergency plan ready for implementation in case of media reports that might implicate recycled water in any crop-contamination cases in the future, either as a result of rumor, intentional misinformation, or an unrelated actual contamination.

There have not been any proactive attempts made to inform the general public specifically about the use of recycled water for irrigation of food crops. Neither was any systematic study conducted to obtain a quantitative measure of public acceptance for this particular use of recycled water in Monterey County. It was argued that produce from any given region is blended with that grown in the rest of the country, and identification of any specific batch as to its origin would be nearly impossible and, in fact, unnecessary. Some major growers, such as Dole—which produces ready-to-eat salad products—conducted their own extensive laboratory analyses to satisfy themselves that the produce irrigated with recycled water was at least as safe as that grown with well water.

Even though the agencies involved in implementing the Monterey County Water Recycling Projects have not advertised the use of recycled water for irrigation of vegetables, they have prepared a number of public educational materials and strategies to avert the possibility of rumors and unfounded fears causing economic harm to the growers.

12.3.4 Current Project Status and Operation

A standing committee was established to maintain communication among the stakeholders. The Water Quality and Operations Committee is composed of representatives of the growers, the local environmental health officer, and the water-recycling agencies. It meets monthly and reviews operational status, any problems encountered, and any issues that need to be addressed. The growers have a major role in day-to-day and long-term decisions affecting irrigation scheduling, use of supplemental wells, and other related matters. The MRWPCA, in collaboration with the county farm advisor, conducts routine soil monitoring in similar fields irrigated with recycled water and those irrigated with well water. Data are regularly shared with the growers and will form a basis for establishing any long-term impact on the soil, over the future years. As a result of the growers' continuing concerns about water quality, MRWPCA has established an active source-control program aimed at reducing salt—especially sodium—input into the recycled water.

12.3.5 Conclusions, Lessons Learned and Recommendations

The first project of a major size using recycled water for irrigation of raw-eaten food crops was, understandably, the most difficult to initiate and implement. Future similar projects should be easier to implement. It is now amply demonstrated, both in pilot and full-scale operations, that this use of recycled water is safe, that it does not impact the soil negatively, and that it does not drive consumers away from purchasing produce irrigated with

Figure 12.3.6 Broccoli grown with recycled water in Monterey County after harvest.

recycled water. As a matter of fact, farmers are experiencing increased yields on fields irrigated with recycled water over those receiving well water alone (Figure 12.3.6).

It would be advisable for farmers in regions planning to provide recycled water to do the following:

Visit farms in Monterey County irrigating with recycled water
Take a tour of a treatment plant producing similar quality recycled water
Smell, taste, and view recycled water in a clear glass container

Tertiary recycled water is the only source of irrigation water that is thoroughly disinfected, is virtually free from pathogens, and has consistent, known chemical and microbiological characteristics.

The safety, long-term benefits, and public acceptance of crops grown with recycled water have now been established in several major locations in California, including Irvine, Orange County, Santa Rosa, Sonoma County, and Napa and Monterey Counties. The long-term direction for recycled water in California appears to be toward more high-value uses of water, consistent with the increased demand, uncertain supplies, and a greater public perception of the scarcity of water resources.

REFERENCES

1. Sheikh, B., et al., Tertiary Reclaimed water for irrigation of raw-eaten vegetables, in *Wastewater Reclamation and Reuse*, Asano, T., ed., CRC Press, Boca Raton, 1998, Chapter 17.
2. Bruvold, W.H., *Public Attitudes Toward Reuse of Reclaimed Water*, contrib. Univ. Calif. Water Resour. Cent., 173, 1972.

13 Conclusions and Summary of Practices for Irrigation With Recycled Water

Valentina Lazarova

CONTENTS

13.1 Assessment of the feasibility of using recycled water for irrigation 383
13.2 Good agronomic practices for irrigation with recycled water 383
13.3 Negative impacts of irrigation water on plants and main corrective actions 388
13.4 Management practices and corrective actions for improvement of the operation of water-reuse treatment schemes 388
13.5 Successful participation programmes improved public acceptance 389
13.6 Successful initiatives to address legal and institutional issues 390

As demonstrated in Chapter 1, as well as by the case studies in Chapter 12, the use of recycled water for irrigation is characterized by numerous benefits. The more apparent and easier to demonstrate is the supply of a cost-competitive drought-proof resource to resolve water-scarcity problems and to conserve the potable water supply. In the case of agricultural reuse, the most valuable contributions are food security and decrease of fertilizer applications. In the

1-56670-649-1/05/$0.00 + $1.50

case of turfgrass and landscape irrigation, the benefits of using recycled water are more obvious, mainly the short distance between treatment plants and irrigation sites, as well as the higher tolerance of turfgrass for some water components such as salinity and nutrients.

The success of water reuse projects, however, depends greatly on the application of appropriate planning and good management practices. As mentioned in Chapters 1, 10 and 11, water reuse planning needs to consider not only engineering issues, but must include also the evaluation of environmental impacts, economic and financial feasibility, institutional framework, and social impacts. In addition, market analysis is a crucial element in identifying and securing final end-users such as farmers and/or cooperatives in agricultural irrigation, as well as public and private users for urban landscape irrigation.

As shown in Figure 13.1, derived from Figure 1.10 (Chapter 1), different institutional, engineering, and agronomic practices could be implemented to improve the efficiency and competitiveness of irrigation with recycled water. Figure 13.1 indicates the chapters where these management practices are described in more detail. In order to aid in decision making and more precisely estimate the benefits of each specific action, the management practices

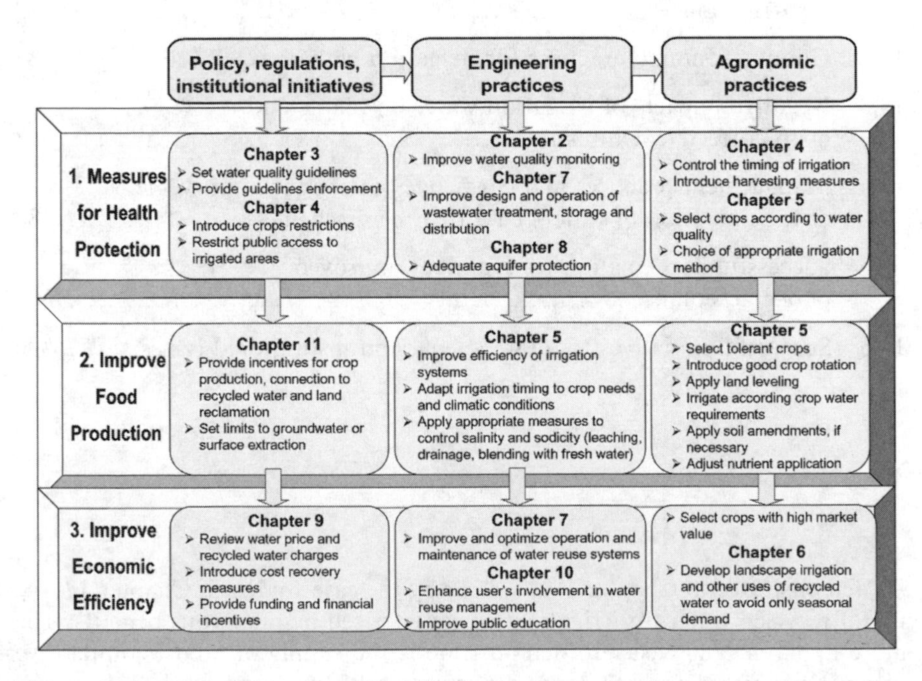

Figure 13.1 Management practices (institutional, engineering, and agronomic) to improve the efficiency and competitiveness of irrigation with recycled water: classification according to the main objective with indication of the chapter number for more details

of irrigation with recycled water are classified in three levels, according to their main objective, listed in increasing order of importance:

1. Health-protection measures
2. Good agronomic practices to improve plant production and to prevent soil and aquifer degradation
3. Improvement of economic competitiveness and public acceptance

Another important criterion in water-reuse systems is the reliability of operation of recycling treatment facilities and the control and monitoring of water quality. In this context, knowledge of the potential operating problems and guidance for corrective actions is of great importance for operators and managers of water-reuse systems.

The quality of irrigation water plays an important role in successful crop production, as well as in maintaining good-quality turfgrass and ornamental plants. In addition to chemical data on water quality, other background information must be taken into account, such as soil characteristics, irrigation method, climatic conditions, etc. Moreover, the application of good management practices allows reducing any potential negative health, agronomic and environmental impacts.

The following sections summarize these issues, offering important guidelines for successful water reuse system management in operation with recommended good agronomic practices, engineering initiatives, and policy and institutional measures.

13.1 ASSESSMENT OF THE FEASIBILITY OF USING RECYCLED WATER FOR IRRIGATION

The first element to be considered when planning water-reuse projects is background data gathering to assess the feasibility of irrigation with recycled water. The feasibility evaluation must be performed by professionals on the basis of site inspection and analysis of the available data on irrigated crops, soil characteristics, irrigation systems, and water-quality considerations. Table 13.1 shows information that should be available for decision making as to the feasibility of water reuse.

13.2 GOOD AGRONOMIC PRACTICES FOR IRRIGATION WITH RECYCLED WATER

Careful considerations and analysis of the background data, as well as market forecast, will allow one to identify and propose the most appropriate management practices for the implementation of irrigation with recycled water or for the retrofit of existing irrigation systems to recycled water (Table 13.2).

Table 13.1 Main Elements to be Considered for Planning of Water Reuse for Irrigation and Factors Influencing Decision Making

Crop type	Soils properties and hydrogeology	Irrigation method	Water quality	Miscellaneous
Agricultural Irrigation				
Water needs (*see Table 5.1*) depending on climate and type of crops	Soil profile and slope	Soil applied:	Coliforms	Storage of recycled water:
	Soil texture (proportions of sand, silt and clay)	surface (flood, border, furrow)	Helminths	surface reservoirs
			pH	aquifers
	Soil structure (aeration heat transfer, etc.)	localized or drip	Total dissolved solids (TDS)	Blending water sources:
Crop tolerance to salinity (*see Tables 5.6–5.10*)		subsurface		groundwater
	Soil chemistry (pH, salinity, SAR, Na, Cl, B, etc.)	Foliar applied (pressurized):	Electrical conductivity	surface water (lakes, rivers)
Crop tolerance to boron (*see Table 5.11*)			Suspended solids	
	Exchangeable cations (Na, Ca, Mg, K, Al)	Sprinklers (central pivot, rolls, stationary, etc.)	BOD	drinking water supply
Crop tolerance to sodicity (*see Table 5.12*)			Forms of nitrogen	Environmental concerns:
	Soil physical characteristics (hydraulic conductivity, bulk density, etc.)		Forms of phosphorus	groundwater
Crop tolerance to chloride (*see Table 5.13*)		Microsprinklers	Boron	surface water (lakes, rivers)
			Cations (Na, K, Ca, Mg)	
Crop growth characteristics	Depth of the water table (ideally >10 m) and seasonal variations	*See Tables 5.3–5.5*	Anions (Ce, NCO_3, SO_4)	wetlands
			Sodium adsorption ratio	Health safety aspects
Crop market price and forecast for future needs			Trace elements	Fertilization practices
	Drainage requirements			Economic and institutional issues (water rights, charges for water supply, eco and other taxes, financial incentives, etc.)
	Quality of groundwater (poor or high quality with or no use for potable supply)		*See Tables 2.1, 2.2, 2.10, 2.12, 5.14–5.17*	
	Proximity to water supply wells (ideally >250 m)			Regulatory issues
				Public acceptance

Landscape and Turfgrass				
Irrigation				
Idem with the following additional considerations: Type of planting: ground covers turf and other grasses trees and shrubs Planting conditions: new or mature plants healthy or stressed Identification of hydrozones for definition of water needs and irrigation scheduling	Idem	Idem with the following additional considerations: Foliar application sprinklers and microsprinklers) may damage some decorative plants (leafs, flowers) Specific infrastructure is needed including cross-connection control, color labeling, couplers, etc.	Idem	Idem with the following additional considerations: Appropriate signaling of sites irrigated with recycled water Irrigation during off-hours

Table 13.2 Summary of Successful Practices of Irrigation with Recycled Water (Agricultural Crops, Turfgrass, and Landscape Ornamentals)

Crop selection and management	Selection of irrigation method	Management of water application
Agricultural irrigation		
Code of practices for salinity management (*see Tables 5.6–5.10*)	**Furrow irrigation** (*see Table 5.4*)	**Leaching and drainage**
Apply source control (industrial wastes, brines, seawater intrusion in sewers)	No foliar injury	Calculate leaching requirement on the basis of water and soil salinity
Reduce evaporation (mulching)	Need for land preparation	Apply preliminary soil flushing by means of 10–20 cm of water before planting
Select salt-tolerant crops	Low cost	Adapt leaching frequency during plant growth
Leaching to remove salts from root zone	Possible water stress	Apply adequate measures to improve soil permeability
Provide adequate drainage	Salt accumulation in the root zone	Apply adequate drainage: surface or sub-surface (shallow water table)
Maintain adequate soil moisture level	Fair to medium suitability for saline water	**Adjusting fertilized application**
Apply more uniform irrigation: localized irrigation, land smoothing	Need of good management and drainage practices	Avoid excess nitrogen application by modification of N fertilizers, crop rotation, denitrification
Increase irrigation frequency to prevent water stress	**Border irrigation** (*see Table 5.4*)	
Irrigate at night or early in the morning at low temperature	Almost no foliar injury (only bottom leaves)	
Apply blending water supply	Need of land preparation	
	Low cost	
	Possible water stress	
	Good salt flushing past the root zone	

Code of practices to overcome B, Na, and Cl toxicity (*see Tables 5.11–5.13*)

- Leaching
- Increase irrigation frequency
- Avoid foliar wetting and injury applying drip irrigation or micro-sprinklers
- Apply soil amendments: organic amendments to improve water infiltration

Code of practices to overcome trace elements toxicity (*see Tables 5.14–5.17*)

- No specific measures for calcareous soils
- Apply soil amendments: liming for acid soils
- Limit the use of acid fertilizers for acid soils
- Select tolerant crops
- Leaching toxic ions past root zone
- Modification of fertilization
- Blending water supplies
- Fair to medium suitability for saline water
- Need of good irrigation and drainage practices

Sprinkler irrigation (*see Tables 5.4 and 5.5*)

- Severe leaf damage can occur
- Good salt flushing past the root zone
- Water stress not possible to maintain high soil water potential
- Poor to fair suitability for saline water
- High capital and O&M costs

Drip irrigation (*see Tables 5.4, 5.18 and 5.19*)

- No foliar injury
- Formation of salt wedge
- Possibility to maintain high soil water potential
- Excellent to good suitability for saline water
- High capital and moderate O&M costs
- Excess phosphorus is not a problem: phosphorus content should be adjusted in first few years by additional fertilizers with further reduction due to P build-up in soils
- Potassium content should be adjusted by additional fertilizers

Management of soil structure

- Deep tillage
- Application of organic amendments
- Chemical amendments such as gypsum
- Indirect calcium amendment by addition of sulfuric acid is not recommended because it needs strict technical supervision

Management of leakage and clogging problems

- Check and repair pipe leakage
- Apply hydraulic flushing with or without chlorination
- Install preliminary filtration of recycled water
- Clean periodically all valves, nozzles and emitters

Successful agronomic practices are divided in three main groups:

Crop selection and management
Selection of irrigation method
Management of water application

Within the first group of practices, salinity control and management is a major issue to be considered when using recycled water for irrigation to avoid any negative impact on crop yield and quality of turfgrass. For some sensitive crops and landscape ornamentals, boron and trace element toxicity could be of major concern. Proper selection of irrigation method and the application of leaching and drainage could reduce negative water-quality impact and prevent salinity build-up. Other cultural practices such as more frequent irrigation, timing of fertilization, and seeding procedures, as well as chemical amendment or blending with other water sources may be needed to deal with temporary increases in recycled water salinity.

Drip irrigation and the use of micro-sprinklers can be used to avoid foliar wetting and leaf injury, with the additional advantage of increased irrigation efficiency and water conservation. In this case, however, adequate engineering-management practices must be applied to reduce clogging problems.

13.3 NEGATIVE IMPACTS OF IRRIGATION WATER ON PLANTS AND MAIN CORRECTIVE ACTIONS

Table 13.3 provides some examples of common negative impacts of water constituents on plant growth with their probable causes and recommended corrective actions. The detection of a probable cause of plant injury requires professional experience. For example, leaf yellowing and necrosis could be due to excessive boron concentration, salt build-up, or high chloride or high fluoride contents in recycled water.

13.4 MANAGEMENT PRACTICES AND CORRECTIVE ACTIONS FOR IMPROVED OPERATION OF WATER-REUSE TREATMENT SCHEMES

Different treatment schemes can be used for the production of recycled water for irrigation. The choice of treatment process depends on numerous factors, including water-quality requirements, plant capacity, climatic and other specific local conditions, land constraints, etc. As a rule, a site-by-site study is to define the most adequate treatment scheme, including recycled water storage for the short or long term.

As mentioned in Chapters 4 and 7, the requirements for the reliability of operation of water-reuse systems are significantly higher compared to municipal wastewater treatment with less stringent discharge restrictions (except for sensitive areas). The main engineering components of water-reuse systems susceptible to failure are power supplies and mechanical equipment. In the case of natural treatment processes such as maturation ponds and long-term storage reservoirs, algae growth is the major constraint leading to degradation of recycled water quality. Table 13.4 provides some examples of good practices for operation of polishing treatment, storage, and distribution.

13.5 SUCCESSFUL PARTICIPATION PROGRAMMES IMPROVED PUBLIC ACCEPTANCE

The development of sustainable water-recycling schemes needs to include an understanding of the social and cultural aspects of water-reuse. The drivers that promote involvement in recycling may vary between households and cultures and will certainly be different for the different reuse applications such as irrigation of agricultural crops, landscape ornamentals, or golf courses.

As a rule, the use of recycled water for irrigation is widely accepted by farmers who believe them to be safer than river waters. However, there are strong consumer concerns about the sale of products that have been irrigated with recovered wastewater, especially vegetables to be eaten raw. Farmers may be able to overcome such resistance through positive evidence from the consumers and the retailers that there will be a market for the products cultivated with recovered water. Also, the establishment of standards for reuse and effective management of monitoring programs promote confidence in water recycling schemes.

In general, there are wide variations in attitudes to water recycling and few robust indicators of positive or negative attitudes that might be used to infer behavior. Current knowledge suggests that neither an individual's gender, age, religious affiliation, nor cultural background predisposes him or her to have a particular attitude towards recycling. Reuse project planners and managers should therefore be prepared to meet with a diversity of responses to water-recycling schemes and focus their efforts on identifying sources of stakeholder concern, promoting education about water-reuse, and achieving consensus among stakeholders on critical project design/operation attributes. Furthermore, the issues of trust and risk need to be managed explicitly within an open and inclusive framework of participation.

Early stages of reuse-project design should involve public participation programs and the involvement of all stakeholders. The development of a successful participation program is characterized by the following main elements:

Is part of a wider decision-making or planning process
Unambiguously defines their aims and objectives
Is clear as to which types of actor or representative are to be consulted

Provides a well-defined and transparent agenda for debate and discussion (including details of frequency of events and how the results of participation are to be used)

Engages those elements of society most likely to be impacted by a decision or project

Involves interested parties in each phase of decision making, not just the final stage

Is sensitive to the concerns of internal stakeholders as well as external stakeholders

Provides a choice of types of participation that reflects the diversity of the individuals and groups being engaged

Provides real opportunities to influence the outcome of project planning or management

13.6 SUCCESSFUL INITIATIVES TO ADDRESS LEGAL AND INSTITUTIONAL ISSUES

In most countries, setting up or adopting a regulatory framework is an essential step for the development and social acceptance of water reuse. Decision makers and politicians need clear, sound, reliable standards to endorse reuse projects. Regulations based on internationally acknowledged guidelines are generally preferred. Although an international effort to reduce discrepancies between current standards is highly desirable, regulations must be adapted to suit each individual country's context, health risk, and affordability. Regulations have a major influence on the choice of treatment technologies and, hence, on the cost of water-reuse projects.

Water-quality issues have a great impact on water-reuse projects because of public perception, liability, and public health concerns. Both overly conservative water-reuse standards and inadequate legislative requirements can affect wastewater-reuse development. The adoption of guidelines in compliance with the recent advance in scientific knowledge, rather than strict conservative standards, would be a reasonable approach, taking into account the cultural and societal issues, the existing infrastructure, and local conditions.

In this context, the development, application, and enforcement of good reuse practices, such as the management practices for irrigation presented in Chapters 4, 5, and 6, could be the critical step for the rational use of recycled water and successful protection of health and environment. Nevertheless, policy, regulatory, and institutional initiatives should play leading roles in the improvement of public health and food safety as well as the cost-efficiency of water-reuse projects for irrigation.

Table 13.5 summarizes the main institutional issues that managers and decision makers must address during the planning and implementation of water-reuse systems.

Table 13.3 Guidelines for Identification of Some Common Negative Impacts of Irrigation Water on Plants and Recommended Corrective Actions

Indicator/observation	Probable cause	Immediate action	Explanation	Corrective actions
Burning or scorched appearance of leaves; Browning and shriveling of fruits	Water stress	Watering operations that avoid water and/or drought stress can reduce or eliminate the problem.	Water stress in crops is caused by drought or poor irrigation. This is because the soil is not allowing enough moisture or fluids to penetrate through the root system to carry enough nutrients to the plants. Other diseases are favored in such conditions.	Urgent watering should be applied using recycled water or other water supply. Appropriate irrigation scheduling should be applied taking into account crop needs, soil moisture and climate conditions. Monitoring soil moisture is commonly used to determine crop water use and to schedule irrigation. Computer-driven irrigation using such sensors is used and could be recommended for landscape and golf course irrigation. While these methods are reasonably reliable, they are costly and give information only for the immediate area in which they are used.
Lodging, excessive foliar growth, weak stalks and stems	Excess nitrogen	Stop addition of N-fertilizer	Very high nitrogen concentrations it can overstimulate plant growth causing problems.	Nitrogen sensitivity varies with the development stage of the crops. It may be beneficial during growth stages, but causes yield losses during flowering/fruiting stages. The immediate corrective action is the modification of N-fertilizers. Crop rotation should be also planned. On-site soil denitrification could be also favored using appropriate measures.
White lime deposits on leaves of plants or on tips of grass	Excess HCO_3 and CO_3		High hardness and alkalinity favor lime precipitation	One possible immediate action is to add polyphosphates. It is also highly recommended to change the irrigation methods and replace sprinkler irrigation by drip irrigation.

(continued)

Table 13.3 Continued

Indicator/observation	Probable cause	Immediate action	Explanation	Corrective actions
Reduced growth (short, thin shoots and weak stems); Brown or yellow patches on greens	Nutrient deficiency (J, P, K)	Check nutrient content and apply adequate fertilizer.	N is part of all amino acids and proteins and is required by chlorophyll molecule. P is necessary for cell division and nucleic acids.	Nutrient deficiency could be easily avoided by preliminary monitoring of water and soil nutrient contents. On the basis of the plant needs (type, grown period), fertilizers should be adequately adjusted.
Yellowing of leaves followed by leaf chlorosis; Sometimes leaf necrosis and burn, premature leaf drop	Boron toxicity	Check B and other toxic ions and elements	Boron primarily regulates the carbohydrate metabolism in plants and is essential for protein synthesis, seed and cell wall formation and germination. Toxicityto crops occurs when sensitive crops were planted and where fertilizers containing boron had been used.	Average boron concentrations in mature leaf tissues can be used to estimate plant boron status as follows: deficient—less than 15 ppm; sufficient—20–100 ppm; and excessive or toxic > 200 ppm. As little as 0.8 mg/L of boron can result in leaf-margin necrosis in many ornamental plants. The immediate measure to avoid further boron toxicity damage is to stop all application of boron fertilizers and good soil leaching (boron is fairly mobile in soils). High levels of calcium may increase the boron tolerance of plants.
Yellowing of leaves followed by leaf chlorosis and necrosis	Fluoride toxicity	Check fluoride concentration in water and soil	Fluoride is a common toxic micronutrient and a number of plants are sensitive under specific conditions.	Fluoride source could be superphosphate fertilizers, water, perlite. Fluoride damage to numerous crops could be effectively reduced by addition of dolomite or calcium hydroxide to increase pH and, thus, to reduce solubility of fluoride. Sensitive plants should be grown in soils with pH > 6.

Yellowing of leaves followed by leaf chlorosis and necrosis 	Iron chlorosis	Check pH	Iron chlorosis is caused by lack of iron compared to plant needs. Iron is needed for the production of chlorophyll and, therefore, a lack of iron results in a loss of the green color in the leaves.	If pH > 7, use sulfur to acidify it and bring pH down. If solid lime is present in the soil, it is not economically possible to decrease the pH. In such soils, consider using raised beds and bringing in good topsoil. Powdered (wettable) sulfur is the most common acidifying agent. The ability of roots to adsorb iron is reduced by inadequate soil aeration (excess soil water).
Sick plants with decline in growth, chlorosis, and leaf tip burn or necrosis 	Salt build up (Cl and Na)	Check water and soil salinity and apply soil leaching	Salinity affects nutrient availability, competitive uptake, transport or partitioning within the plant, phosphate uptake, etc., affecting the quality of both vegetative and reproductive organs. Salinity can cause a combination of complex interactions affecting plant metabolism or susceptibility to injury. Many ornamental plants, especially woody plants (roses, azaleas, camellias, rhododendrons, etc.), are sensitive to chlorine accumulation.	The most efficient immediate action to deal with crop damage due to salt build up is to apply soil leaching in order to decrease salinity in the root zone. An additional measure should be the increase of irrigation frequency to maintain high soil moisture. Water blending should be applied when recycled water is characterized by high salt content. In the long term, salinity problems could be overcome by selection of salt tolerant crops and source control measures. In addition to salinity, it is important to check also the exchangeable sodium percentage of the soil: the higher the measured value, the higher would be water salinity that will not lead to damage of soil structure. Investigate the reasons for the presence of salts: salinity may be indicative of other limitations such as seasonal waterlogging.

Table 13.4 Examples of Good Operational Practices and Corrective Actions to Improve the Reliability of Wastewater Treatment, Storage, and Distribution

Advanced disinfection processes	Natural treatment processes	Storage and distribution of recycled water
Tertiary filtration	**Maturation ponds**	**Short-term storage reservoirs**
Adequate choice of filter media and chemicals	Periodic inspection and removal of weeds	Use covered reservoirs to prevent bacterial regrowth and complementary contamination
Regular filter backwashing	Odors indicate anaerobic conditions and/or overloading	Add recirculation pump to avoid water stagnation and odor generation
Periodically adjust coagulant dosage by jar tests	Algae growth could be controlled by increased dissolved oxygen, addition of plankton and/or algae eating fishes	**Deep long-term storage surface reservoirs**
Chlorination	Add baffling or rock filter to prevent algae loss with treated effluents	Algae growth can be controlled by improvement of oxygenation and other measures
Monitor chlorine dosage and demand	**Soil-aquifer treatment**	Destratification devices for oxygenation and recirculation
Adequate safety measures	Provide adequate flooding-drying cycle in recharge basins	Hypolimnetic oxygenation of bottom layers
Regular inspection and cleaning	Regular removal of the upper sediment layer of the recharge zone	Chemical addition
UV Disinfection	Water-quality monitoring in observation wells	Regular dredging
Adequate design and operation of downstream processes		Biomanipulation (plankton, specific fishes)
Adequate grounding and regular check of electronic signals		
Regular lamp and sensor cleaning		
Ballast card protection by adequate ventilation		

Table 13.5 Main Institutional Issues to be Addressed for Successful Development of Water-reuse Projects for Irrigation

Main element or action	Issues to be addressed
Legal and institutional drivers for water reuse	Understand the laws governing water supply and wastewater treatment
	Identify which parties are responsible for each aspect of water supply, distribution, wastewater treatment, and disposal
	Define any special constraints that support water reuse
Ownership of recycled water	Identify responsibilities to downstream users
	Identify users that can claim rights to the effluent and include them as stakeholders in the planning process with appropriate compensation
	Check if stopping wastewater discharge has negative environmental impacts (necessity to maintain minimum flow)
Relevant institutions and participating stakeholders	Identify the institutions affected by a water-recycling program
	Assess the need for alternative institutional structure for operation of the water-reuse system
	Identify as early as possible any legislative changes that might be required to create the necessary institutions and the level of governance at which the legislation must be enacted
	Establish and maintain contacts with the federal, state, and local agencies involved, as well as customer groups, public interest organizations, and others
Reuse technology	Assess the best available reuse technology, taking into account all the requirements for water quality and sampling conformity
	Evaluate the needs for operation and maintenance (power, labor, chemicals, etc.)
	Identify alternative treatment technologies, in particular for developing countries and rural areas
	Assess the needs for short- or long-term storage

(*continued*)

Table 13.5 Continued

Main element or action	Issues to be addressed
Project funding and cost recovery	Identify sources for funding (local, national, or international)
	Formulate contract provisions
	Propose cost-recovery techniques and revenue plan with short-term and long-term business goals for the program
	Assess the cash flow needed to anticipate funding requirements
Contracts	Establish formal contracts with farmers, cooperatives, or other users with provisions relating to the quality and quantity of the reclaimed water
	Identify the responsibility for any storage facilities and/or supplemental sources of water
	Identify the responsibility for ownership and maintenance of the recycling facilities, as well as those of distribution network
Project follow-through	Long-term monitoring responsibility must be specified, especially if a monitoring program is required as a condition to use reclaimed water for irrigation purposes
	Specific compliance with environmental regulations must be assigned to each party, along with a schedule of performance and consequences for failing to abide by the terms of the agreement
Public education program	Develop an educational outreach program providing accurate information about the nature of recycled water and the benefits and costs of water reuse

Index

A

Aarhus Directive, 288
Abu Dhabi, 20
Acanthamoeba, 213
Activated sludge, *173*, 173–174, *see also* Sludge treatment and disposal
Actopan River, 345
Advantages, *see* Benefits
Adverse effects of sewage irrigation
  behavior of compounds, 238–253
  compounds with potential adverse effects, *237–238*
  disinfection by-products, *38–39*, 247–248
  endocrine disruptors, 248
  groundwater water-table management, 253–260
  historical developments, 236–237
  hormones, 248
  nutrients, 244–247
  organic containments, 248–253
  pharmaceuticals, 248–253
  salt, *110–111*, *128–130*, 239–244, *241*, *245*, 253–260
  south-central Arizona, 259–260
Aerobic biological treatment, *169*
Africa, *see also* South Africa; specific country
  helminthiasis, 350
  lagooning, 183
  program management case study, *336*
  public perception, 289
  unplanned aquifer recharge, 357
  wastewater reuse effects, 348
Agronomic codes of practices
  alternating water sources, 145
  basics, 104
  blending of water sources, 145
  border irrigation, 114
  boron toxicity, 135–137, *136*, *138–139*
  chloride toxicity, 135–137, *136*, *138–139*
  clogging, 147–148, *147–148*
  comparisons of methods, 113–122, *114–115*
  criteria for selection, 108–113, *109–113*, 122–123
  crop selection and management, 123–141, *126–135*, *140–141*
  drainage, 142–145
  drip irrigation, 118–121, *119–120*
  fertilizer applications, 145–146
  furrow irrigation, 114–115, *115*
  groundwater, 144–145
  leaching, 142–145
  management strategy, 106–108, *107*
  maximum crop production, 105–106
  method selection, 108–123
  micro-sprayer irrigation, 118, *118–119*
  quantity of water, 104–105, *105*
  salinity hazards, 123–135
  sodium toxicity, 135–137, *138–139*
  soil structure management, 146–147
  source control, 124
  sprinkler irrigation methods, *116–117*, 116–118
  storage systems, 148–149
  subirrigation, 115–116
  subsurface irrigation, 121–122, *122*
  surface irrigation methods, 113–116, *114–115*
  trace elements toxicity, 137, 140–141
  water application, 142–149
Agronomic significant parameters, 45–55
Al Samra, Jordan, 183
Alternating water sources, 145
Alternative sources of irrigation water, 4

Anaerobic biological treatment, *169*
Andalusia, 75
Annual equivalent fixed cost, 273–274
Annual operating costs, 273
Application control, 94–96, *95–96*
Aqueduct, Central Arizona Project, 254
Aquifers, 223–224, 356–357, *357–360*
Argentina, South America
  lagooning, 181, 183
  water reuse, *16*, 26
Arizona, United States, *see also* specific city
  adverse effects of sewage irrigation, 254, 259–260
  international health guidelines and regulations, *74*
  water reuse, *17*, *22–23*, 25
  water rights, 318
Arizona Public Service vs. Long, 319
Arnstein studies, *299*, 299–300
Article 14 Water Framework Directive, 288
Asano studies, 1–29
Asia, 26–27, 348, 357
Athletes, risks, 100
Atlantic coast, 181
Australia
  international health guidelines and regulations, 73
  membrane filtration, 215
  participative planning processes, 302
  UV disinfection, 208
  water reuse, 4, *16*, 26
  wetlands, 183

B

Bacon, Francis, 253
Bacteria removal, 353–354, *354*, *see also* specific method
Bahri studies
  agronomic codes of practices, 103–149
  health protection codes of practices, 83–100
  quality considerations, 31–58
Balearic Islands, 75
Bathing zone comparison, *77–78*
Behavior of compounds, 238–253
Beijing, 27
Belgium, *16*, 18
Benefits
  economics of water recycling, *9–10*, 267–268
  factors influencing, 268–269
  landscape and golf course codes of practices, 152
  water reuse, 8, *9–10*
BGS, *see* British Geological Survey (BGS)
Bicarbonate, 53–54
Biofilm technologies, 176–178, *177*
Biological treatment, 173–176, 188–219
Biomanipulation, 225
Blending of water sources, 145
Border irrigation, 114
Boron toxicity, 135–137, *136*, *138–139*
Bouwer studies, 31–58, 235–260
Brackish water, *126*
Brevard County, Florida, 318
British Geological Survey (BGS), 356
Bubbler irrigation, 121

C

California, United States, *see also* specific city or county
  adverse effects of sewage irrigation, 238
  construction issues, 328
  international health guidelines and regulations, 64, 66–67, *71–72*, 73, *74*, 75
  landscape and golf course codes of practices, 155
  membranes bioreactors, 219
  ozonation, 92
  UV disinfection, 198–199, 205, 208
  wastewater treatment, 172
  water recycling economics, 278, 280
  water reuse, 4, *17*, 21, *22–23*, 24–25
  water rights, 319
California Water Code, 316
California Water Recycling Criteria
  adverse effects of sewage irrigation, 238
  basics, 66–67, *71–72*, 73
  health protection codes of practices, 85
  international health guidelines and regulations, 62, 64
  ozonation, 212–214
Campo Espejo, 26
Canada
  international health guidelines and regulations, 73
  ozonation, 211
  water reuse, 15, *16*
Cape Verde, 181
Carbonate, 53–54
Case studies
  Costa Brava, Spain case study, 363–373
  El Mezquital, Mexico case study, 345–357
  food crop irrigation, 374–379
  France, *313–314*
  golf course, 363–373

Irvine Ranch Water District, *331*
Israel, *314*
Japan, *317–318*
largest irrigation district, 345–357
Los Angeles, *323–324*
Middle East, *315–316*
Monterey County, California case study, 374–379
Sardinia, *291–292*
Silicon Valley (San Jose), *321*
Sydney, NSW, *324–325*
Tunisia, *336*
United Kingdom, *296–297*
Cash flow needs, 338
Castell-Platja d'Aro wastewater treatment plant (WWTP), *see* Costa Brava, Spain
Central America
helminthiasis, 350
public perception, 289
water reuse, 26
Central Arizona Project Aqueduct, 254
Challenges of water reuse
Asia, 26–27
basics, 4–6
benefits and constraints, 8–10
Central America, 26
Europe, *18*, 18–20
international experience, 14–27, *15–17*
management actions, *28*, 28–29
Mediterranean region, *18*, 18–20
Middle East, *18*, 18–20
Oceania, 26–27
planning specifics, 11–12, *12–13*, 14
role, 6–8, *7*
South America, 26
United States, 20–21, *22–25*, 24–25
water security, 1–2, *3*, 4
Chemicals, water quality, 36–37, *37–39*, 40
Chemical treatment, management strategies, 225
Chiba Prefecture Kobe City, Japan, *317*
Chicago metropolitan area, 377
Chile, South America, *16*, 160
China
chlorine dioxide, 198
wastewater regulations, 320
water quality, 36, 53
water reuse, 2, 4, *16*, 27
Chloride toxicity, 135–137, *136*, *138–139*
Chlorination
basics, 194–195, *195*
design, 195–197
disinfection, 89, *90*, 91
efficiency, 195–197
residual of chlorine, 197–198
Chlorine dioxide, 198
Choice of treatment, 166, *166*, *see also* specific treatment
Citizens' juries, 301
Citizens' panels, 301
Clean Water Act (CWA), 319–320
Clermont-Ferrand, France, 19, 181
Clogging, 147–148, *147–148*
CNA, *see* National Water Commission (CNA)
Coagulation, *169*, 171–172, *172*
Colorado River, 254, 259
Community and institutional perceptions and barriers
basics, 285–286, *287*, 288–289
institutional barriers, 292–296, *296–297*
models for participative planning, 298–301, *299*
participative planning processes, 301–305, *303–304*
public perceptions, 289–291, *291–292*
Community liaison groups, 300
Compound behavior, 238–253
Compounds with potential adverse effects, 237–238
Constraints, 8, *9–10*, 152
Construction issues, 328–329, *329*
Contracting power, 331
Contracts, 338–339
Corrective actions, *394*
Costa Brava, Spain
basics, 344, 372–373
electrical conductivity, 364–365, *365–366*
historical developments, 363, *364*
maturation ponds, 368–372, *372*
nutrients, 365–368, *367–371*
water management, 364
Costs, 226–228, 355–356, *356*, *see also* Economics
Coullons, 181
Crook studies, 61–80
Crop consumers, risks, 99
Crops, *see also* Adverse effects of sewage irrigation; Plants
restrictions, 96–98, *97*
selection and management, 123–141, *126–135*, *140–141*
Cryptosporidium, 213, 231
Customers, 160–161, 329, 332–333
CWA, *see* Clean Water Act (CWA)
Cyprus, 18

D

Dalian, 27
Dan Region, Israel
  lagooning, 183
  nonconventional natural systems, 178
  recharge basin, *187*
  soil-aquifer treatment, 188
  water reuse, 19
Dead Sea, 257
Design parameters, wastewater treatment, 214–215
Destratification, 225
Disinfection
  basics, *169*, 188–189
  chlorination, 194–198, *195*
  health protection codes of practices, 89, *90*, 91–92
  membrane filtration, 215–216
  membranes bioreactors, *216–218*, 216–219
  ozonation, 208–215, *212–214*
  selection, 228, *229–230*, 231
  tertiary treatment, 188–193, *190*, *192–195*
  UV disinfection, 198–208, *199–203*, *205*, *207*, *209–210*
Disinfection by-products, *38–39*, 247–248
Dissolved inorganic matter, 204
Dissolved organic matter, 204
Distribution systems, 225–226, *226*, *see also* Storage and distribution
Doha, 20
Drainage, agronomic codes of practices, 142–145
Dredging, 225
Drinking water standards, *37–39*
Drip irrigation
  agronomic codes of practices, 118–121, *119–120*
  clogging, 147
  methods, *95*
Dubai, 20

E

Economics, *see also* Costs
  analysis, 267
  basics, 266, 282
  benefits of recycled water, *9–10*, 267–268
  economic analysis, 267
  examples, water prices, 279–280, *280–281*
  factors influencing benefits, 268–269
  financial analysis, 266
  financing power, 331
  function of water prices, 275–277
  incentives/disincentives, 334
  landscape and golf course codes of practices, 158–160
  prices for recycled water, 275
  pricing instruments, 278–279
  setting prices, criteria, 277–278
  supply options, *269*, 269–271
  system components, 269, *270*
  water-recycling options, *269*, 271–274, *272*, *274*
Effluent regulations, 320–322
Egg removal, 351–352, *351–352*
Egypt
  water recycling economics, 279
  water reuse, 15, *16*
  wetlands, 183
Electrical conductivity, 364–365, *365–366*
El Kantaoui Golf Course, 153, *154*
El Mezquital, Mexico, *see also* Mexico City
  aquifer recharge, unplanned, 356–357, *357–360*
  bacteria removal, 353–354, *354*
  basics, 344–346, *346*
  costs of recycled water, 355–356, *356*
  egg removal, 351–352, *351–352*
  filtration step, 352–353, *353*
  helminthiasis, 349–352, *350–352*
  legislation, 348–349
  reuse effects, 348, *348–349*
  sludge treatment and disposal, 355, *355*
  unplanned aquifer recharge, 356
  wastewater quality, 347, *347*
El Pasa, Texas, 259
Emilia Romagna, 19
Endocrine disruptors, 248
England, 18, *see also* United Kingdom
Ensure follow-through, 339
Environmental regulations, 326–328
EPA, *see* U.S. EPA
Epidemiological studies, *44*
Escherichia coli, 219, 231, 270
Euphrates River, 315, *315*
Europe, *see also* specific country
  additional treatment and reuse cost, 227
  public perception, 289
  water reuse, *18*, 18–20
  wetlands, 183
European Union, 320
Exposure control, human, 98–100

F

Far East, 350
Feasibility

main considerations, 11–12, *12–13*, 14
practices, 383, *384–385*
program management, 333–334
Federal Ministry for the Environment, 320
Federal Water Act, 320
Federal Water Pollution Control Act (Clean Water Act), 319
Fertilizer applications, 145–146
Field workers, risks, 99
Filtration, *169*, 352–353, *353*, *see also* specific type of filtration
Financial aspects and analysis, 158–160, 266, *see also* Economics
Flocculation, *169*, 171–172, *172*
Flood irrigation, *95*
Florida, United States, *see also* specific city or county
construction issues, 328
UV disinfection, 208
water recycling economics, 278
water reuse, 4, *17*, 21, *22–23*, 24
water rights, 318
Flotation, 172–173
Focus groups, 301
Food crop irrigation, *see* Monterey County, California
France
chlorination, 231
chlorine dioxide, 198
helminthiasis, 351
infiltration-percolation, 186
international health guidelines and regulations, *74*, 75
lagooning, 181
membrane filtration, 215
public perception, 290
short-term storage, 220
UV disinfection, 208
water reuse, 15, *16*, 19
water rights, 313
water rights case study, *313–314*
Free chlorine, 55
Fukuoka City, Japan, *318*
Furrow irrigation, *95*, 114–115, *115*

G

Germany, 18, 313, 320
Ghana, 279
Giardia, 213, 231
Golf Course Costa Brava, Spain, 366, 372, *see also* Costa Brava, Spain
Golf Course d'Aro, 363, 367, *see also* Costa Brava, Spain
Golf Course L'Angel, 366, *see also* Costa Brava, Spain
Golf Course Less Serres de Pals, 365, 367, *see also* Costa Brava, Spain
Golf courses, *152–153*, 153–155, *155*, *see also* Costa Brava, Spain; Landscape and golf course codes of practices
Golfers, 100
Good practices, 383, *386–387*, 388
Granular filtration, 208
Greater Tel Aviv area, 19
Groundwater, 144–145, *see also* Adverse effects of sewage irrigation
Groundwater water-table management, 253–260
Guidelines, 337–339
Gujarat, India, 327

H

Haifa, Israel, 19, 183
Handan, 27
Hawaii, United States, 21, *22–23*, 24
Health protection codes of practices
application control, 94–96, *95–96*
appropriate action flow chart, *86*
basics, 83–85, *85–86*
crop restrictions, 96–98, *97*
disinfection processes, 89–92, *90*
human exposure control, 98–100, *99–100*
infection risks, *96–97*, *99*
public access restrictions, 96–98, *97*
reliability of operation requirements, 93–94
specific wastewater treatments, 86–94
storage and distribution requirement, 92–93, *93*
treatment schemes, *87–88*, 87–89
Health significant parameters, 36–45
Helminthiasis, 349–352, *350–352*
Hohokam civilization, 254
Hormones, 248
Human exposure control, 98–100, *99–100*
Hungary, 36
Hydraulic residence time, 213–214
Hypolimnetic aeration, 225

I

Implementation issues, 324–333
India, 2, 4, *16*, 27, 327
Indicators, biological and microbiological, *41–42*, 62, *182*, *see also* Pathogens
Infiltration-percolation, 184–186, *185*
Initial capital costs, 273

Institutional drivers, 337
Institutional issues, *see also* Community and institutional perceptions and barriers
  assess cash flow needs, 338
  basics, 310
  construction issues, 328–329, *329*
  customer agreements, 332–333
  effluent regulations, 320–322
  ensure follow-through, 339
  environmental regulations, 326–328
  France case study, *313–314*
  guidelines, 337–339
  institutional drivers, 337
  integrated planning, 333
  Irvine Ranch Water District case study, *331*
  Israel case study, *314*
  Japan case study, *317–318*
  land use planning, 324–326
  legal drivers, 337
  Los Angeles case study, *323–324*
  matrix analysis, 333–335, *334–335*, *337*
  Middle East case study, *315–316*
  Northern California CalFed Program, *327*
  ownership, 311–319, 337
  participating stakeholders and institutions, 338
  planning and implementation issues, 324–333
  practices, 390, *395–396*
  prepare contracts, 338–339
  pretreatment for protection, 322–323
  program management, 333–339
  public education program, 339
  retailer issues, 329–332, *330*
  reuse technology, 338
  rights to recycled water, 318–319
  Silicon Valley (San Jose) case study, *321*
  Sydney, NSW, case study, *324–325*
  Tunisia case study, *336*
  wastewater regulations, 319–323
  water rights, 311–315
  water use limits, 316–318
  wholesaler issues, 329–332, *330*
Integrated planning, 333
Interactive web sites, 300
International health guidelines and regulations, *see also* specific country
  Arizona, *74*
  Australia, 73
  basics, 61–62, *63*, 64
  California criteria, 66–67, *71–72*, 73, *74*
  Canada, 73
  France, *74*
  Israel, 73, *74*
  Japan, 73
  Jordan, 73
  Mediterranean area, 73
  South Africa, 73
  Spain, *74*
  standard enforcement and perspectives, 79–80
  urban use standards, 76, *77–78*, 79
  USEPA, 66, *68–70*, *74*
  WHO guidelines, 64, *65*, *74*
International water reuse, 14–27, *15–17*
Iraq, 315
Irrigation and irrigation systems, *see also* specific type of system
  basics, *3*
  features, *110–111*
  methods, *95*, *112*, 113
  surface irrigation comparison, *117*
  water application efficiency, 109, *113*
Irvine, California, 379
Irvine Ranch Water District
  landscape and golf course codes of practices, 160
  plug-flow chlorine-contact basins, *195*
  water reuse, 24
  wholesaler/retailer issues, *331*
Israel
  adverse effects of sewage irrigation, 257
  chlorine dioxide, 198
  infiltration-percolation, 186
  international health guidelines and regulations, 73, *74*
  lagooning, 183
  short-term storage, 220
  soil-aquifer treatment, 188
  surface reservoirs, 221
  water reuse, 4, *16*, 18–19
  water rights, 313, 315–316
  water rights case studies, *314–316*
Israel case study, *314*
Italy, *16*, 19, 290

J

Japan
  institutional issues case study, *317–318*
  international health guidelines and regulations, 73
  membrane filtration, 215
  program management, 333
  water reuse, *16*, 26
  water rights case studies, *317–318*
Jeffrey studies, 285–305
Jimenez studies, 345–357

Jordan
adverse effects of sewage irrigation, 257
international health guidelines and regulations, 73
lagooning, 183
water reuse, 4, 15, *16*, 18–19
water rights, 315
water rights case study, *315*
Jordan River, 315, *315–316*
Jubail, 20

K

Kenya, 181, 183
Kesterson (Lake), *see* Lake Kesterson
King County, 21
Kishon Complex, Greater Haifa, 19
Kishon reservoir, 221
Kuwait, *16*, 216
Kuwait City, Kuwait, 20
Kyoto, Japan (World Water Forum), 333

L

Lagooning
applications, 181, 183
basics, 178–179
maturation ponds, 179–181, *180*, *182*
Lake Kesterson, 238
Landscape and golf course codes of practices, *see also* Golf courses
basics, 151
benefits and constraints, 152
customer acceptance, 160–161
economic and financial aspects, 158–160
golf courses, *152–153*, 153–155, *155*
turfgrass, effects on, 153, *153–154*, *156*, 156–157
water quality criteria, *77–78*
Land use planning, 324–326
Largest irrigation district, *see* El Mezquital, Mexico
Las Vegas, 21
Latin America, 320, 348, 357
Lazarova studies
agronomic codes of practices, 103–149
challenges of water reuse, 1–29
economics of water recycling, 265–282
health protection codes of practices, 83–100
international guidelines and regulations, 61–80
practices, 381–390
wastewater treatment, 163–231
water quality, 31–58
Leaching, 142–145
Legal drivers, 337
Legal issues, 390, *395–396*
Legislation, 348–349
Lime clarification, 172, *172*
Liquid media, *170*
Lloret de Mar wastewater treatment plant (WWTP), 367, *see also* Costa Brava, Spain
Loading, *140–141*
Local residents, risks, 99–100
Long-term storage, 92–93, 220–224
Los Angeles, California, *323–324*, 377
Low-rate infiltration, *170*

M

Management
agronomic codes of practices, 106–108, *107*
practices, *382*, 388–389, *394*
water reuse, *28*, 28–29
Mas Nou Golf Course, 363, *see also* Costa Brava, Spain
Matrix analysis, 333–335, *334–335*, *337*
Maturation ponds, *see also* Polishing process
characteristics, 179–181, *180*
Costa Brava, Spain case study, 368–372, *372*
disinfection, *90*
operational problems, 181, *182*
Maximum crop production, 105–106
Mediterranean area
additional treatment and reuse cost, 227
international health guidelines and regulations, 73
lagooning, 181
water reuse, 15, *18*, 18–20
Medjerda catchment area, *336*
Melle, 181
Membrane bioreactors
basics, *169–170*
tertiary treatment, *216*, 216–217, *218*, 219
Membrane filtration, 208, 215–216
Mendoza, Argentina, 26, 183
Mendoza City Basin, 26
Mequital Valley, 26
Mesa, Arizona, 318
Mesopotamia, 254
Metropolitan Water District of Southern California, *323–324*
Mexico, 4, *17*, 26
Mexico City, 26, *see also* El Mezquital, Mexico
Mezquital, *see* El Mezquital, Mexico

Microbiological indicators, *41–42*, *see also* Pathogens; specific type of indicator
Microorganisms influence, 206–207, *207*
Micro-sprayer irrigation, 118, *118–119*
Middle East, *see also* specific country
institutional issues case study, *315–316*
unplanned aquifer recharge, 357
UV disinfection, 208
water reuse, 2, 15, *18*, 18–20
water rights, 315
water rights case study, *315–316*
Milan, 18
Mindelo, Cape Verde, 181
Models for participative planning, 298–301, *299*
Monitoring strategies, 55–58, *56–57*
Montana, 270
Monterey County, California
conclusions, 378–379, *379*
current project status and operation, 378
historical developments, 374, *375*
lessons learned, 378–379, *379*
Monterey County Water Recycling projects, 379
project overview, 374, 376, *376–377*
public perception, 344, 377–378
recommendations, 378–379, *379*
Monterey Regional Water Pollution Control Agency, 374
Monterey Wastewater Reclamation Study for Agriculture (MWRSA), 24, 43, 377, *see also* Monterey County, California
Montreal, Canada, 211
Mont-Saint-Michel, 181
Morocco, *17*, 181, 186
Morris studies, 265–282

N

Naegleria, 213
Nairobi, Kenya, 183
Namibia, *17*, 181
Napa County, California, 379
Narmada River Valley, 327
National Water Commission (CNA), 356
Natural sources of irrigation water, 2, 4
Nevada, 21, *22*
New Calendonia, 220
New York metropolitan area, 377
Nitrogen, 54
Nogales, *see* Prescott-Nogales corridor (Arizona)
Noirmoutier, 181
Nonconventional natural systems, 178, *178*
North Africa, 2, *see also* Africa; specific country
Northern California CalFed Program, *327*
Norway, 351
Nutrients
adverse effects of sewage irrigation, 244–247
Costa Brava, Spain case study, 365–368, *367–371*
water quality, 54–55

O

Oak Hill Cemetery, San Jose, California, *158*
Oceania, 26–27
Oman, 15, *17*
Open house and exhibition, 301
Operational problems, 193, *194*
Opinion polls and surveys, 300
Orange County, California, 155, 379
Organic containments, 248–253
Orlando, Florida, 21
Ownership, 311–319, 337
Ozonation
basics, 91–92, 208, 211–215
design parameters, 214–215
disinfection, *90*
hydraulic resistance time, 213–214
wastewater quality influence, *203*, 212–213, *213–214*

P

Pakistan, 2, 27
Papadopoulos studies, 103–149
Paris, 18
Participation, 298–305, *299*, *303–304*, 338
Particles, 204
Pathogens, 40, *41–44*, 43, 45
Pebble Beach golf course, *156*
Pénestin, France, *185*
Petra, Jordan, *180*
pH, 53, *140*
Pharmaceuticals, 248–253
Phoenix, Arizona, 242, 254, 257, 260, 319
Phosphorus, 146
Photobacterium phosphoreum, 357
Physiochemical treatment, 168–173
Pinal Active Management Area, 254
Planning issues, 324–333
Planning specifics, 11–12, *12–13*, 14

Plants, *see also* Adverse effects of sewage irrigation; Crops
  practices, negative impact on, 388, *391–393*
  salt tolerance, *126*, *128–134*, *153*, *241*, 255–256
Polishing process, *see also* Maturation ponds
  disinfection, 89, *90*
  secondary effluents, 20
  wastewater treatment, 226
Political motivation, 334
Pomona Virus Study, 24
Porquerolles, 181
Potassium, 54
Practices
  basics, 381–383
  corrective actions, *394*
  feasibility assessment, 383, *384–385*
  good practices, 383, *386–387*, 388
  institutional issues, 390, *395–396*
  legal issues, 390, *395–396*
  management practices, 388–389, *394*
  negative impact on plants, 388, *391–393*
  public acceptance, 389–390
Prescott-Nogales corridor (Arizona), 259
Pretreatment for protection, 322–323
Prices, *see* Economics
Primary sedimentation, 171
Process design, 191–193, *192–193*
Processes used, 166, *167*, 168, *169–170*
Program management, 333–339
Public acceptance, 389–390
Public access restrictions, 96–98, *97*
Public education program, 339
Public meetings, 300
Public perceptions, 289–291, *291–292*, 377–378

Q

Qingdao, 27
Quality control, 225–226, *226*
Quantity of water, 104–105, *105*
Queensland Water Recycling Strategy, 302

R

Reims, 18
Ré (island), 181
Reliability of operation requirements, 93–94
Replacement costs, 273
Reservoir management, 224–225
Retailer issues, 329–332, *330*
Reuse effects, 348, *348–349*
Reuse technology, 338
Revenue programs, 329
Reverse osmosis, 208
Rights to recycled water, 318–319
Riyadh, 20
Role of water reuse, 6–8, *7*
Rosenblum studies, 309–339
Rotating biological contactors, 176
Rouse Hill area, NSW, *324–325*

S

Sacramento, California, 219
Sacramento River, 315
Salado River, 345
Sala studies, 363–373
Salinity
  classification, *47*
  crop selection, *127*
  hazards, agronomic codes of practices, 123–135
  management, 259–260
  south-central Arizona, 259–260
  water quality, 45–46, *47*
Salt
  adverse effects of sewage irrigation, *110–111*, *128–130*, 239–244, *241*, *245*, 253–260
  groundwater water-table management, 253–260
  loadings, 234–235
  management, 256–259, *258*
  plant tolerance, *126*, *128–134*, *241*, 255–256
  water relations, *239–244*, *240–241*, *245*
Salton Sea, California, 257
Salt River, 255, 259
Salt River Valley, 242, 255, 257
Sampling strategies, water quality, 55, *56–57*, 58
San Francisco Bay, California, 315, 377
San Joaquin River, 315
Santa Rosa, California, 379
Sardar Sarovar Project, 327
Sardinia case, *291–292*
Saudi Arabia, 4, 15, *17*
Screening, physiochemical treatment, 168–171
Sedimentation, *169*
Setting prices, criteria, *see* Economics
Sewer configuration, 168, *171*
Sheikh studies, 151–161, 374–379
Shenyang, 27
Shenzhen, 27
Shijiazhuang, 27
Short-term storage, 92, 220
Sidi Salem Dam, *336*

Silicon Valley (San Jose), California, *321*, 335
Sludge treatment and disposal, 355, *355*, *see also* Activated sludge
Sodium adsorption ratio, 48
Sodium toxicity, 135–137, *138–139*
Soil-aquifer treatment, *186–187*, 186–188
Soil structure management, 146–147
Sonoma County, California, 379
Soreq treatment line, 188
Source control, 124
South Africa, *see also* Africa; specific country
  international health guidelines and regulations, 73
  UV disinfection, 208
  water reuse, 2, *17*
  wetlands, 183
South America, 26, 350
South-central Arizona, 254, 259–260, *see also* Arizona, United States
Spain
  helminthiasis, 351
  infiltration-percolation, 186
  international health guidelines and regulations, *74*, 75
  UV disinfection, 208
  water reuse, *17*, 19
  wetlands, 183
Specific wastewater treatments, 86–94
Sprinkler irrigation
  agronomic codes of practices, *116–117*, 116–118
  basics, 95
  clogging, 147
  methods, *95*
St. Petersburg, Florida, 24
Standard enforcement and perspectives, 79–80
Storage and distribution
  aquifers, 223–224
  basics, 219–220
  health protection codes of practices, 92–93, *93*
  long-term storage, 220–224
  reservoir management, 224–225
  short-term storage, 220
  surface reservoirs, 220–223, *222*
  water quality control, 225–226, *226*
Storage systems, 148–149
Subirrigation, 115–116
Subsurface irrigation, *95*, 121–122, *122*
Sulaibiya, Kuwait, 20, 216
Supply options, *269*, 269–271
Surface irrigation methods, 113–116, *114–115*
Surface reservoirs, 220–223, *222*
Sweden, 15, *17*, 351
Sydney, NSW, Australia, 27, *324–325*
Syria, 315, *315*
System components, 269, *270*

T

Tabqa Dam, *315*
Tacoma, 21
Taif, 20
Taiyan, 27
Tallahassee, Florida, 21
Tanzania, 181
Technologies, UV disinfection, 199–200, *200–201*
Tel Aviv, Israel, 183
Temperature, 202
Tertiary treatment
  basics, 188–189
  chlorination, 194–198, *195*
  chlorine dioxide, 198
  efficiency of treatment, *190*, 190–191
  filtration, 189–193
  operational problems, 193, *194*
  ozonation, 208–215, *212*
  process design, 191–193, *192–193*
  UV disinfection, 198–208, *199–203*, 199–208, *205*, *207–210*
Texas, *22–23*, 259
Texcoco, 26
Thayer vs. City of Rawlins, Wyoming, 319
Thessaloniki, Greece, *193*
Tianjin, 27
Tigris River, 315
Tokyo, Japan, *317*
Total average annual costs, 274
Toxic ions, 46–47
Trace elements
  toxicity, 137, 140–141
  water quality, *33*, *38–39*, 49, *50–52*, 51–53
Transmittance, UV disinfection, 202–204
Treatment schemes, *87–88*, 87–89
Trickling filters, 174–176, *175*
Tucson, Arizona, 254, 260
Tula River and Tula Valley, 345, 356
Tunisia
  lagooning, 181
  program management case study, *336*
  turfgrass, 153, *154*
  water reuse, 4, *17*, 18–20
Turfgrass, *see also* Landscape and golf course codes of practices
  effects of recycled water, 153, *153–154*
  prevention of effects, *156*, 156–157
Turkey, 315, *315*
Tyrrel studies, 265–282

U

Uganda, 183
United Arab Emirates, 15, *17*, *296–297*
United Kingdom
  chlorination, 231
  institutional perspectives, *296–297*
  public perception, 290
  salad growers, 270–271
  water reuse, 15, *17*
United States, *see also* specific state
  adverse effects of sewage irrigation, 251
  agricultural irrigation projects, *23*
  chlorine dioxide, 198
  helminthiasis, 351
  infiltration-percolation, 186
  international health guidelines
    and regulations, 79
  membrane filtration, 215
  participative planning processes, 302
  public perception, 289, 377
  UV disinfection, 199
  wastewater regulations, 319–320
  water quality, 40
  water reuse, 4, *17*, 20–21, *22–25*, 24–25
  water rights, 312, 315
  wetlands, 183
Upstream treatment influence, *205*, 205–206
Urban use standards, 76, *77–78*, 79
U.S. EPA
  international health guidelines
    and regulations, 64, 66, *68–70*, *74*
  Mexico City wastewater, 347
  Monterey County Water Recycling
    projects, 374
  wastewater regulations, 320
U.S. EPA Reuse Guidelines, 205, 212
U.S. EPA/USAID Guidelines for Water
  Reuse, 311
Uses, recycled water, *7*
UV irradiation
  advantages and disadvantages, *199*
  basics, *70*, 91, 198–199
  design and operation, 207–208, *209–210*
  disinfection, *90*
  dissolved inorganic matter, 204
  dissolved organic matter, 204
  microorganisms influence, 206–207, *207*
  particles, 204
  technologies used, 199–200, *200–201*
  temperature, 202
  transmittance, 202–204
  upstream treatment influence, *205*,
    205–206
  water quality influence, 201–205, *202–203*

V

Value engineering, 329
Vitoria, 19

W

Wadi Mousa treatment plant, *180*
Washington, 21, 25
Wastewater
  quality, case study, 347, *347*
  regulations, 319–323
  treatment, *203*, 212–213, *213–214*
Wastewater Charges Act, 320
Wastewater treatment
  activated sludge, 173–174
  aquifers, 223–224
  basics, 164–168, *165*
  biofilm technologies, 176–178, *177*
  biological treatment, 173–176
  chlorination, 194–198, *195*
  chlorine dioxide, 198
  choice of treatment, 166, *166*
  coagulation, 171–172, *172*
  costs, additional treatment and reuse,
    226–228
  design parameters, 214–215
  disinfection, 188–219, 228, *229–230*, 231
  flocculation, 171–172, *172*
  flotation, 172–173
  hydraulic residence time, 213–214
  infiltration-percolation, 184–186, *185*
  lagooning, 178–183
  long-term storage, 220–224
  maturation ponds, 179–181, *180*, *182*
  mechanisms, *169–170*
  membrane bioreactors, *216*, 216–217,
    218, 219
  membrane filtration, 215–216
  nonconventional natural systems, 178, *178*
  operational problems, 193, *194*
  ozonation, 208, 211–215
  physiochemical treatment, 168–173
  polishing process selection, 226
  primary sedimentation, 171
  process design, 191–193, *192–193*
  processes used, 166, *167*, 168, *169–170*
  quality control, 225–226, *226*
  reservoir management, 224–225
  rotating biological contactors, 176
  screening, 168–171
  sewer configuration, 168, *171*
  short-term storage, 220

Wastewater treatment (*Continued*)
soil-aquifer treatment, *186–187*, 186–188
storage and distribution, 219–226
surface reservoirs, 220–223, *222*
trickling filters, 174–176, *175*
UV disinfection, *70*, 198–208, *199*
wastewater quality, *203*, 212–213, *213–214*
wetlands, 183–184, *184*
Water application, 142–149
Water availability as benchmark, 14
Water Environment Research Foundation, 302
Water Framework Directive, Article 14, 288
Water management, 364
Water quality
agronomic significant parameters, 45–55
basics, 31–36, *32–35*
bicarbonate, 53–54
carbonate, 53–54
chemicals, 36–37, *37–39*, 40
criteria, *32*
distribution systems, 225–226, *226*
drinking water standards, *37–39*
free chlorine, 55
health significant parameters, 36–45
influence, 201–205, *202–203*
monitoring strategies, 55–58, *56–57*
nutrients, 54–55
parameters, *33–35*
pathogens, 40, *41–44*, 43, 45
pH, 53
salinity, 45–46, *47*
sampling strategies, 55, *56–57*, 58
sodium adsorption ratio, 48
toxic ions, 46–47
trace elements, *33*, *38–39*, 49, *50–52*, 51–53
Water-recycling options, *269*, 271–274, *272*, *274*
Water reuse
advantages and disadvantages, *9–10*
Asia, 26–27
basics, 4–6
benefits and constraints, 8–10
Central America, 26
decision factors, *384*
Europe, *18*, 18–20
international experience, 14–27, *15–17*
main characteristics, *22*
management actions, *28*, 28–29
Mediterranean region, *18*, 18–20
Middle East, *18*, 18–20
Oceania, 26–27
planning, 11–12, *12–13*, 14
role, 6–8, *7*
South America, 26
United States, 20–21, *22–25*, 24–25
water security, 1–2, *3*, 4
worldwide projects, *15–18*
Water rights, 311–315
Water security, 1–2, *3*, 4
Water use limits, 316–318
Web sites, interactive, 300
West Basin Water Recycling Plant, 172
West Coast Basin Barrier, California, *172*
Western Europe, 227, *see also* Europe
Wetlands, 183–184, *184*, 225
Wholesaler issues, 329–332, *330*
Windhoek, Namibia, 181
World Bank, 14–15, 327
World Health Organization (WHO)
adverse effects of sewage irrigation, 238
international health guidelines and regulations, 62, 64, *65*, *74*, 75–76
lagooning, 181
maturation ponds, 179
ozonation, 212, 214
UV disinfection, 205, 208
water quality, 36, 40
water recycling economics, 268, 271
wetlands, 183
World Water Council, 310, 333
World Water Forum (Kyoto, Japan), 333

Y

Yarmouk River, *316*
Yavne treatment line, 188

Z

Zhaozhuang, 27